3rd Materials Research Society of Indonesia Meeting (MRS-Id 2018)

AA006923

Materials for Energy and Environment Sustainability

IOP Conference Series: Materials Science and Engineering
Volume 622

Bali, Indonesia
31 July - 2 August 2018

Editors:

Khairurrijal Khairurrijal
Rino R. Mukti
Hutomo Suryo Wasisto
Ariando
Satria Z. Bisri

ISBN: 978-1-7138-0902-9
ISSN: 1757-8981

Printed from e-media with permission by:

Curran Associates, Inc.
57 Morehouse Lane
Red Hook, NY 12571

Some format issues inherent in the e-media version may also appear in this print version.

This work is licensed under a Creative Commons Attribution 3.0 International Licence.
Licence details: http://creativecommons.org/licenses/by/3.0/.

No changes have been made to the content of these proceedings. There may be changes to pagination and minor adjustments for aesthetics.

Printed with permission by Curran Associates, Inc. (2020)

For permission requests, please contact the Institute of Physics
at the address below.

Institute of Physics
Dirac House, Temple Back
Bristol BS1 6BE UK

Phone: 44 1 17 929 7481
Fax: 44 1 17 920 0979

techtracking@iop.org

Additional copies of this publication are available from:

Curran Associates, Inc.
57 Morehouse Lane
Red Hook, NY 12571 USA
Phone: 845-758-0400
Fax: 845-758-2633
Email: curran@proceedings.com
Web: www.proceedings.com

TABLE OF CONTENTS

IOP Conf. Series: Materials Science and Engineering **622** (2019) 011001 doi:10.1088/1757-899X/622/1/011001

Preface

3rd MRS-id Meeting is the continuation of successful biannual series including 1st MRS-id Meeting in Bali (2014) and 2nd MRS-id Meeting in Bandung (2016) organized by Materials Research Society of Indonesia (MRS-id). The conference was held on July 31st to 2 August 2nd, 2018 at Aston Denpasar Hotel & Convention Center, Bali, Indonesia. The theme of current meeting is "Materials for Energy and Environment Sustainability". The focus of the MRS-id meeting is to promote materials science and technology in Indonesia, regional area (such as ASEAN and Asia) as well as global. The MRS-id meeting is a platform for the gathering of researchers, lecturers and students who always seek for collaboration, exchanging ideas across all scientific disciplines.

The 3rd MRS-id Meeting was attended by 7 plenary speakers, 13 keynote speakers, 95 oral speakers and 45 poster presenters. These are the contribution from 11 countries across the world. The scope of meeting covers various aspects of advanced materials science and technology such as photonic materials, optical materials, biomaterials, biomedical materials, electronic materials, magnetic materials, porous materials and membrane. The submitted manuscripts from this meeting have been peer-reviewed and to the selected manuscripts, those has been published on Materials Research Express. Eventually, other 34 accepted manuscripts are for the publication on IOP Conference Series: Materials Science and Engineering.

The 3rd MRS-id Meeting organizing committee would like to acknowledge the enthusiasm of the participants, support from the sponsors, contribution from the international advisory boards and support from the heads of universities that lead to the success of this event.

Rino R. Mukti

Chair of 3rd MRS-id Meeting

Editors:

Khairurrijal Khairurrijal, Institut Teknologi Bandung, Indonesia (krijal@fi.itb.ac.id)

Rino R. Mukti, Institut Teknologi Bandung, Indonesia (rino@chem.itb.ac.id)

Hutomo Suryo Wasisto, Technische Universitaet Braunschweig (h.wasisto@tu-braunschweig.de)

Ariando, National University of Singapore, Singapore (ariando@nus.edu.sg)

Satria Z. Bisri, RIKEN Center for Emergent Matter Science, Japan (satria.bisri@riken.jp)

Content from this work may be used under the terms of the Creative Commons Attribution 3.0 licence. Any further distribution of this work must maintain attribution to the author(s) and the title of the work, journal citation and DOI.
Published under licence by IOP Publishing Ltd

Peer review statement

All papers published in this volume of *IOP Conference Series: Materials Science and Engineering* have been peer reviewed through processes administered by the proceedings Editors. Reviews were conducted by expert referees to the professional and scientific standards expected of a proceedings journal published by IOP Publishing.

Content from this work may be used under the terms of the Creative Commons Attribution 3.0 licence. Any further distribution of this work must maintain attribution to the author(s) and the title of the work, journal citation and DOI.
Published under licence by IOP Publishing Ltd

The Properties of Bolus Material using Silicone Rubber

Heri Sutanto[1,*], Indras Marhaendrajaya[1], Gede Wiratma Jaya[1], Eko Hidayanto[1], Astri Suppa Supratman[1], Santi Yuli Astuti[1], Tris Budiono[2] and Muhammad Agus Firmansyah[2]

[1]Departement of Physics, Faculty of Science and Mathematics, Diponegoro University, Jalan Prof. H. Soedarto, SH, Tembalang, Semarang City, Central Java, Indonesia
[2]Radiotherapy Installation, RSUP Dr. Sardjito, Jalan Kesehatan, Sleman District, Yogyakarta, Indonesia

[*]Corresponding author: herisutanto@fisika.undip.ac.id

Abstract. Bolus is a material equivalent to tissue and used in radiotherapy process to increase a dose surface using electron beam. The bolus synthesis from a material equivalent to tissue is not easy, one of the alternative materials used is silicone rubber (SR). In this research bolus was synthesized with dimension of length x width x thickness is (17 x 17 x 1) cm^3. Bolus has been characterized by CT-Scan to find relative electron density (RED) and linear accelerator (LINAC) to investigate percentage of surface dose (PSD) with two energy (8 MeV and 10 MeV). The RED value for bolus is 1.176, these results show the RED value for bolus between soft tissue and solid tissue. The PSD value at 8 MeV and 10 MeV are 102.32% and 101.32%, respectively. These results indicate that the silicone rubber material can be used as an alternative bolus material because it corresponds to the bolus function in radiotherapy.

Keyword: Bolus, Soft Tissue, Silicone Rubber, Electron Density, Percentage of Surface Dose.

1. Introduction

Bolus is a material equivalent to tissue and used in radiotherapy process [1]. In radiotherapy, bolus has several function such as increasing the surface dose, leveling the body surface of the irregular patient in order to obtain radiation dose uniformity, and reduce a possible electron penetration to soft tissue around the area of cancer [2,3]. Bolus synthesis with material equivalent to tissue is not easy, but some studies have used some materials that are equivalent to the tissue such as parafin granules, elasto-gel pad, superflab, thermoplastic sheets, dental wax, polypropylene, and rayon blend [4-6]. The use of these material as bolus, especially in Indonesia still very limited and must import to be used in radiotherapy. One of the alternative material that can be used for bolus fabrication is silicone rubber (SR).

SR is a synthetic polymer material derived from polydimethyl siloxane (PDMS) [7,8]. SR is easy to find in Indonesia because it is a product derived from petrochemicals. SR has several advantages such as has an excellent level of elasticity at very low temperature, excellent properites against heat, resistance to ultraviolet light, has a high level of resistance with various types irradiation, and low toxicity [9-12]. In addition, SR has been used as radioprotection material in radiotherapy with electron beam and can reduce electron penetration by 42.9% at 6 MeV [13]. In medical field, SR has been used

Content from this work may be used under the terms of the Creative Commons Attribution 3.0 licence. Any further distribution of this work must maintain attribution to the author(s) and the title of the work, journal citation and DOI.
Published under licence by IOP Publishing Ltd

to fabrication artificial organs such as rectum, obturator musculus internus and levator ani musculus as a group organ for prostate brachytherapy application [14].

In this work, SR is used as a bolus material because SR has material characteristics similar to tissue. To investigate SR characteristics, some characterizations have been done, among others relative electron density (RED) measurement and percentage of surface dose (PSD) using an electron beam radiation. This characterization is a basic test to show SR material can be used as an alternative bolus material.

2. Materials and Methods

2.1. Materials

Bolus used in this research was synthesized using silicone rubber (SR) RTV-5 and bluesil catalyst 60 R as hardener (Indrasari Chemical Store, Semarang, Indonesia). Bolus dimension (length x width x thickness) is (17 x 17 x 1) cm^3. For synthesis process using a simple molded method, which that SR with volme of 28 ml and hardener with volume of 11 ml have been mixed by mixer (Maspion, Indonesia) for 6 minutes. After mixed process, the sample (SR and hardener) poured into wooden cast and wait for the sample to dry completely. Then, sample has been removed from wood mold and wiped by paper, for the result can be seen in Figure 1.

Figure 1. Bolus from silicone rubber material

2.2. Relative Electron Density

Measurement of relative electron density (RED) on bolus by scanning process using CT-Scan (Thosiba, Japan). The scanning method used axial scanning and the tomography image was send to the computer for determine CT-Number. To calculate RED value from CT-Number using equation as follows:

$$\rho_e = 1.052 + 0.00048N_{CT} \quad (1)$$

$$\rho_e = 1.000 + 0.001N_{CT} \quad (2)$$

with ρ_e is electron density and N_{CT} is CT-Number. If the CT-Number is greater than 100 then use an equation (1) and CT-Number less thann 100 using equation (2) [15].

2.3. Percentage of Surface Dose (PSD)

The measurement of the percentage of surface dose (PSD) on the bolus has been done by giving the electron beam radiation from the Linear Accelerator (LINAC) device (Elekta,Sweden). The energy used are 8 MeV and 10 MeV with applicator field of 10 x 10 cm^2, and source to surface dose (SSD) has been set up in a position of 100 cm. The plan parallel chamber detector (Iba Dosimetry, Sweden) is used to measure electron beam radiation in surface and depth dose maximum position (d_{max}) on solid water phantom shown in Figure 2. For d_{max} value of 8 MeV and 10 MeV is 1.7 cm and 2.2 cm, respectively. The data can read by an electrometer (Iba Dosimetry, Sweden) in units of nanocoulomb (nC). To calculate PSD value using equation as follows:

$$P_A = \frac{D_S}{D_M} \times 100\ \% \quad (3)$$

$$P_B = \frac{D_{SB}}{D_M} \times 100\ \% \quad (4)$$

with P_A is PSD value without bolus, P_B is PSD value using bolus, D_S is radiation dose measured on the surface of solid water phantom without bolus, D_{SB} is radiation dose measured on the surface of solid water phantom using bolus, and D_M is radiation dose measured on a d_{max} position of solid water phantom without using bolus [2].

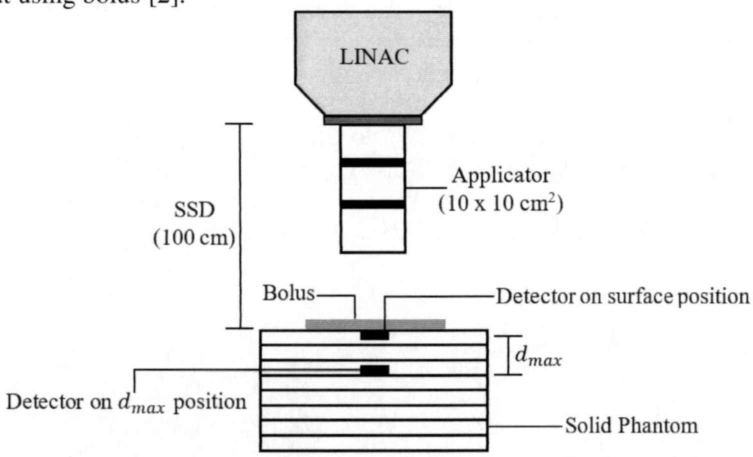

Figure 2. Schematic to measurement percentage of surface dose (PSD)

3. Results and Discussion

3.1. Relative Electron Density (RED)

In this results, the relative electron density (RED) from bolus made of silicone rubber (SR) is 1.176, for the RED value of some tissue can be seen in Tabel 1 below.

Tabel 1. Relative Electron Density of Various Tissues/Materials

Tissue/Material	Relative Electron Density	Reference
Water	1.000	[16]
Muscle	1.043	[16]
Breast	0.976	[17]
Dense Bone	1.512	[17]

Based on Table 1, the RED produced by the bolus is above the RED value of water and soft tissue such as muscle and breast. This is because of different material compositions between soft tissues and boluses. Soft tissue consists of hydrogen (H), carbon (C), oxygen (O), nitrogen (N), phosphor (P), sulphur (S), chlorine (Cl), and potassium (K) [18,19]. On the other hand, bolus with SR material consists of inorganic polymer bond composed by polysiloxan (Si-O) bonds and methyl (CH_3) bonds [20]. When compared with a solid tissue such as dense bone material, the electron density of bolus is below the value of electron density dense bone. If viewed from the composing dense bone material, the dense bone consists of mineral phase, hydroxyapatite ($Ca_{10}(PO_4)_6(OH)_2$), an organic phase, and water [21]. So it has different material density both with bolus material and soft tissue. When compared with alternative materials such as ethyl methacrylate [15] and durian seed [17], RED value is much smaller than SR. This is because the two materials have a material composition similar with water. However, the results of this study support the results of research conducted by Li et al [14] who have used SR as an ingredient in prostate group organs synthesis. Thus, SR can be an alternative material that is equivalent to a tissue.

3.2. Relative Electron Density (RED)

For PSD results on solid phantom surface can be seen in Figure 4. Based on Figure 4, the PSD value without using bolus on 8 MeV and 10 MeV are 89.18% and 90.55%, respectively. At the time of using bolus, the PSD value for 8 MeV and 10 MeV has increased to 102.32% and 101.32%, respectively. These results indicate the use of bolus wtih SR material can provide an increase in PSD value with a percentage above 100%, when using radiation energy greater than 8 MeV, the value of PSD was decreased. Supposedly, a greater radiation energy can increase the value of PSD [1]. This is because the amount of radiation energy can provide an increase in kinetic energy of electron particles so that electrons can penetrate deeper solid water phantom. When compared with elasto gel, the resulting PSD value is 98.2% at 9 MeV with a thickness of 1.25 cm [2]. This result shows that SR with a thickness of 1 cm can provide a higher PSD value than using elasto gel. However, based on the aim of used a bolus, bolus with SR material has been succeeded in increasing PSD value on solid water phantom surface. The use of SR can be an alternative material that can be used as a bolus. For the future, a further research is needed to investigate of varying composition between SR and hardener to measure an RED and PSD for each composition variation.

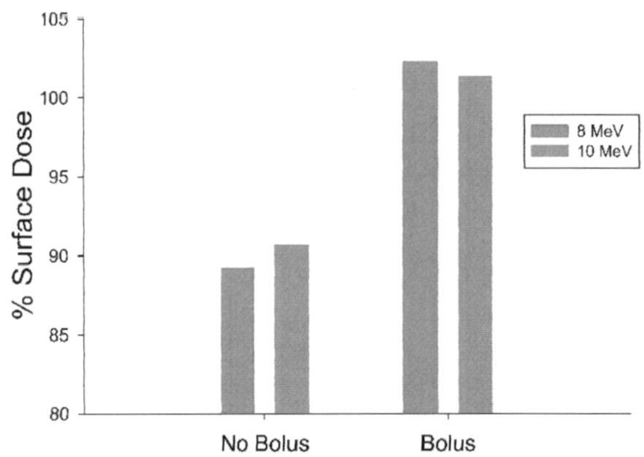

Figure 3. Percentage of surface dose (PSD) graph

4. Conclusion

Based on the results of this research, bolus with silicone rubber material has electron density value of 1.176. These results show the value of RED bolus between soft tissue and solid tissue. The PSD value at 8 MeV and 10 MeV are 102.32% and 101.32%, respectively. These results indicate that the silicone rubber material can be used as an alternative bolus material because it corresponds to the bolus function in radiotherapy.

5. Conclusion

The authors thank to Physics Material Laboratory, Department of Physics, Diponegoro University and Radioteherapy Instalation, RSUP dr. Sardjito Yogyakarta for supporting our research.

References

[1] Podgorsak, E.B. 2005 *Radiation Oncology Physics: A Handbook for Teacher and Students*. Vienna: Internastional Atomic Energy Agency.

[2] Günhan, B., Kemikler, G., and Koca, A. 2003. .Determination of Surface Dose and The Effect of Bolus To Surface Dose In Electron Beams. *Medical Dosimetry* **28**, pp. 193-198.

[3] Demir, B., Okutan, M., Cakir, A., Goksel, E., and Bilge, H. 2008. The Effect of Oblique Electrom Beams to the Surface Dose Under the Bolus. *Medical Dosimetry* **34**, pp. 311-316.

[4] Seppala, T., Collan, J., Auterinen, I., Serén, T., Salli, E., Kotiluoto, P., Kortesniemi, M., Leemput, K.V., Kankaanranta, L., Joensu, H., Savolainen, S. 2004. A dosimetric study on the use of bolus material for treatment of superficial tumors with BNCT. *Applied Radiation and Isotopes* **61**, pp. 787-91.

[5] Chang, F., Chang, P., Benson., K., and Share, F. 1992. Study of Elasto-Gel Pads Used as Surface Bolus Material in High Energy Photon and Electron Therapy. *J. Radiation Oncology. Biology. Physics* **22**, pp. 191-193.

[6] Visscher, S., and Barnett, E. 2016. Comparison of Bolus Material to Highly Absorbent Polypropylene and Crayon Cloth. *Journal of Medical Imaging and Radiation Sciences* **48**, pp. 55-60.

[7] Braun, U., Lorenz, E., Weimann, C., Sturm, H., Karimov, I., Ettl, J., Meier, R., Walter, A., Wohlgemuth., Berger, H., and Wildgruber, M. 2016. Mechanic and surface properties of central-venous port catheters after removal: A comparison of polyurethane and silicon rubber materials. *Journal of The Mechanical Behavior Of Biomedical Material* **64**, pp. 281-291.

[8] Jaya, G.W., and Sutanto, H. 2018. Fabrication and characterization of bolus material using polydimethyl-siloxane. *Material Research Express* **5**, pp. 1-7.

[9] Seyedmehdi, S.A., Zhang, H., and Zhu, J. 2015. Influence of production method, silicone type and thickness on silicon rubber superhydrophobic coatings. *Progress in Organic Coatings* **90**, pp. 291-295.

[10] Zhou, Y., Liu, R., Hou, F., Zhang, Xu., and Xue, W. 2013. Morphology of electrical trees in silicon rubber. *Journal of Electrostatics* **71**, pp. 440-448.

[11] Shaojian, He., Jianbin, Hu., Chong, Z., Jiaqi, W, Lin, C., Xingming, Bi., Jun, L., and Xiaoze, Du. 2018. Performance improvement in nano-alumina filled silicone rubber composite by using vinyl tri-methoxsilane. *Polymer Testing* **67**, pp. 295-301.

[12] Yan, Xu., Qun, Gao., Hongqin, L., and Kangsheng, Z. 2016. Effect of functional graphene oxide on properties of phenyl silicone rubber composite. *Polymer Testing* **54**, pp. 168-175.

[13] Spunei, M., Malaescu, I., Mihai, M., and Marin, C.N. 2014. Absorbing Materials with Apllications In Radiotherapy and Radioproctection. *Radiation Protection Dosimetry* **162**, pp. 167-170.

[14] Li, P., Jiang, S., Yu, Y., Yang, J., and Yang, Z. 2015. Biomaterail characteristics and applicatoion of silicone rubber and PVA hydrogels mimicked in organ groups for prostate brachytherapy. *Journal of The Mechanical Behavior of Biomedical Material* **49**, pp. 220-234.

[15] Montaseri, A., Alinaghizadeh, M., and Mahdavi, S.R. 2012. Physical Properties of Ethyl Methacrylate as a Bolus in Radiotherapy. *Iranian Journal of Medical Physics* **9**, pp. 127-134.

[16] Saw, C.B., Loper, A., Komanduri, K., Combine, T., Huq, S., Scicutella, C. 2005. Determination of Ct-To-Density Conversion Relationship for Image-Based Treatment Planning Systems. *Medical Dosimetry* **30**, pp. 145-148.

[17] Lothfy, F.A., Mustafa, I.S., Yahya, Z.S., Ishak, S.A., and Daud, N. M. 2015. To Study The Durian Seed As A New Substrate For Bolus In Radiotherapy. *Proceedings of Sixteenth The IIER International Conference, Kuala Lumpur, Malaysia*, pp. 110-113.

[18] International Commission on Radiation Units and Measurements (ICRU) Report 44 1989 *Tissue substitutes in radiation dosimetry and measurement* Betheda , Maryland, USA.

[19] Torikoshi, M., Tsunoo, T., Sasaki, M., Endo, M., Noda, Y., Ohno, Y., Kohno, T., Hyodo, K., Uesugi, K., and Yagi, N. 2003. Electron density measurement with dual-energy x-ray CT using

synchrotron radiation. *Physics In Medicine and Biology* **48**, pp. 673-685
[20] Segura, T., and Burillo, G. 2013. Radiation modification of silicone rubber with glycidylmethacrylate. *Radiation Physics and Chemistry* **91**, pp. 101–107
[21] Boskey, A.L. 2013. Bone composition: relations to bone fragility and antiosteoporotic drug effects. *BoneKEy Report* **2**, pp. 1-11

Effect polyethylene glycol (PEG 400) to the physical properties of gadolinium doped cerium ($Ce_{0.9}Gd_{0.1}O_{1.95}$) nanoparticles synthesized by co-precipitation method

Damisih[1], J. Raharjo[1], H. Yuliani[1], A.U. Hapsari[1], Masmui[1], R.D. Pravitasari[1] and Grandevi[2]

[1] Center for Materials Technology, Agency for the Assessment and Application of Technology, Bldg 224, Puspiptek Area, South Tangerang, Indonesia 15314
[2] Department of Physic-Energy Engineering, Surya University, Jl. M.H. Thamrin KM 2.7, Tangerang, Indonesia 15143

*Corresponding author: damisih@bppt.go.id

Abstract. Cerium oxide base materials have been attracting great attention as a promising electrolyte for intermediate temperature of solid oxide fuel cell (IT-SOFC) due to its excellent conductivity at a lower temperature. In this works, cerium from Indonesia local raw material was developed as a cheaper alternative precursor for preparing gadolinium doped cerium ($Ce_{0.9}Gd_{0.1}O_{1.95}$ or GDC10) electrolyte. The effects of polyethylene glycol 400 (PEG 400) as a surfactant on to physical properties of GDC10 electrolyte were studied. GDC10 powders were synthesized using co-precipitation method with the addition of various PEG 400 concentration *i.e* 0,1,2 and 3v/v%. Synthesized powders were characterized by using X-Ray Diffraction (XRD), Particle Size Analyzer (PSA), Scanning Electron Microscopy (SEM), Energy Dispersive Spectroscopy (EDS) and Fourier Transform Infrared (FTIR) Spectroscopy. The XRD analysis indicates that crystallinity was enhanced and all of the peaks on samples correspond to the fluorite crystal structure of single phase CeO_2. The average crystallite size is about 11.37, 7.27, 6.75 and 7.02 nm for PEG 400 concentration of 0, 1, 2 and 3v/v%, respectively. SEM images show different morphology of particle regarding with the addition of surfactant. Particle size analysis exhibits decreasing of particle diameter as the addition of PEG surfactant. The smallest particle size was about 1.47 μm for 1v/v% of PEG concentration. The results of this works confirm that the addition of PEG 400 surfactant strongly affects particle size and morphology of GDC10 powders. However, addition PEG 400 as surfactant should be delivered in a certain amount to give optimum effects where according to this works it is about 1-2v/v%.

Keyword: Gadolinium doped cerium, solid oxide fuel cell, polyethylene glycol, surfactant, co-precipitation

1. Introduction

In recent years, cerium based ceramic oxide materials have been attracting great attention of many researchers. It has many applications such as oxygen sensor, fluorescence material, catalyst, UV absorbents and solid oxide fuel cell (SOFC) [1], [2], [3]. However, the most application of cerium base ceramic oxide was focused on SOFC as solid electrolyte due to its excellent ionic conductivity in the intermediate temperature range of 500-800°C [4]. Gadolinium doped cerium (GDC) denotes as one of

Content from this work may be used under the terms of the Creative Commons Attribution 3.0 licence. Any further distribution of this work must maintain attribution to the author(s) and the title of the work, journal citation and DOI.
Published under licence by IOP Publishing Ltd

cerium based ceramic oxide that excessively used for solid electrolyte. GDC with the composition of 10%mol of Gd dopant as known GDC10 is reported to possess better stability in reducing atmosphere at an intermediate temperature [5]. Moreover, GDC10 is also reported for having higher ionic conductivity than other compositions [6]. From the point of view fuel cell technology, intermediate temperature SOFC has some advantages such as the smaller thermal mismatch between its components and rapid startup with less energy consumption [7].

Some different methods of synthesis and processing have been reported to prepare GDC such as sol-gel [8], sol-combustion [9], hydrothermal [10], solid state [11] and co-precipitation. Among all of them, the co-precipitation method is widely employed due to low preparation cost, relatively simple process and high yield [4]. Furthermore, the wet-chemical method through co-precipitation provides powders with better reactivity as well as purity than solid-state method [12]. It has been reported the addition of PEG as a non-ionic surfactant to synthesize cerium base particle by chemical precipitation. Thus, the results demonstrate that it plays a significant role to prevent particle agglomeration and obtain finer particle. It can strongly bond on the particle surface and produce the particle with smaller size and lower PDI [13]. It is also actively absorbed monolayer of organic or inorganic molecules to prevent agglomeration in nanoparticles synthesis and then stabilize them [14].

Polyethylene glycol 400 (PEG 400) is reported to have low toxicity and soluble in water, acetone, alcohols and benzene [15]. It is a low molecular weight grade of polyethylene glycol and frequently used as the surfactant. The molecular weight of PEG governs the surface attraction or repulsion property [16]. It could be responsible for either inducing aggregation or stabilizing the dispersion. The rate of particle aggregation becomes a major factor that controls the crystallinity and morphology of products. So that, by adjusting the amount of PEG as the surfactant, particle size and morphology of powders can be modified. Some researchers have reported the use of PEG for powders synthesis where it was proved to decrease particle size and improve its notable properties [17], [18], [19], [20]. It is important information of nanoemulsion or microemulsion where it lowers the interfacial tension so that different tension between inside and outside of the droplets is reduced. Hence, stress needed for breaking up a drop is reduced and smaller droplets are formed. Smaller emulsion droplets tend to provide smaller particles. So that why PEG is important for particle synthesis due to its restricting effect of the microemulsion and the PEG molecules at the surface embrace the crystal and control their excess growth [21].

In this works, GDC10 powders were synthesized by co-precipitation methods using local raw materials of cerium precursor from Indonesia. The effects of PEG 400 addition as surfactant particularly its concentration on to properties of GDC10 was investigated.

2. Materials and Methods

2.1. GDC10 synthesis by co-precipitation method

GDC10 powders were synthesized through co-precipitation method as like previously performed by Ramasamy *et al.* [22]. Slight modification on that method was implemented with the addition of gadolinium as a dopant. $Gd(NO_3)_3.6H_2O$ as precursor was commercially purchased from Sigma Aldrich, while $Ce(OH)_4$ was obtained from Center of Science and Accelerator Technology, BATAN, Indonesia. Monazite sand was leached in strong sulfuric acid for 1 hour. It was added with water for quenching and then it was filtered. Ammonia was added into the filtrate so that mixed precipitant of REOH (rare-earth hydroxide) and REP_2O_7 (rare earth pyrophosphate) were formed. REP_2O_7 was mixed with NaOH at 140°C for 1 hour to form REOH. It was filtered and washed until neutral pH was obtained. After that, it was dried in oven at110°C. REOH concentrate was dissolved in nitric acid and heated at 100°C. After oxidation, ammonia was added until pH of 1 where Ce compound including $Ce(OH)_4$ was precipitated. $Ce(OH)_4$ precipitant was calcined at 900°C for 4 hours.

For the synthesis of GDC10 powders, first, a stoichiometric amount of $Ce(OH)_4$ was diluted in 25v/v% of distilled water and then added with 0.1887v/v% of nitric acid (HNO_3 65%) and to form cerium nitrate hexahydrate ($Ce(NO_3)_3.6H_2O$).

Gadolinium nitrate hexahydrate [$Gd(NO_3)_3.6H_2O$] was separately dissolved in 25v/v% of distilled water. Both of precursor solution that completely dissolved in distilled water were mixed and stirred for around 5 minutes. After that, PEG 400 as the surfactant was slowly added with various concentration of 0, 1, 2, 3v/v%. The solution was stirred for 10 minutes until a homogenous solution was obtained.

Then, NH_4OH 32% was added drop by drop to the solution to control pH. Prior to the addition, NH_4OH was dissolved in 25v/v% distilled water. After addition of NH_4OH was completely performed, solution was stirred for 2 hours at room temperature to mix and then precipitate it. The solution was filtered and washed with distilled water and acetone subsequently. Distilled water was used to remove absorbed salt ion on resultant precipitant, while acetone was useful for eliminating absorbed water. The precipitant was dried at a temperature of 120°C for 24 hours to remove water content. It was subsequently powdered in an agate mortar and then the powders produced were calcined at 600°C for 24 hours. During calcination, the surfactant was removed and the yellow colored powder was obtained.

2.2. GD10 powders characterization

The GDC10 powders obtained from synthesis were characterized by means of X-Ray Diffractometer (XRD) (PANalytical EMPYREAN) with Cu-Kα operating at 40 kV and 30 mA. Step size used was 0.0260° 2θ in the range of 10-90° at scan step time of 22.44 s with continuous mode and 0.5° of slit size. Particle size was revealed through Particle Size Analyzer (PSA) (NanoPlus HD) based on dynamic light scattering (DLS). Before measurement, the sample was performed sonication until completely dissolved. Deionized water was used as medium and analysis was carried out for 3 (three) times of repetition. The morphology of GDC10 powders was observed by Scanning Electron Microscope (SEM) (JEOL JSM-6510LA) with energy dispersive spectroscopy (EDS) attachment. The SEM observation didn't perform for all of the samples but only for without PEG 400 and with 3v/v% addition of PEG 400. The Fourier transform infrared (FTIR) spectrum was measured using FTIR spectrometer (Thermo Scientific Nicolet iS50). FTIR measurement was done using at room temperature ranging from wavenumber of 4000 to 500 cm^{-1}. The spectrum was collected at a resolution of 0.964 cm^{-1}.

3. Materials and Methods

Figure 1 shows XRD pattern data of GDC10 prepared by co-precipitation method with different addition of PEG concentration starting of 0, 1, 2 and 3v/v%. XRD analysis was performed after GDC10 powders were calcined at 600°C for 2 hours. Several peaks were found for all of the samples regarding with addition of PEG 400 as the surfactant. The diffraction peaks were indexed to (111), (200), (220), (311), (222), (400), (331), (420) and (422) crystal plane with (111) as a dominant peak. All of the peaks correspond to the fluorite structure of CeO_2 which demonstrated a single phase. This is in good agreement with the powder database from ICSD-98-002-8795. It was also observed no other diffraction peaks beside CeO_2 demonstrating high purity of GDC10 powders were obtained from synthesis. XRD crystalline peak corresponding to Gd_2O_3 wasn't found for all of the samples which indicated well mixing of Ce^{3+} and Gd^{3+} cations in the precursor [23].

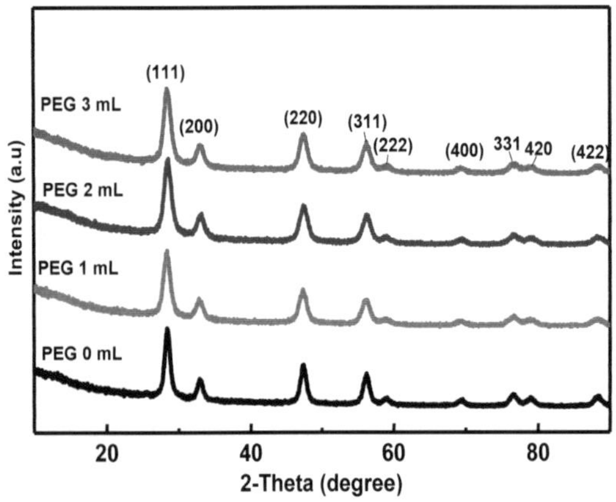

Figure 1. Schematic to measurement percentage of surface dose (PSD)

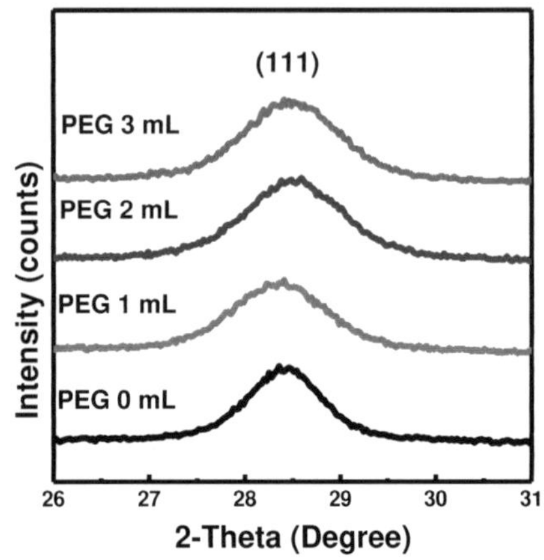

Figure 2. A single peak of (111) XRD pattern

The peak intensity for XRD pattern shows changes, where the intensity of (111) peak for 2v/v% of 400 is higher than other concentration indicating crystallinity is enhanced. This peak becomes the highest among other concentration. It could also indicate the change of grain size possibly happened [24]. In addition, shifting of the peak is also slightly observed to higher angle as shown by Figure 2, which means crystal quality is improved. This might be due to the strain effect. The (111) peak that noted as preferred orientation is seen broadening as a result of the addition of PEG 400 concentration. The peak broadening is attributed to changes in crystallite size as well as chemical composition. Similar results are also reported by other researchers where they discovered if crystallinity enhancement of powders could be obtained by the addition of surfactant [25], [26].

From XRD analysis, the crystallite size that denoted by D is calculated using Scherrer's formula as follows [22]:

$$D = \frac{0.9\,\lambda}{\beta \cos \theta} \tag{1}$$

Where λ is X-ray wavelength (λ = 0.15406 nm), β is full width at half maximum (FWHM) and θ expresses Bragg's diffraction angle. The estimated crystallite size was calculated for all of the peaks found and then taken its average values, as shown in Table 1 and Figure 3.

Tabel 1. The average crystallite size of GDC10 powders with different PEG 400 concentration

PEG Conc.	2θ (deg.)	Planes	FWHM	D (nm)	D Average (nm)
0v/v%	28.39	(111)	0.902	9.49	11.37
	32.90	(200)	0.880	9.83	
	47.09	(220)	0.610	14.83	
	55.84	(311)	0.500	18.78	
	58.94	(222)	0.950	10.03	
	69.26	(400)	1.080	9.34	
	76.54	(331)	1.220	8.66	
	78.90	(420)	1.250	8.60	
	88.00	(422)	0.900	12.82	

IOP Conf. Series: Materials Science and Engineering **622** (2019) 012003 doi:10.1088/1757-899X/622/1/012003

1v/v%	28.41 32.86 47.27 55.98 58.91 69.21 76.50 78.90 88.26	(111) (200) (220) (311) (222) (400) (331) (420) (422)	1.08 1.14 1.30 1.11 1.35 1.30 1.54 1.64 1.84	7.92 7.59 6.96 8.46 7.06 7.75 6.86 6.55 6.28	7.27
2v/v%	28.49 33.02 47.40 56.27 59.03 69.31 76.62 78.99 88.36	(111) (200) (220) (311) (222) (400) (331) (420) (422)	1.21 1.18 1.37 1.37 1.33 1.51 1.63 1.67 1.89	7.07 7.33 6.61 6.87 7.17 6.68 6.49 6.44 6.12	6.75
3v/v%	28.28 32.92 47.38 56.25 58.99 69.30 76.60 78.94 88.33	(111) (200) (220) (311) (222) (400) (331) (420) (422)	1.01 1.05 1.31 1.36 1.47 1.44 1.63 1.64 1.89	8.47 8.24 6.91 6.92 6.48 7.00 6.49 6.55 6.12	7.02

The average crystallite size is about 11.37, 7.27, 6.75 and 7.02 nm for PEG 400 concentration of 0, 1, 2 and 3v/v%, respectively. With the addition of 1v/v% of PEG 400 surfactant, the crystallite size clearly decreases from 11.37 nm to 7.27 nm. It further decreases with the addition of 2v/v% PEG became 6.75 nm and noted as the lowest average crystallite size obtained in this work. After 3v/v% of PEG 400 was added, it slightly increased to be 7.02 nm. It is noted that the addition of a small amount of PEG 400 surfactant decreases the crystallite size of powders. From these results, it can be said the optimum addition of PEG 400 as surfactant on the co-precipitation method for GDC10 synthesis is around 2v/v%.

The lattice constant ($\boldsymbol{a}$) is estimated by using the following equation [22]:

$$\boldsymbol{a} = \mathrm{d_{hkl}}\sqrt{\mathrm{h}^2 + \mathrm{k}^2 + \mathrm{l}^2} \tag{2}$$

Where d is d-spacing and *hkl* is the Miller indices. Meanwhile lattice strain ($\boldsymbol{\varepsilon}$) and dislocation density are obtained with formulas below [22]:

$$\varepsilon = \frac{\beta \cos \theta}{4} \tag{3}$$

$$\delta = \frac{1}{\mathrm{D}^2} \tag{4}$$

The lattice constant, lattice strain and dislocation density are tabulated on Table 2 and shown in Figure 3.

Tabel 2. The lattice constant and lattice strain of GDC10 powders with different PEG 400 concentration

PEG 400 Conc. [v/v%]	Lattice constant (a) [nm]	Micro strain (ε) x 10^{16}	Dislocation density (δ) x 10^{16} [lines/m^2]
0	1.000	1.9460	0.9247
1	1.043	2.8760	2.2159
2	0.976	3.0804	2.2131
3	1.512	2.9862	2.0940

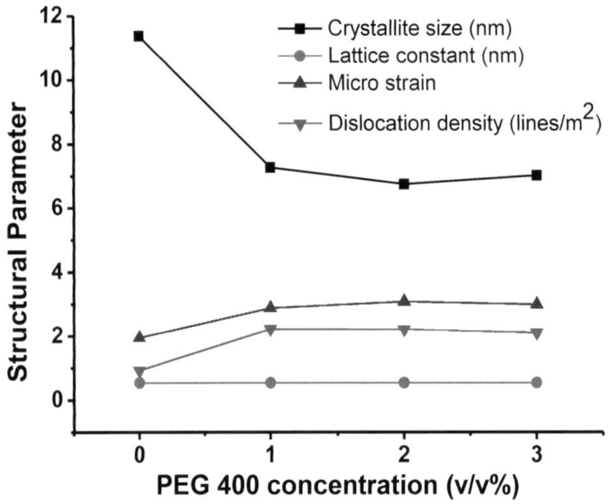

Figure 3. Schematic to measurement percentage of surface dose (PSD)

The average lattice constant for PEG 400 concentration of 0, 1, 2, and 3v/v% is 0.5435, 0.5431, 0.5418 and 0.5427 nm, respectively. The pure CeO_2 is recorded to have theoretical lattice parameter of about 0.5411 nm [24]. However, lattice constant value found in this work is higher than pure CeO_2 because of the presence of Gd^{3+} that replacing Ce^{4+} lattice in the fluorite structure of CeO_2 [12]. With the addition of PEG 400 surfactant, the lattice constant is observed to decrease and reach the lowest value at 2v/v% of PEG 400. After that, it slightly increases. On the other hand, microstrain is noted to gradually increase as PEG 400 concentration an increase. Similarly, dislocation density also tends to increase with surfactant addition.

The particle size distributions of GDC10 powders regardless with the addition of PEG 400 as a surfactant at different concentration are shown by Figure 4. Meanwhile, Table 3 shows the average diameter and polydispersity index (PDI) number. As seen from Figure 4, the addition of PEG 400 obviously affects particle size distribution as well as particle size. Without the addition of PEG 400, multiple peaks were initially observed on the particle size distribution graph, as shown by Figure 4(a). After the addition of PEG 400 as the surfactant, it became only a single peak appear for 1v/v% and 2v/v% concentration where PDI number was also noted to decrease. The presence of a single definite peak suggests the monodispersed distribution of particle [27]. PDI number was 1.154 for the sample without PEG 400 addition, and it decreased to 0.716 after addition of 1v/v% of PEG 400. But then, it slightly increased as the value of 0.823 and 0.918 for 2v/v% and 3v/v% of PEG 400, respectively. However, these numbers are still much lower compared to sample with no addition of PEG 400. PDI is used to describe the non-uniformity degree of size distribution [27]. The lowering of PDI number after addition of PEG 400 could be related with the homogeneity of particle where it became more uniform in size. When $Ce(NO_3)_3.6H_2O$ and $Gd(NO_3)_3.6H_2O$ were dissolved in water, microemulsion was formed. PEG surfactant molecules played a role in creating the interfacial layer which separates the aqueous layer and the inorganic phases. It also reduced the interfacial tension between the microemulsion and the excess phase and act as a steric barrier layer for preventing the coalescence of droplets [28]. This would allow particle size to be controlled just by controlling the droplet size where smaller droplets would produce the smaller size of the particle.

With the addition of PEG 400 surfactant at a certain concentration, the average diameter of GDC powders was reduced. All samples regardless of surfactant addition demonstrate lower particle size diameter compared to sample without surfactant addition. Yao *et al.* mentioned that indention-like amphiphilic molecular of PEG 400 could provide a steric hindrance effect with hydrophobic ends adsorbed onto particle surface [4]. As a result, the neighboring particle is prevented from getting close to each other. It means re-agglomerated is also prevented so that particle size becomes smaller. However, as seen in Table 3, particle size decrease until reach optimum value and then increase again. Here, the smallest particle size was about 1.84 μm for 1v/v% of PEG 400 concentration. This result matched well with crystallite size calculation from XRD result even though the smallest crystallite size was given by a sample with 2v/v% addition of PEG 400. It was acceptable since the term of crystallite size is considered for individual crystalline calculation, and particle is constructed from several different crystallites.

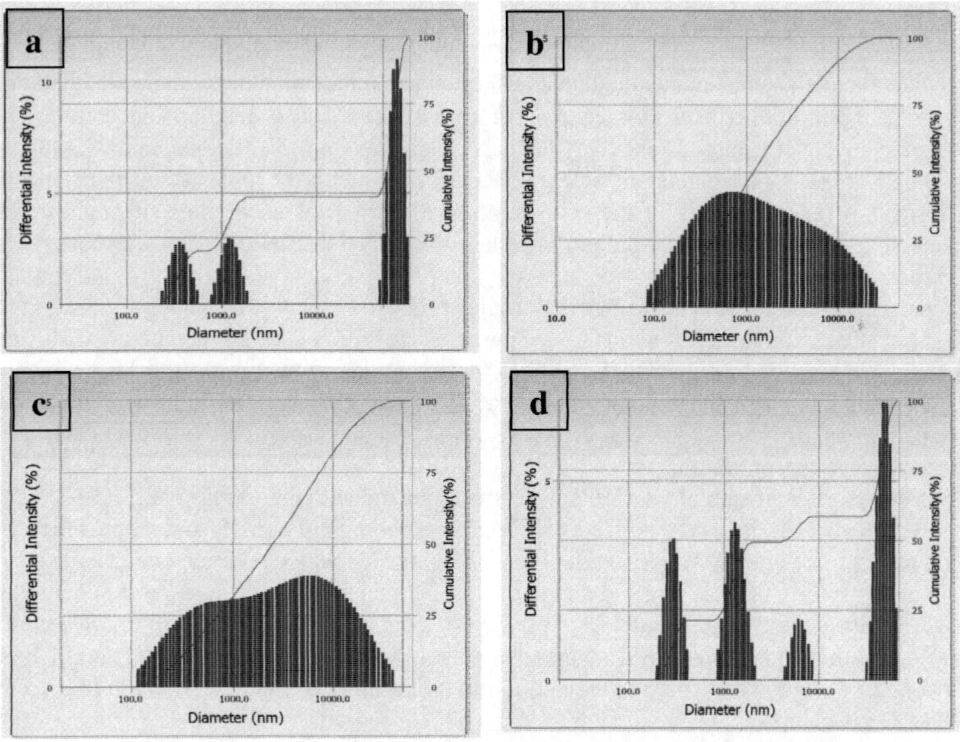

Figure 4. . The particle size distribution of GDC10 powder with different concentration of PEG 400 addition (a) 0v/v% (b) 1v/v% (c) 2v/v% (d) 3v/v%

Yao *et al.* reported that de-agglomeration treatment to the precursor by both ultrasonic and surfactant was effective to improve particle size distribution of doped-ceria nanoparticle synthesized by co-precipitation route [4]. PEG 400 was also used as a surfactant, however, with higher the concentration of 15% as optimal dosage. In this case, a small amount of PEG 400 was used and it demonstrated enough for providing particle size reduction effects.

It was also reported by Ketzial *et al.* the effect of surfactant on the particle size distribution of CeO_2 by chemical precipitation [3]. By the addition of 1v/v% PEG 400, the particle size of powder was in the range 0.510 – 39.23 µm with 53% of a particle smaller than 13 µm and average diameter of 12.8 µm. Furthermore, particle size in the range of 0.339-51.47 µm was observed for 2v/v% of PEG 400 addition with 60% of particle smaller than 13 µm and average diameter of 9.9 µm. Our results exhibited smaller value where the average diameter of 1v/v% of PEG 400 was 1.47 µm and it was 1.84 µm for 2v/v%.

The effect of PEG as the surfactant was also studied by Cholan *et al.* on cerium doped nanoparticle ZnS prepared by precipitation method [18]. They suggested 2.5% of PEG as optimum concentration to control particle size. Introduction of PEG into during synthesis process will not only stabilize small particle but also restrict agglomeration. It was observed in this work that at 3v/v% of PEG 400, particle size somewhat increased suggesting that there was a certain amount of surfactant addition to give optimum effect on particle size reduction.

Tabel 3. The average diameter and polydispersity index (PDI) of GDC10 powders with different PEG 400 concentration

PEG 400 concentration (v/v%)	Average Diameter (nm)	Polydispersity index (PDI)
0	2410	1.154
1	1472	0.716
2	1842	0.823
3	2015	0.918

Surfactant concentration affects the size of particle related to Critical Micellar Concentration (CMC) value. If surfactant concentration is above CMC, particle size will be smaller than below CMC. Particle size has to be increased with higher surfactant concentration due to the presence of more micelles for particle formation [18]. However, if the concentration of surfactant is too high, it might destabilize micelle and then lead particle to grow bigger. Micelle is a colloidal dispersion that spontaneously formed at certain surfactant concentration.

Surfactant concentration affects the size of particle related to Critical Micellar Concentration (CMC) value. If surfactant concentration is above CMC, particle size will be smaller than below CMC. Particle size has to be increased with higher surfactant concentration due to the presence of more micelles for particle formation [18]. However, if the concentration of surfactant is too high, it might destabilize micelle and then lead particle to grow bigger. Micelle is a colloidal dispersion that spontaneously formed at certain surfactant concentration.

Figure 5 shows SEM images of GDC10 powders without and with the addition of PEG 400 surfactant at 500X magnification. It definitely indicates different morphology of particle and shape where without surfactant, particles are likely to agglomerate. It is also observed that the addition of PEG 400 surfactant resulted in more uniform particle size and distribution compared with no surfactant was added, as shown in Figure 5(b). This confirms the results from particle size analysis and also crystallite calculation from XRD data. It should be noted that SEM images show aggregated particles which made up of crystallite as a primary particle with an average size of 11.37 nm (without PEG 400) and 7.02 nm (with 3v/v% of PEG 400 as a surfactant).

Figure 5. SEM images of GDC10 powders synthesized by co-precipitation method at 500X magnification (a) without addition PEG 400 (b) with the addition of 3v/v% of PEG 400

Figure 6. SEM images of GDC10 powders synthesized by co-precipitation method at 5000X magnification (a) without the addition PEG 400 (b) with the addition of 3v/v% of PEG 400

The use of surfactants has been many reported having a strong effect on morphology of synthesized metal oxide powders and nanoparticles. Erten-Ela *at al.* reported the presence of PEG 400 led ZnO to form rod-like structure and without PEG 400 flower-like structure was revealed [29]. Cholan *et al.* also observed flower-like nanostructure of cerium doped ZnS without PEG, while spherical nanoparticles were found mixed with some rod-like structure [18].

At a higher magnification of SEM images, shape and structure of particle are clearly observed of difference as shown by Figure 6. The irregular shape of the particle was revealed for the sample without PEG 400, whereas with the addition of PEG 400 as a surfactant resulted in sub-angular particle shape resembling rock like grain. The change of particle shape can be related to surfactant role that it acts as a capping agent and a strong influence on the formed particle [15]. By the addition of PEG 400, the growth rate of various crystallographic is controlled and orientation in crystal formation is created. As a result, more uniform shape and size of particles could be obtained.

Elemental analysis of GDC10 powders was also performed by using EDS and the result is shown in Table 4 and Figure 7. It was detected only oxygen (O), cerium (Ce) and gadolinium (Gd) element suggesting that synthesized product was gadolinium doped cerium (GDC10). As expected, no other elements were detected indicating high purity of GDC10 powders.

4. Results and Discussion

4.1. Relative Electron Density (RED)

In this results, the relative electron density (RED) from bolus made of silicone rubber (SR) is 1.176, for the RED value of some tissue can be seen in Tabel 1 below.

Tabel 4. Relative Electron Density of Various Tissues/Materials

Tissue/Material	Relative Electron Density	Reference
Water	1.000	[16]
Muscle	1.043	[16]
Breast	0.976	[17]
Dense Bone	1.512	[17]

Based on Table 1, the RED produced by the bolus is above the RED value of water and soft tissue such as muscle and breast. This is because of different material compositions between soft tissues and boluses. Soft tissue consists of hydrogen (H), carbon (C), oxygen (O), nitrogen (N), phosphor (P), sulphur (S), chlorine (Cl), and potassium (K) [18,19]. On the other hand, bolus with SR material consists of inorganic polymer bond composed by polysiloxan (Si-O) bonds and methyl (CH_3) bonds [20]. When compared with a solid tissue such as dense bone material, the electron density of bolus is below the value of electron density dense bone. If viewed from the composing dense bone material, the dense bone consists of mineral phase, hydroxyapatite ($Ca_{10}(PO_4)_6(OH)_2$), an organic phase, and water [21]. So it has different material density both with bolus material and soft tissue. When compared with alternative materials such as ethyl methacrylate [15] and durian seed [17], RED value is much smaller than SR. This is because the two materials have a material composition similar with water. However, the results of this study support the results of research conducted by Li et al [14] who have used SR as an ingredient in prostate group organs synthesis. Thus, SR can be an alternative material that is equivalent to a tissue.

4.2. Relative Electron Density (RED)

For PSD results on solid phantom surface can be seen in Figure 4. Based on Figure 4, the PSD value without using bolus on 8 MeV and 10 MeV are 89.18% and 90.55%, respectively. At the time of using bolus, the PSD value for 8 MeV and 10 MeV has increased to 102.32% and 101.32%, respectively. These results indicate the use of bolus wtih SR material can provide an increase in PSD value with a percentage above 100%, when using radiation energy greater than 8 MeV, the value of PSD was decreased. Supposedly, a greater radiation energy can increase the value of PSD [1]. This is because the amount of radiation energy can provide an increase in kinetic energy of electron particles so that electrons can penetrate deeper solid water phantom. When compared with elasto gel, the resulting PSD value is 98.2% at 9 MeV with a thickness of 1.25 cm [2]. This result shows that SR with a thickness of 1 cm can provide a higher PSD value than using elasto gel. However, based on the aim of used a bolus, bolus with SR material has been succeeded in increasing PSD value on solid water phantom surface. The use of SR can be an alternative material that can be used as a bolus. For the future, a further research is needed to investigate of varying composition between SR and hardener to measure an RED and PSD for each composition variation.

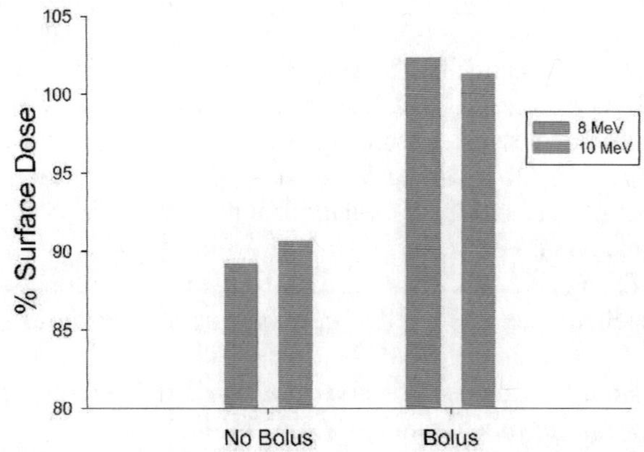

Figure 7. Percentage of surface dose (PSD) graph

5. Conclusion

Based on the results of this research, bolus with silicone rubber material has electron density value of 1.176. These results show the value of RED bolus between soft tissue and solid tissue. The PSD value at 8 MeV and 10 MeV are 102.32% and 101.32%, respectively. These results indicate that the silicone rubber material can be used as an alternative bolus material because it corresponds to the bolus function in radiotherapy.

6. Conclusion

The authors thank to Physics Material Laboratory, Department of Physics, Diponegoro University and Radioteherapy Instalation, RSUP dr. Sardjito Yogyakarta for supporting our research.

References

[1] Podgorsak, E.B. 2005 *Radiation Oncology Physics: A Handbook for Teacher and Students*. Vienna: Internastional Atomic Energy Agency.

[2] Günhan, B., Kemikler, G., and Koca, A. 2003. .Determination of Surface Dose and The Effect of Bolus To Surface Dose In Electron Beams. *Medical Dosimetry* **28**, pp. 193-198.

[3] Demir, B., Okutan, M., Cakir, A., Goksel, E., and Bilge, H. 2008. The Effect of Oblique Electrom Beams to the Surface Dose Under the Bolus. *Medical Dosimetry* **34**, pp. 311-316.

[4] Seppala, T., Collan, J., Auterinen, I., Serén, T., Salli, E., Kotiluoto, P., Kortesniemi, M., Leemput, K.V., Kankaanranta, L., Joensu, H., Savolainen, S. 2004. A dosimetric study on the use of bolus material for treatment of superficial tumors with BNCT. *Applied Radiation and Isotopes* **61**, pp. 787-91.

[5] Chang, F., Chang, P., Benson., K., and Share, F. 1992. Study of Elasto-Gel Pads Used as Surface Bolus Material in High Energy Photon and Electron Therapy. *J. Radiation Oncology. Biology. Physics* **22**, pp. 191-193.

[6] Visscher, S., and Barnett, E. 2016. Comparison of Bolus Material to Highly Absorbent Polypropylene and Crayon Cloth. *Journal of Medical Imaging and Radiation Sciences* **48**, pp. 55-60.

[7] Braun, U., Lorenz, E., Weimann, C., Sturm, H., Karimov, I., Ettl, J., Meier, R., Walter, A., Wohlgemuth., Berger, H., and Wildgruber, M. 2016. Mechanic and surface properties of central-venous port catheters after removal: A comparison of polyurethane and silicon rubber materials. *Journal of The Mechanical Behavior Of Biomedical Material* **64**, pp. 281-291.

[8] Jaya, G.W., and Sutanto, H. 2018. Fabrication and characterization of bolus material using polydimethyl-siloxane. *Material Research Express* **5**, pp. 1-7.

[9] Seyedmehdi, S.A., Zhang, H., and Zhu, J. 2015. Influence of production method, silicone type and thickness on silicon rubber superhydrophobic coatings. *Progress in Organic Coatings* **90**, pp. 291-295.

[10] Zhou, Y., Liu, R., Hou, F., Zhang, Xu., and Xue, W. 2013. Morphology of electrical trees in silicon rubber. *Journal of Electrostatics* **71**, pp. 440-448.

[11] Shaojian, He., Jianbin, Hu., Chong, Z., Jiaqi, W, Lin, C., Xingming, Bi., Jun, L., and Xiaoze, Du. 2018. Performance improvement in nano-alumina filled silicone rubber composite by using vinyl tri-methoxsilane. *Polymer Testing* **67**, pp. 295-301.

[12] Yan, Xu., Qun, Gao., Hongqin, L., and Kangsheng, Z. 2016. Effect of functional graphene oxide on properties of phenyl silicone rubber composite. *Polymer Testing* **54**, pp. 168-175.

[13] Spunei, M., Malaescu, I., Mihai, M., and Marin, C.N. 2014. Absorbing Materials with Apllications In Radiotherapy and Radioproctection. *Radiation Protection Dosimetry* **162**, pp. 167-170.

[14] Li, P., Jiang, S., Yu, Y., Yang, J., and Yang, Z. 2015. Biomaterail characteristics and applicatoion of silicone rubber and PVA hydrogels mimicked in organ groups for prostate brachytherapy. *Journal of The Mechanical Behavior of Biomedical Material* **49**, pp. 220-234.

[15] Montaseri, A., Alinaghizadeh, M., and Mahdavi, S.R. 2012. Physical Properties of Ethyl Methacrylate as a Bolus in Radiotherapy. *Iranian Journal of Medical Physics* **9**, pp. 127-134.

[16] Saw, C.B., Loper, A., Komanduri, K., Combine, T., Huq, S., Scicutella, C. 2005. Determination of Ct-To-Density Conversion Relationship for Image-Based Treatment Planning Systems. *Medical Dosimetry* **30**, pp. 145-148.

[17] Lothfy, F.A., Mustafa, I.S., Yahya, Z.S., Ishak, S.A., and Daud, N. M. 2015. To Study The Durian Seed As A New Substrate For Bolus In Radiotherapy. *Proceedings of Sixteenth The IIER International Conference, Kuala Lumpur, Malaysia*, pp. 110-113.

[18] International Commission on Radiation Units and Measurements (ICRU) Report 44 1989 *Tissue substitutes in radiation dosimetry and measurement* Betheda , Maryland, USA.

[19] Torikoshi, M., Tsunoo, T., Sasaki, M., Endo, M., Noda, Y., Ohno, Y., Kohno, T., Hyodo, K., Uesugi, K., and Yagi, N. 2003. Electron density measurement with dual-energy x-ray CT using

synchrotron radiation. *Physics In Medicine and Biology* **48**, pp. 673-685
[20] Segura, T., and Burillo, G. 2013. Radiation modification of silicone rubber with glycidylmethacrylate. *Radiation Physics and Chemistry* **91**, pp. 101–107
[21] Boskey, A.L. 2013. Bone composition: relations to bone fragility and antiosteoporotic drug effects. *BoneKEy Report* **2**, pp. 1-11

Metamaterial Antenna on Electro-Optic Modulator for Wireless Terra-Hertz Detection Through Radio-Over-Fibre Technology

Yusuf Nur Wijayanto[1], Ashif Aminulloh Fathnan[1,2], Atsushi Kanno[3], Dadin Mahmudin[1], and Pamungkas Daud[1]

[1] Research Centre for Electronics and Telecommunication, Indonesian Institute of Sciences, Bandung, Indonesia
[2] School of Engineering and Information Technology, University of New South Wales Canberra, Australia
[3] National Institute of Information and Communication Technology (NICT), Tokyo, Japan

*Corresponding author: yusuf.nur.wijayanto@lipi.go.id

Abstract. We propose a new metamaterial antenna on electro-optic (EO) modulator for wireless terra-hertz detection through radio-over-fibre (ROF) technology. By wireless terra-hertz signal irradiation to the proposed device, strong terra-hertz electric field can be induced on the electric-LC metamaterial resonator. The induced terra-hertz electric field can be used for optical modulation through EO effects when a light-wave propagates into an optical waveguide located under the capacitive gap which the strongest induced terra-hertz electric field. Analysis of optical modulation is presented in details for operational frequency of 0.1 THz. The device fabrication process and the results of its measured characteristics are also reported.

Keywords: metamaterial antenna, electro-optic modulator, terra-hertz, radio-over-fibre

1. Introduction

Recently, wireless radio-wave are used widely with their advantages of easy in operation for many applications such us communication, sensing, imaging, and so on. In communication application based on Cisco's report, global mobile data traffic will increase sevenfold between 2016 and 2021. Mobile data traffic will grow at a compound annual growth rate (CAGR) of 47 percent from 2016 to 2021, reaching 49.0 exabytes per month by 2021 [1]. Therefore, high-speed communication is required in the future to avoid traffic bottleneck. In sensing application, users always require precise and accurate measurement to make decision quickly [2]. In imaging application, high-resolution image are required and easy to analyse [3]. In order to meet the requirements, wireless radio-wave with high frequency operation is the candidate to anticipate the future problems.

Radio-wave has widely frequency operation bands include terra-hertz bands. The terra-hertz has a frequency ranges in the electromagnetic spectrum which in between microwave and infrared [4]. Terra-hertz wave has large propagation loss in air as shown in Fig. 1 due to environmental condition such as

Content from this work may be used under the terms of the Creative Commons Attribution 3.0 licence. Any further distribution of this work must maintain attribution to the author(s) and the title of the work, journal citation and DOI.
Published under licence by IOP Publishing Ltd

humidity, rain, snow, smoke, fog, an soon [5]. The propagation loss characteristics of the terra-hertz is also depending on specific operational frequency, some specific frequency in terra-hertz bands has low propagation loss compare the others. Therefore, the specific low propagation loss terra-hertz wave can be considered for communication applications. Furthermore, it can be used for specific short area in applications by considering the drawbacks. The terra-hertz bands are promising for high-speed communication since large bandwidth characteristics and high-resolution imaging with small wavelength operation. In wireless terra-hertz, an important device is a detector of the terra-hertz wave. It can be realized by using an antenna with small size since the terra-hertz wavelength is small.

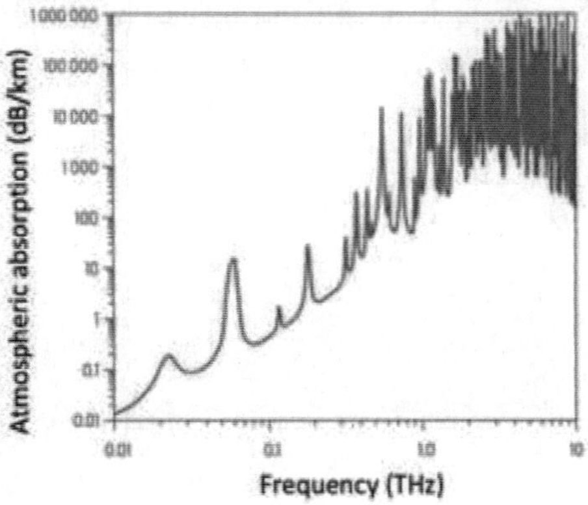

Figure 1. Wireless terra-hertz propagation loss in the air.

The main drawback of the wireless terra-hertz is large propagation loss in air. In order to overcome the drawback, optical fibre networks can be used to connect between the wireless terra-hertz cells by adopting Radio-Over-Fibre (ROF) technology [6,7]. As we know, optical fibre has extremely low propagation loss for carrying light-wave in several kilometre length. As a result, coverage areas of wireless terra-hertz can be improved furthermore. In this configuration requires a device to convert between wireless terra-hertz and light-wave.

Conversion devices using an optical modulator with an antenna for converting wireless radio-wave to light-wave were proposed and reported [8,9]. The conversion devices using discrete, integrated, and fusion configurations with patch antenna and planar Yagi antenna were realized [10-13]. The realized conversion devices were successfully characterized. Basic operations of the conversion devices for converting wireless radio-wave to light-wave were performed. The reported conversion devices are operated in microwave bands by using an antenna with quite large size. In terra-hertz bands, a small antenna is required to detect the wireless terra-hertz wave.

In this paper, we propose and discuss a conversion device from wireless terra-hertz wave to light-wave using metamaterial antenna on electro-optic (EO) modulator for wireless terra-hertz detection through ROF technology. The proposed conversion device is composed of a straight optical waveguide and metamaterial structure with electric-field-coupled resonators on an EO crystal. Basically, the resonators consist of planar inductance and capacitance for generating strong terra-hertz electric field. EO modulation using the induced electric field can be obtained. Meandering-gaps are also adopted to compensate for optical modulation degradation due to transit time effects. Based on that, the conversion device from wireless terra-hertz wave to light-wave can be obtained effectively.

In the following sections, we will discuss the proposed conversion device structure and analyse its characteristics in detail for 0.1 THz operational frequency. The experiments and measurements of the fabricated prototype conversion device are also presented.

2. Device Structure

Figure 2 shows the proposed conversion device from wireless terra-hertz wave to light-wave using metamaterial antennas on an electro-optic modulator. The proposed conversion device is fabricated on a substrate with an EO crystal such as LiNbO3, LiTaO3, EO polymer, and so on. The proposed conversion device is composed of a straight optical waveguide and planar antennas with metamaterial structures. The optical waveguide is fabricated using titanium diffusion method on the EO substrate. The metamaterial antennas are composed mainly with inductive-capacitive (LC) circuits in planar structures. By considering values of inductance and capacitance in the circuits, resonant operational frequency can be designed [14]. The metamaterial antennas using an array of electric-LC planar resonators is fabricated on the EO substrate using a metal material. A buffer layer with thin silicon dioxide (SiO2) film is inserted between the metal antennas and EO substrate for avoiding large optical propagation loss and protection. There is no ground metal electrode on the reverse side of the EO substrate. Additionally, polarization reversal with meandering gaps is also considered in this proposed conversion device for effective operation.

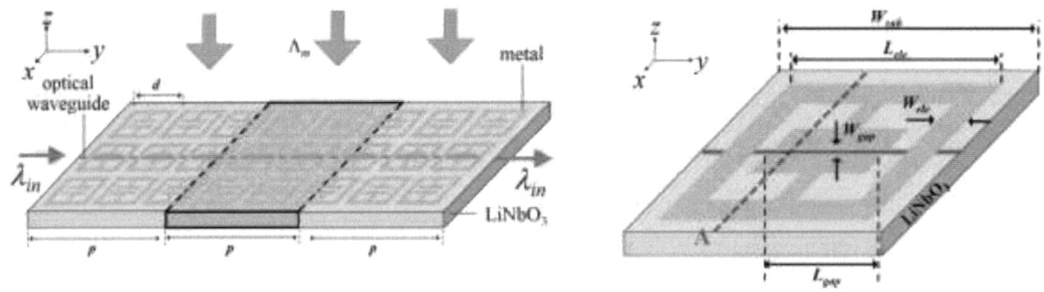

(a) (b)

Figure 2. Basic structure of the metamaterial antenna on optical modulator: (a) whole structure and (b) a unit cell of the metamaterial structure.

The principle operation of the proposed conversion device for detecting and converting wireless terra-hertz wave to light-wave is discussed. When wireless terra-hertz wave is irradiated to the proposed conversion device. The LC circuits of the metamaterial antennas will be interacted. Effective performance can be achieved at the resonant operational frequency by considering parameters of inductance (L) and capacitance (C). As a result, effective reflection and transmission characteristics of the resonator can be obtained. Therefore, large terra-hertz electric field is induced along the capacitive gaps. This terra-hertz electric field can be used for EO modulation. When light-wave propagates to the optical waveguide located along to the capacitive gaps, the interaction between light-wave and terra-hertz electric fields can be occurred. During the light-wave propagation, the light-wave and terra-hertz electric field interaction should be controlled due to transit-time effect [15]. Effective EO modulation can be obtained by considering the light and terra-hertz electric field interaction. Therefore, the proposed conversion device can be used for detecting wireless terra-hertz wave and converting it to light-wave. The detail analysis of those phenomena will be discussed in next section.

3. Design and Analysis

3.1. Microwave Analysis

The characteristics of the proposed conversion device in the terra-hertz frequency operation were analysed in detail by use of electromagnetic field analysis software. To understand the work of metamaterial electric-LC resonator as an antenna receiver for EO modulators, we show the simulated scattering parameters result of the structure in Fig. 3, with the corresponding design parameters are shown in Table 1. It is clearly seen that when a transverse electric (TE) plane wave is irradiated to the structure, a high transmission at specific frequency is detected, in which the EO modulation is expected to work. This transmission is a result of pure electric resonance due to the fact that the capacitive element of the structure corresponds with parameter of the capacitor gap width couples strongly to the incident electric field.

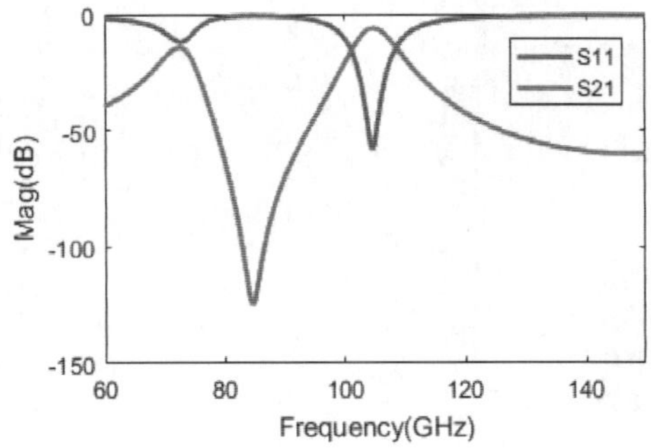

Figure 3. S-parameters result of simulated Electric-LC Resonators of the metamaterial antenna.

Furthermore, the symmetry configuration of the resonators revokes the collective magnetic flux, making the structure theoretically having neither magnetic nor magneto-electric responses [16,17]. For the purpose of modulation using Pockels effects, this strong electric resonance without magnetic involvement is most enticing to realize, due to the fact that only local electric fields are needed in changing refractive index of the optical crystal.

Table 1. Design Parameters of the proposed conversion device

Parameters	Value (μm)
Substrate Width (W_{sub})	210
Electrode Length (L_{ele})	190
Electrode Width (W_{ele})	30
Capacitor Gap (g)	10
Capacitor Length (L_{gap})	110
Substrate Depth (s)	200

We further analyzed the proposed structure using 3D electromagnetic simulation and observe that the electric field at the capacitor gap is the higher field across the entire structure. As shown in Fig. 4, the electric field is captured in 2D (x-y and x-z plane). The strong electric field confinements between the capacitor gaps, resulted at the resonance frequency, are the main mechanism to gain modulation effect onto LiNbO3 crystal through the Pockels effects. In Fig. 5, 1D cut of x-y plane at z = 100 μm just below the capacitor gap (y = 0) is also captured, showing the electric field profile across the x-axis.

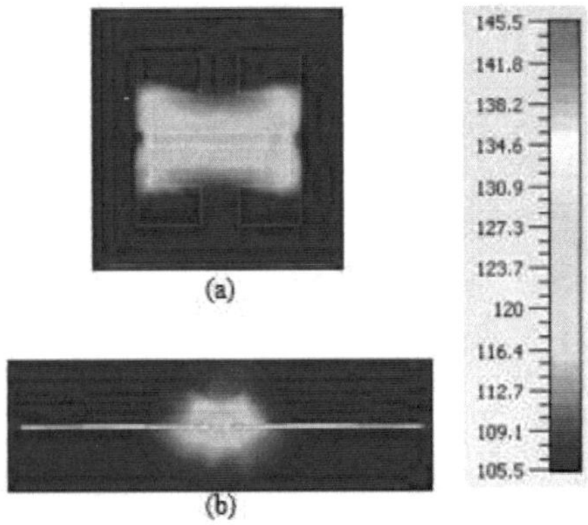

Figure 4. Electric field distributions (V/m) on the simulated electric-LC resonator at (a) x-y plane (z = 100 μm or just below the buffer layer), (b) x-z plane (y = 0 μm or at the centre of the structure).

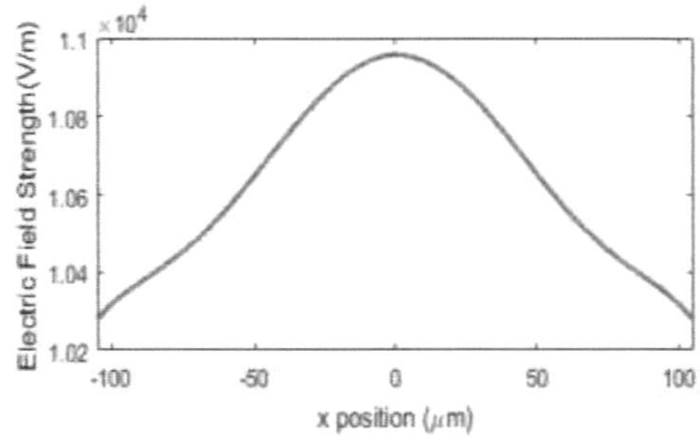

Figure 5. Electric profile along x position where y = 0 μm, z = 100 μm.

In order to properly design the frequency operation of the EO modulator, we use a qualitative approach to model the electric-LC resonator by its equivalent circuit. As shown in Fig. 6(a), the electric-LC resonator is described as capacitive element of plate with gap equivalent to a capacitor, connected in parallel to two loops inductive element, equivalent to inductors. This configuration leads to an electric field driven LC-resonance, it can be expressed as following equation.

$$f_0 = \frac{1}{2\pi\sqrt{\mathrm{LC}}} \tag{3.1}$$

Accordingly, the resonance frequency or working range of optical modulation can be tuned by ariation to the design parameters. Particularly, by only investigating the role of capacitive element, an effective adaptation to the working frequency can be done by altering the value of L_{gap} and W_{gap}. Adding the value of the width of capacitor gap (W_{gap}) can lower its capacitance, thus making the resonance frequency higher. Contrarily, adding the value of its length (L_{gap}) can increase its capacitance thus decreases resonance frequency. The corresponding graph can be seen from Fig. 6(b).

In the proposed electric-LC resonator, the terra-hertz electric field across the capacitive gap can be expressed as

$$E_{THz} = E_0 \sin(\omega_{THz} t) \tag{3.2}$$

Where, ω_{THz} is the wireless terra-hertz signal angular frequency.

3.2. Microwave Analysis

The EO modulation characteristics driven by the terra-hertz electric fields across the capacitive gaps on the metamaterial antennas are discussed also. In order to calculate conversion efficiency (sensitivity) of the proposed conversion device, the terra-hertz electric fields observed by the light-wave propagating in the optical waveguides should be considered by EO effects and by taking into account of the transit-time effect [15].

IOP Conf. Series: Materials Science and Engineering **622** (2019) 012004 doi:10.1088/1757-899X/622/1/012004

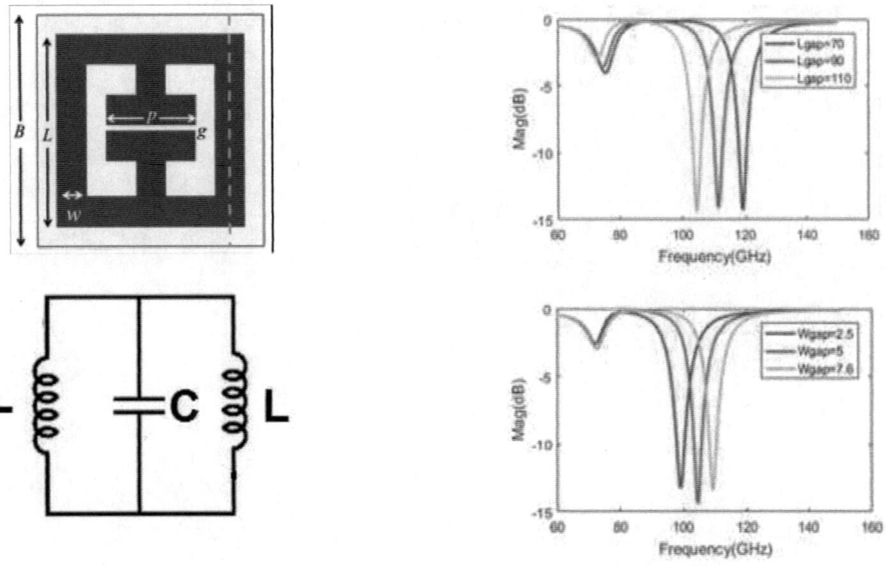

(a) (b)

Figure 6. Equivalent circuit of ELC resonator (a) and S11 of the simulated structure (b) where the value of Lgap and Wgap are varied.

The transformation due the transit-time effect during the light-wave propagation in the optical waveguide can be expressed by y^'=y-v_g t, where y' denotes the point of the light-wave in the coordinate system moving with the light-wave, and vg is the group velocity of the light-wave. Therefore, the terra-hertz electric field observed by the light-wave along the optical waveguide become as

$$\begin{aligned} E_{THz-Light} &= E_{THz}\left(t = \frac{y-y'}{v_g}\right) \\ &= E_0 \sin(\omega_{THz}\frac{y-y'}{v_g}) \\ &= E_0 \sin(k_{THz} n_g y + \varphi \end{aligned} \quad (3.3)$$

where kTHz is the wave number of the terra-hertz in vacuum (kTHz = □THz/c), ng is the group index of the light-wave propagating in the optical waveguide (ng = c/vg), φ is an initial phase of the light-wave in the optical waveguide (φ = kTHz ng y'), and c is the light velocity in vacuum.

The light-wave propagating in the optical waveguide is modulated by the induced terra-hertz electric field across the capacitive gap through the EO effect. The obtained light-wave from the optical waveguide is optical phase modulation by the wireless terra-hertz wave. The modulation index, □□x can be calculated by taking into account of the overlapping between of the induced terra-hertz and light-wave electric fields. It is expressed as follows,

$$\Delta\phi_\lambda = \frac{\pi r_{33} n_e^3}{\lambda}\Gamma \int_0^P E_{THz-Light}(y)dy \quad (3.4)$$

where λ is the light-wave wavelength propagating in the optical waveguide, r33 is the EO coefficient, ne is the extraordinary refractive index of the substrate, Γ is a factor expressing the overlapping between the induced terra-hertz and the light-wave electric field, and P is the length of the electric field interaction in a one period for effective optical modulation. When the modulation index is smaller than unity, the

IOP Conf. Series: Materials Science and Engineering **622** (2019) 012004 doi:10.1088/1757-899X/622/1/012004

conversion efficiency (sensitivity) from the wireless terra-hertz wave to the light-wave corresponds to the square of the calculated modulation index of the optical phase modulation [18].

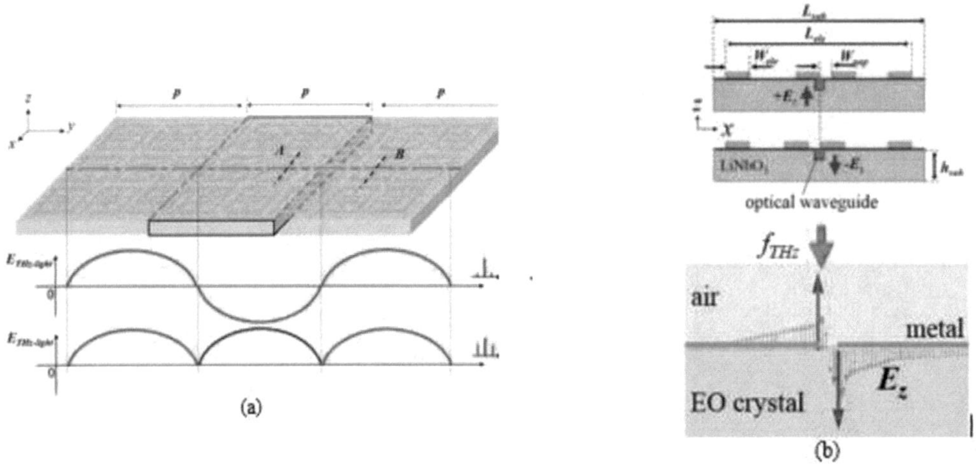

Figure 7. Illustration of interaction between terra-hertz wave and light-wave (a) The polarization reversal technique and electric field vector across the capacitive gap (b)

Furthermore, the conversion efficiency (sensitivity) can be improved using n- period electric field interactions (nP) as shown in Fig. 7. In order to compensate degradation of optical modulation due to transit–time effect, the interaction of terra-hertz and light-wave electric fields should be manipulated. The polarization reversal technique is a candidate to solve the problem. In here, the polarization reversal with meandering gap is proposed. The modulation index can be calculated with considering the polarization reversal for n- period electric field interaction. It can be expressed using following equation

$$\Delta\phi_n = \frac{\pi r_{33} n_e^3}{\lambda}\Gamma\left(\sum_{n=0}^{n(even)}\int_0^P E_{THz-Light}(y)dy - \sum_{n=1}^{n(odd)}\int_0^P E_{THz-Light}(y)dy\right) \quad (3.5)$$

where n denotes n-period electric field interaction, P is the length of the electric field interaction in a one period effective optical modulation.

3.3. *Experiment*

The proposed conversion device was designed at about 0.1THz operational frequency using a 0.25 mm-thick z-cut LiNbO3 crystal. The size of the electric-LC resonators was calculated using electromagnetic analysis software as shown in the previous section.

In the fabrication process as shown in Fig. 8(a), an EO crystal of z-cut LiNbO3 was prepared as the substrate of the proposed conversion device. A straight optical waveguide was fabricated on the substrate by using titanium diffusion method and effective for operation in 1.55 □m wavelength [19]. A thin SiO2 buffer layer was deposited on the surface of the substrate after the optical waveguide fabrication. Next, electric–LC resonators with a polarization reversal pattern were fabricated using a 1 μm-thick gold metal film on the buffer layer by use of thermal vapor deposition, a standard photolithography technique, and a lift-off process. The edges of the gaps were aligned onto the optical waveguides for effective conversion. The typical pictures of the fabricated conversion device are shown in Fig. 8(a) using microscope. After device fabrication process, the fabricated conversion device was coupled with optical fibres for easy in measurement. The photograph of the fabricated conversion device is shown in Fig. 8(b).

IOP Conf. Series: Materials Science and Engineering **622** (2019) 012004 doi:10.1088/1757-899X/622/1/012004

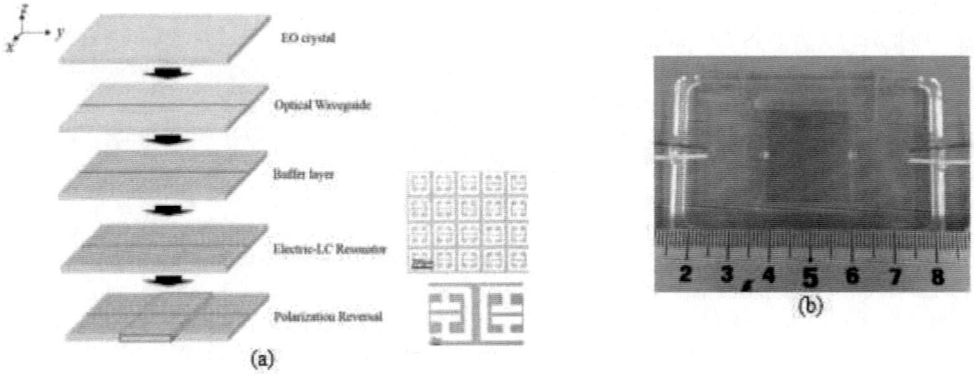

Figure 8. Fabrication process of the proposed conversion device (a) and the photograph of the fabricated conversion device (b).

4. Result and Discussion

The characteristics of the fabricated conversion device were measured experimentally with a measurement setup as shown in Fig. 9. A 0.1 THz wireless signal was generated by use of multiplying several time of the microwave signal. The terra-hertz power was improved using an amplifier. Finally, the terra-hertz signal was irradiated to the air using a horn antenna. As a result, the terra-hertz wireless signal can be radiated to the air. Then, light-wave with operational wavelength of 1.55μm from laser were propagated to optical fibres and coupled to the fabricated device. Optical polarizer was inserted between the lasers and fabricated device and set to the z-axis by considering EO crystal orientation. In order to measure the light-wave output, an optical spectrum analyser (OSA) was used to identify the optical carrier and sidebands due to terra-hertz radiation.

The optical waveguides in the fabricated conversion device were coupled with optical fibres with measured insertion loss of about -8dB. The basic operation of the proposed conversion device was measured. Typical of the measured light-wave spectra are shown in Fig. 10 without and with wireless terra-hertz irradiation. We can see that clear optical sidebands can be obtained using the fabricated conversion device by 0.1 THz wireless irradiation. The power ratio between optical carrier and sidebands of about 60 dB were obtained using the fabricated conversion device.

Fig. 11 shows the measured frequency dependence of the terra-hertz wireless irradiation for characteristics of the fabricated conversion device. The experiment result almost coincided with the calculation result. The operational frequency can be tuned furthermore by adjusting the electric-LC resonator of the metamaterial antenna. Based on the experiment results, basic operations of the proposed conversion device were successfully demonstrated. The wireless terra-hertz signal can be received and converted to the light-wave signal directly using the proposed conversion device.

The conversion efficiency or sensitivity can be improved furthermore by considering longer interaction length of terra-hertz wave and light-wave electric fields. In this proposed device, an optical phase modulation was obtained. Therefore, single-sideband modulated light-wave with high power is required for detecting to a photo-detector. It can be realized using a a sharp-cut optical filter and an optical amplifier [20, 21]. The power levels of the optical sideband and the carrier can be set almost the same by tuning the filter characteristics to improve conversion efficiency or sensitivity.

5. Result and Discussion

The conversion device from wireless terra-hertz to light-wave using metamaterial antenna on EO modulator was proposed for terra-hertz wireless detection through ROF technology. The proposed conversion device was analysed, designed, and optimized for 0.1 THz operational frequency. The designed device was fabricated successfully. The fabricated device was measured also for its basic operation of detection and conversion of wireless terra-hertz to light-wave. Clear optical sidebands were obtained by irradiating 0.1 THz operational frequency. Therefore, the basic operations of the proposed device for the detection and conversion were demonstrated successfully. The proposed device has a compact and simple structure, passive operation, and low terra-hertz loss. The results show that this device is promising for future high-speed wireless communication and extremely low loss electromagnetic compatibility measurement using ROF technology.

Acknowledgments

This research activity is conducted as part of the research projects on the "Research and Development of Radio and Optical Components Based-on Planar Structure for Communication and Sensing" funded by the Indonesian Institute of Sciences (LIPI), Indonesia, the "Research on Electromagnetic Applications for Food Quality Control" funded by Ministry of Research, Technology, and High Education, Indonesia and the "Research and Development on Short Distance Communication and Imaging for Application in ASEAN Region" funded by National Institute of Information and Communication Technology (NICT), Japan.

Acknowledgments

This research activity is conducted as part of the research projects on the "Research and Development of Radio and Optical Components Based-on Planar Structure for Communication and Sensing" funded by the Indonesian Institute of Sciences (LIPI), Indonesia, the "Research on Electromagnetic Applications for Food Quality Control" funded by Ministry of Research, Technology, and High Education, Indonesia and and the "Research and Development on Short Distance Communication and Imaging for Application in ASEAN Region" funded by National Institute of Information and Communication Technology (NICT), Japan.

References

[1] "Cisco Visual Networking Index: Global Mobile Data Traffic Forecast Update, 2016 - 2021," Cisco, 2017.

[2] Z. Popovic and E.N. Grossman, "THz Metrology and Instrumentation," in IEEE Transactions on Terahertz Science and Technology, vol. 1, no. 1, pp. 133-144, Sept. 2011.

[3] K.B. Cooper, R.J. Dengler, N. Llombart, B. Thomas, G. Chattopadhyay and P.H. Siegel, "THz Imaging Radar for Standoff Personnel Screening," in IEEE Transactions on Terahertz Science and Technology, vol. 1, no. 1, pp. 169-182, Sept. 2011.

[4] J. Choi, W.S. Kwon, K.S. Kim and S. Kim, "Improvement of terahertz spectrum accuracy by selective extraction of sample-induced echo signals," 2014 39th International Conference on Infrared, Millimeter, and Terahertz waves (IRMMW-THz), Tucson, AZ, 2014, pp. 1-2.

[5] Mark J. Rosker and H. Bruce Wallace, "Imaging Through the Atmosphere at Terahertz Frequencies," IEEE MTT-S International Symposium, June 2007

[6] J.-W. Shi, C.-B. Huang and C.-L. Pan, "Millimeter-Wave Photonic Wireless Links for Very High Data Rate Communication," NPG Asia Materials, vol. 3, pp. 41-48, 2011.

[7] A. Kanno; T. Umezawa; T. Kuri; N. Yamamoto; T. Kawanishi; Y.N. Wijayanto "Key technologies for millimeter-wave distributed radar system over a radio over fibre network," 2016 International Conference on Radar, Antenna, Microwave, Electronics, and Telecommunications 2016, Tangerang, Indonesia.

[8] F.T. Sheehy, W.B. Bridges, and J.H. Schaffner, 60 GHz and 94 GHz Antenna-Coupled LiNbO3 Electrooptic Modulators, IEEE Photonics Technology Letters, vol. 5, no. 3, pp. 307-310 (1993).

[9] D.H. Park, V.R. Pagán, T.E. Murphy, J. Luo, A.K.-Y. Jen, and W.N. Herman, "Free space millimeter wave-coupled electro-optic high speed nonlinear polymer phase modulator with in-plane slotted patch antennas," Opt. Express 23, 9464-9476 (2015)

[10] S. Shinada, T. Kawanishi, and M. Izutsu, A 10-GHz Resonant-Type LiNbO3 Optical Modulator Array, IEEE Transaction Electronics, vol. E90-C, no. 5, pp. 1090-1095 (2007).
[11] H. Murata, N. Kohmu, Y.N. Wijayanto, Y. Okamura, "Integration of Patch Antenna on Optical Modulators," IEEE Photonics Society News, vol. 28, no. 2, April 2014.
[12] Y.N. Wijayanto, H. Murata, and Y. Okamura, "Electro-Optic Millimeter-Wave-Light-wave Signal Converters Suspended to Gap-Embedded Patch Antennas on Low-k Dielectric Materials," IEEE Journal of Selected Topics in Quantum Electronics, vol.19, no.6, November/ December 2013.
[13] Y.N. Wijayanto, H. Murata, and Y. Okamura, "Electro-Optic Microwave-Light-wave Converters Utilizing Patch Antennas with Orthogonal Gaps," Journal of Nonlinear Optical Physics and Material, vol. 21, no. 1 (2012) 1250001 (12 pages)
[14] K. Chang, RF and Microwave Wireless Systems, New York: John Wiley & Sons, Inc, 2000.
[15] A. Yariv, Quantum electronics, (Wiley, New York, 1989).
[16] D. Schurig, J.J. Mock, and D.R. Smith, "Electric-field-coupled resonators for negative permittivity metamaterials," Appl. Phys. Lett. 88(4), 041109 (2006).
[17] W. Withayachumnankul, C. Fumeaux, and D. Abbott, "Compact electric-resonators for metamaterials," Optical Express, vol. 18, no. 25, pp. 25 912–25 921, 2010.
[18] Y.N. Wijayanto, A. Kanno, H. Murata, T. Kawanishi, N. Yamamoto, and Y. Okamura, "Array of Patch-Antennas with Meandering-Gaps on Optical Modulator for Wireless Millimeter-Wave Beam-Steering ," International Journal of Microwave and Wireless Technologies, Vol. 8, Special Issue 4-5, pp 759-765, June 2016,
[19] L.W. Stulz, "Titanium In-Diffused LiNbO3 Optical Waveguide Fabrication," Applied Optics, vol. 18, no. 12, pp. 2041-2044, 1979.
[20] D. Mahmudin, T.T. Estu, A. A. Fathnan, Y.Y. Maulana, P. Daud, G. Sugandhi, Y.N. Wijayanto, "Optimization of optical filter using triple coupler ring resonators structure based on polyimide substrate", Proc. SPIE 10150, Second International Seminar on Photonics, Optics, and Its Applications (ISPhOA 2016).
[21] F. Li, A.S. Helmy, "Gigahertz to terahertz tunable all-optical single-side-band microwave generation via semiconductor optical amplifier gain engineering," Optical Letters. 2013 Nov 15;38(22):4542-5.

Antimicrobial hollow fiber polypropylene/ZnO membrane for effective air filtration

Anak Agung Istri Agung Sri Komaladewi [1], K. Khoiruddin[2], I Dewa Gede Ary Subagia[1], Dea Indriani Astuti[3] and I Gede Wenten[2*,4]

[1] Mechanical Engineering Department, Udayana University, Jl. Raya Kampus UNUD, Bukit Jimbaran, Kuta Selatan, Bali, 803611, Indonesia
[2] Chemical Engineering Department, Institut Teknologi Bandung, Jl. Ganesha 10, Bandung 40132, Indonesia
[3] Department of Microbiology, School of Life Science and Technology, Institut Teknologi Bandung, Jl. Ganesha 10, Bandung 40132, Indonesia
[4] Research Center for Nanosciences and Nanotechnology, Institut Teknologi Bandung, Jl. Ganesha 10, Bandung 40132, Indonesia

*Corresponding author: igw@che.itb.ac.id

Abstract. Air pollution has become an environmental problem since it poses a serious effect on human health. In addition, circulation of air containing airborne particles including pathogens in an enclosed building like a hospital may lead to more severe health effects. Therefore, an air purifier is needed to prevent the spreading of those particles through air circulation. To address the issue, an antimicrobial membrane was prepared by embedding ZnO nanoparticles onto hollow fiber polypropylene (PP) membrane. Results showed that the prepared PP/ZnO membrane had a high air filtration performance as well as antibacterial properties. The membrane also showed a high filtration capacity or permeability with a relatively low-pressure drop. With those interesting features, the newly developed PP/ZnO membrane can be applied in air filtration, in particular for indoor spaces and other medical applications.

Keywords: air filter, air pollution, clean air, fine particles, membrane

1. Introduction

Worsening air quality has become one of the environmental problems due to their adverse effects on human health. This is associated with various activities, such as industry, transportation, forest fire, and fuel usage [1]. Airborne particles such as fine particles and bio-aerosols are among the important pollutants which gain increasing attention due to their negative impacts on human health. Since they have a small size, those pollutants may from one place to another easily and can be suspended in the atmosphere for a long time due to its stability [2–4].

Those airborne particles can be removed for producing clean air by employing technologies, such as electrostatic precipitation, cold plasma, wet scrubbing, cyclonic air filtration, and filters [5–7]. Among the technologies, filtration is the most widely used because it is easy to operate, less energy consumption, and low cost.

Content from this work may be used under the terms of the Creative Commons Attribution 3.0 licence. Any further distribution of this work must maintain attribution to the author(s) and the title of the work, journal citation and DOI.
Published under licence by IOP Publishing Ltd

Recently, membrane technology has been applied in almost industrial sectors which were driven by interesting features of the smaller footprint, easy to scale-up, and lower operating cost [8–11]. In air filtration, membranes show superior filtration efficiency than conventional filter due to the ability to perform a filtration in the molecular level. Membranes allow the passes of clean air while rejects fine particles via sieving mechanism. With the high removal efficiency of fine particles, the membrane may be used as an effective air filtration replacing the conventional ones [12,13].

Even though the airborne particles can be effectively retained, the airborne microorganism may be accumulated on the membrane surface becoming secondary bioaerosols polluting source [14]. For a long-term application, a membrane with antimicrobial properties should be used. An antimicrobial membrane can be prepared by introducing antimicrobial agents in its structure. Metal oxide particles, silver, and natural extracts are usually employed as antimicrobial agents [15–19]. ZnO is one of the materials known to have antimicrobial functionality. ZnO antimicrobial activities are derived from its chemical and physical interaction with bacterial cells [20]. The antibacterial functionality of ZnO in an air filter has been demonstrated by Zhong et al. [21] by growing ZnO nanorods on expanded polytetrafluoroethylene. In this study, the excellent antimicrobial activity of ZnO is utilized to prepare the antimicrobial polypropylene (PP) membrane for effective air filtration. PP hollow fiber membrane was chosen since it has high mechanical strength, high chemical and thermal stability, and low cost [22]. Hollow fiber membrane also provides a high packing density and is easy to fabricate. To introduce ZnO nanoparticles, a simple permeation method is used. Combining simple fabrication and the low-cost material is expected to produce membrane air filter with low cost.

2. Materials and methods

PP membrane module and ZnO particles used in this study were obtained from GDP Filter Indonesia. Properties of PP membrane were summarized in Table 1. Technical grade ethanol (obtained from a local supplier, 96%) was used for the preparation of ZnO suspension.

ZnO particles were dispersed in ethanol with a concentration of 4 g/L (2.5 L). The suspension was mixed until optically homogeneous. The suspension was pumped into the shell side of the membrane module. The solvent was permeated through the membrane in a dead-end mode (at 0.2 bar operating pressure; 9 L/h flow rate) while the permeate was recirculated into the feed tank. Permeation was conducted for 30 minutes. The membrane was then allowed to dry for 24 hours. To remove the remaining ethanol, air scouring was applied for 30 minutes.

Table 1. Properties of PP membrane

Parameter	Value
Outside diameter (mm)	2.2
Inside diameter (mm)	1.8
Pore size (nm)	~50
Membrane surface area (m^2)	3

An air filtration was performed by filtering synthetic polluted air prepared by combusting papers in a 100 L combustion chamber. The chamber was connected to a blower. The polluted air was transferred into the module in a dead-end mode and outside-in flow configuration. Permeate quality was measured immediately by laser-based light scattering detector (Origins Technology, 1-999 µm/m3). Microbial filtration test was conducted by using a mixed culture solution (EM4; PT. Songgolangit Persada, Indonesia; total plate counts = 2.8 x 106; Lactobacillus = 3.0 x 105; Phosphate dissolving bacteria = 3.4 x 105; Yeast = 1.95 x 103). A 20 mL mixed culture was diluted into demineralized water until the final volume of 2 L was obtained. The solution was then filtered by PP membrane (at 1 bar operating pressure). The feed and permeate were analyzed by microscope. For the antimicrobial test, fibers of PP and PP/ZnO membrane were cut and wound into a circle with a diameter of 2 cm. The fiber was placed on nutrient agar which was previously added microbes obtained from the mixed culture. After 5 days, the membrane was observed.

IOP Conf. Series: Materials Science and Engineering **622** (2019) 012005 doi:10.1088/1757-899X/622/1/012005

FTIR analysis of the membrane was carried out by The PerkinElmer SpectrumTwo™ while SEM analysis was performed using scanning electron microscopy (JEOL JSM 6510 LA, Tokyo, Japan).

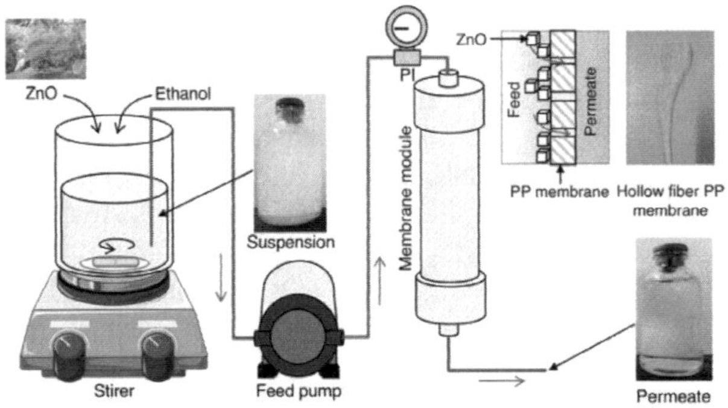

Figure 1. Experimental set-up for ZnO deposition.

3. Results and discussion

3.1. SEM and FTIR analysis

SEM images of PP and PP/ZnO membrane are shown in Figure 2 a and b. It can be seen that ZnO particles were successfully deposited on the membrane indicated by bright spots. It was obvious since the ZnO particles have an average size (about 300 nm, Figure 2 c) which was larger than the membrane pore size. Consequently, ZnO particles were retained on the membrane due to the sieving mechanism.

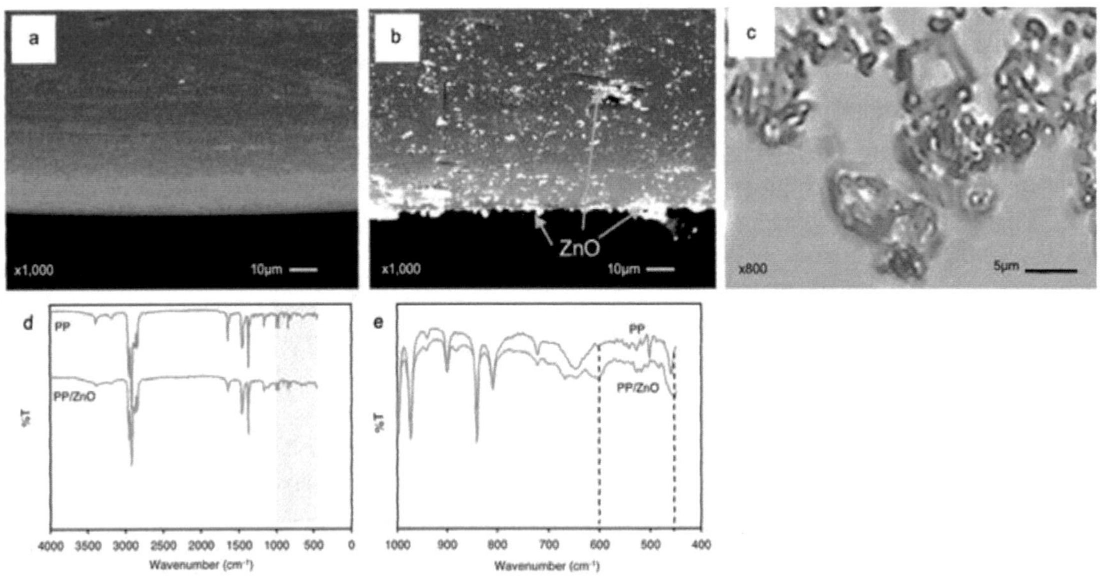

Figure 2. SEM Image of (a) PP and (b) PP/ZnO membranes, (b) digital microscope image of ZnO particles, (d), (e) IR spectra of PP and PP/ZnO membranes.

FTIR analysis was also conducted to ensure the presence of ZnO on the membrane surface. Wang et al. [23] observed a peak of around 450 cm-1 in their FTIR analysis of ZnO particles. The peak was attributed with the stretching mode of crystal ZnO. Meanwhile, Zhu et al. [24] found the stretching mode of Zn-O bond spanning at the wavenumber of 540-570 cm-1. FTIR analysis of PP and PP/ZnO membranes are shown in Figure 2 d and e. Two peaks were found in PP/ZnO, i.e. wavenumber of ~450 cm-1 and ~600 cm-1 which may indicate the presence of Zn-O bond.

3.2. Membrane performance

The performance of PP membrane during filtration of microbial containing solution is depicted in Figure 3a and b. As shown in Figure 3 b, complete removal of the microbes was observed. This was due to the small pore size of the PP membrane (see Table 1).

Figure 3 c shows that the membrane could remove PM2.5 particles (particulate matters with the size of less than 2.5 □m) almost completely (~99.9%). In addition, the membrane exhibited relatively low-pressure drops which are beneficial for their application. The lower pressure drop will result in lower energy consumption.

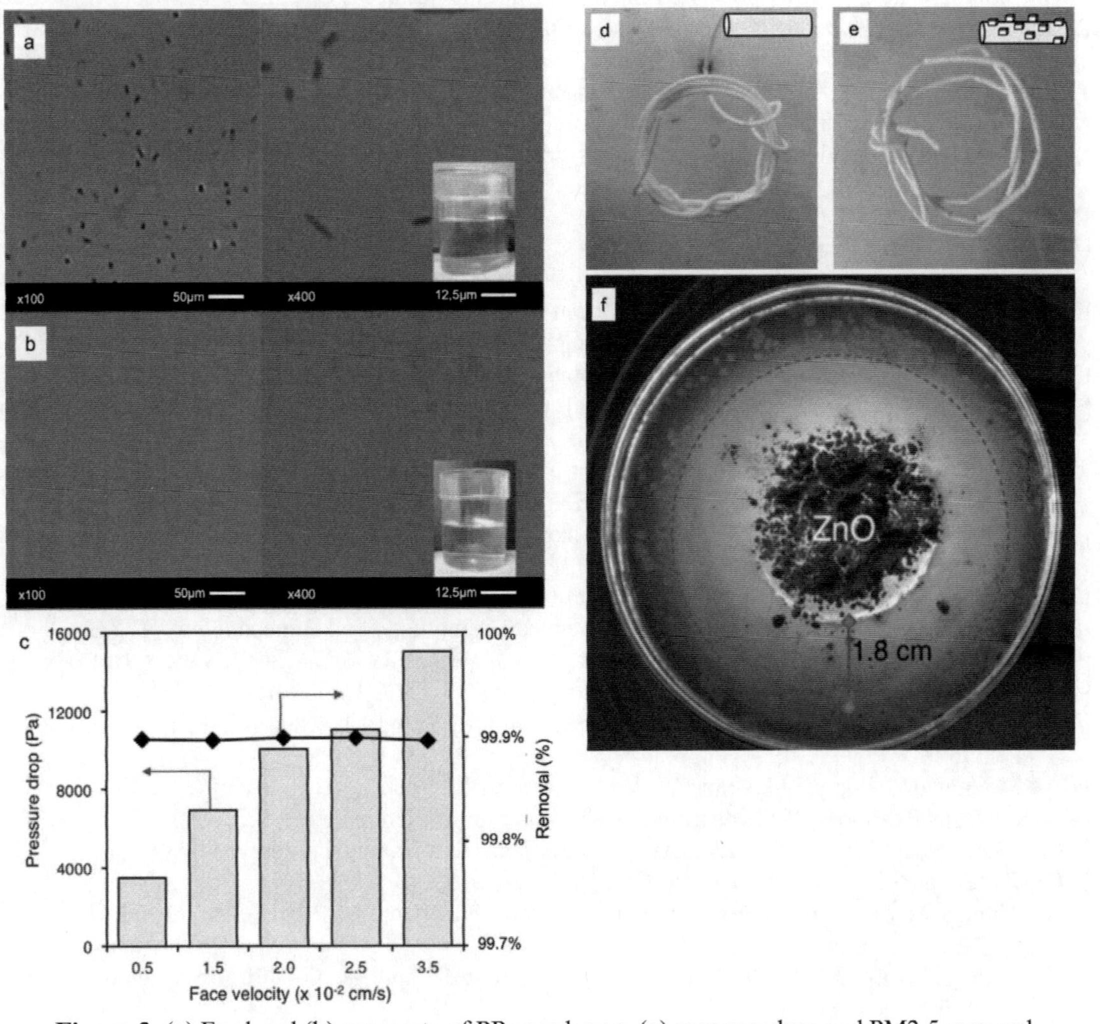

Figure 3. (a) Feed and (b) permeate of PP membrane, (c) pressure drop and PM2.5 removal efficiency (feed: 990 µm/m3), (d) PP membrane, (e) PP/ZnO, and (f) ZnO on microbial culture.

3.3. Antimicrobial activity of PP/ZnO membrane

Results of the antimicrobial test of PP, PP/ZnO, and ZnO particles were shown in Figure 3 d-e. As can be seen in Figure 3 f, the inhibition zone of ZnO (0.06 g ZnO/cm2) when it is placed on a medium with microbial culture. ZnO particles are known as the antimicrobial agent. Antimicrobial activity of ZnO particles has been demonstrated in several studies [20,25–27]. Antimicrobial activity of ZnO in composite membranes was also reported in literature [25,28]. The antimicrobial property of ZnO was derived from its oxidative activity. It was found that the death of bacteria due to the presence of ZnO particles was associated with the high generation rate of oxygen on ZnO surface and its activity increased with the decreasing size of ZnO [26]. The antimicrobial activity of the prepared PP/ZnO membrane was also observed in this study. As shown in Figure 3 d and e, the presence of ZnO (estimated to be 0.33 mg ZnO/cm2 membrane) resulted in lower growth of the microbial indicated by the cleaner area surrounding the PP/ZnO membrane. This may imply that the antimicrobial membrane is successfully prepared.

4. Conclusion

In this study, an antimicrobial air filter was prepared by depositing ZnO nanoparticles onto hollow fiber PP membrane. The ZnO was deposited on the membrane surface by a simple permeation method. Results showed that ZnO particles were successfully coated on the membrane surface showed by SEM and FTIR analysis. The prepared PP/ZnO membrane had a high air filtration performance as well as antimicrobial properties. The membrane also showed a high filtration capacity or permeability with a relatively low-pressure drop. With those interesting features, the newly developed PP/ZnO membrane can be potentially applied in air filtration, particularly for indoor spaces and other medical applications.

References

[1] Fugiel A, Burchart-Korol D, Czaplicka-Kolarz K and Smoliński A 2017 Journal of cleaner production 143 159
[2] Smith GJD, Vijaykrishna D, Bahl J, Lycett SJ, Worobey M, Pybus OG, Ma SK, Cheung CL, Raghwani J, Bhatt S, Peiris JSM, Guan Y and Rambaut A 2009 Nature 459 1122
[3] Beggs CB 2003 Indoor and Built Environment 12 9
[4] Kang JS, Kim H, Choi J, Yi H, Seo SC, Bae G-N and Jung JH 2016 Aerosol and Air Quality Research 16 2059
[5] Brincat J-P, Sardella D, Muscat A, Decelis S, Grima JN, Valdramidis V and Gatt R 2016 Trends in Food Science and Technology 50 175
[6] Joubert A, Laborde J-C, Bouilloux L, Chazelet S and Thomas D 2011 Chemical engineering journal 166 616
[7] Hung C-H and Leung WW-F 2011 Separation and purification technology 79 34
[8] Drioli E and Curcio E 2007 Journal of Chemical Technology and Biotechnology 82 223
[9] Sianipar M, Kim SH, Khoiruddin, Iskandar F and Wenten IG 2017 RSC Advances 7 51175
[10] Wenten IG, Khoiruddin K, Hakim AN and Himma NF 2017 (Elsevier) pp 199–218
[11] Wenten IG, Dharmawijaya PT, Aryanti PTP, Mukti RR and Khoiruddin 2017 RSC Advances 7 29520
[12] Miaomiao Z, Jingquan H, Fang W, Wei S, Ranhua X, Qilu Z, Hui P, Yong Y, Keshari SS, Feng Z and Chaobo H 2017 Macromolecular Materials and Engineering 302 1600353
[13] Bae G-N and Jung JH 2016 Journal of Nanoscience and Nanotechnology 16 4487
[14] Batterman SA and Burge H 1995 HVAC&R Research 1 61
[15] Thunyasirinon C, Sribenjalux P, Supothina S and Chuaybamroon P 2015 Aerosol and Air Quality Research 15 600
[16] Hwang GB, Lee JE, Nho CW, Lee BU, Lee SJ, Jung JH and Bae G-N 2012 Science of the Total Environment 421–422 273
[17] Jung JH, Hwang GB, Lee JE and Bae GN 2011 Langmuir 27 10256
[18] Woo CG, Kang J-S, Kim H-J, Kim Y-J and Han B 2015 Aerosol Science and Technology 49 611
[19] Detsri E, Kamhom K and Ruen-ngam D 2016 Journal of Experimental Nanoscience 11 930

[20] Kumar Rajesh, Umar Ahmad, Kumar Girish and Nalwa HS 2017 Ceramics International 43 3940
[21] Zhong Z, Xu Z, Sheng T, Yao J, Xing W and Wang Y 2015 ACS Applied Materials and Interfaces 7 21538
[22] Himma NF, Wardani AK and Wenten IG 2017 Materials Research Express 4 054001
[23] Wang Z, Zhang H, Zhang L, Yuan J, Yan S and Wang C 2003 Nanotechnology 14 11
[24] Zhu Y, Apostoluk A, Gautier P, Valette A, Omar L, Cornier T, Bluet JM, Masenelli-Varlot K, Daniele S and Masenelli B 2016 Scientific Reports 6 23557
[25] Wang Y, Zhang Q, Zhang C and Li P 2012 Food Chemistry 132 419
[26] Padmavathy N and Vijayaraghavan R 2008 Science and Technology of Advanced Materials 9 35004
[27] Espitia PJP, Soares N de FF, Coimbra JS dos R, de Andrade NJ, Cruz RS and Medeiros EAA 2012 Food and Bioprocess Technology 5 1447
[28] Al-Hinai MH, Sathe P, Al-Abri MZ, Dobretsov S, Al-Hinai AT and Dutta J 2017 ACS Omega 2 3157

Evaluation Wettability of nanofluid on the Surface of Zircalloy 4 Before and after Oxidation

Djoko Hadi Prajitno*, Very Trisnawan, and Jupiter Pane Sitorus

PSTNT-BATAN Jl. Tamansari 71 Bandung 40132, Indonesia

*Corresponding author: djokohp@batan.go.id

Abstract. The solid surface tension plays an important role in the heat and mass transfer system for heat exchanger equipment. In the nuclear power plant industry, Zircalloy 4 has been used for long time as structure materials. The purpose of the experimental is to study solid state surface tension behavior by measure contact angle of aquades and Nano fluid contain nano particle alumina on metal surface of Zircalloy 4 before and after oxidation at 700o C by sessile drop method. The experiment is to measure the static contact angle and drop of aquades and nano fluid contains nano particle alumina on zircalloy 4 with different spreading time from 1 to 30 minute. It was observed that zircalloy 4 after oxidation lose their hydrophobic properties with increasing elapsed time during drop of a aquades and nano fluid on the surface of alloy compared with zircalloy 4 before oxidation. As a result the contact angle of aquades and nano fluid on surface of zircalloy 4 before and after oxidation is decrease grdually during increasing elapsed time. While the magnitude diameter drops of aquades and nano fluid and wetting surface are increase with increasing elapsed time on the surface of Zircalloy 4 before and after oxidation. It was observed that oxidised zircalloy4 good wettability properties compare with non oxidised zircaloy 4.

Keywords: zircalloy 4,contact angle,nanofluid,wettability,oxidised

Content from this work may be used under the terms of the Creative Commons Attribution 3.0 licence. Any further distribution of this work must maintain attribution to the author(s) and the title of the work, journal citation and DOI.
Published under licence by IOP Publishing Ltd

1. Introduction

Zirconium alloys such as zircalloy 1,2,3,4 , zirlo, excel, and ZrNb alloys are used in nuclear industries due to of their low neutron cross section absorption and their excellent mechanical and corrosion properties. The application of zircaloy-4 with Zr-1.5%Sn-0.2%Fe-0.1%Cr. alloying element in the alloy is used for pressurized water reactors (PWR) nuclear power plant. The zircalloy 4 is used for structural material of nuclear fuel rod for cladding, control rod guide tubes, and grid spacers. (1,2). During accident such as loss of Flowing Accident (LOFA) and loss of Coolant accident (LOCA), fuel rods made of a zirconium alloy were oxidized at high temperatures producing H gas that leading to the core underwent degradation and then directly to meltdown. Currently, the nuclear community tried to evaluate and improve fuel systems with the main purpose to preventing and mitigating possible deleterious effects in nuclear fuel materials under high risk accident [3,4]. One of the methods to improved performance of zirconium alloy is to modified zirconium alloys by alloying [5]. The other method to improve zirconium alloy is used multiple coating with excellent corrosion resistance, chemical inertness [6,7]. Unfortunately, their applicability of ceramic material is limited due to low ductility, low heat transfer and crack resistance. The second method to mitigating structural nuclear materials meltdown during LOCA is to use nanofluid as emergency core cooling system (ECCS) media during accident. Nanofluid is suspension of nano particles in a base fluid. The properties of nanofluid are improved the heat transfer coefficient and conductivity compared with the original base fluid. These nano particles usually metal or metal oxide, increase the conduction and convection coefficients of base fluid (8). The studies of nanofluid for nuclear power plants application such as secondary cooling system and emergency core cooling system (ECCSs) should be compatible with conventional systems during normal operations to make nanofluid technologies practical in Nuclear Power Plants. One of the methods to evaluation performance nanofluid as media cooling system is wettability. Wettability can be defined as interaction of a fluid with a solid surface of material such as metal. The wetting phenomena plays an important role in the mass and heat transfer for engineering structure such as heat exchanger, steam generation, waste heat recovery and emergency cooling system. An index of wetting properties is defined by the magnitude of contact angle between the liquid and solid surface of metal (9, 10). Solid metal provides planar surface that allow the geometric measurement of a contact angle on the metal. If the contact angle is lower than 90° the material is hydrophilic otherwise it is hydrophobic.

The zircalloy 4 as an engineering material is used as nuclear material structure in the primary cooling system. The nominal chemical composition of Zircaloy-4 is Zr e 1.5 wt%Sn e0.2 wt%Fe e 0.1 wt%Cr [(11). The contact angle is highly influenced by surface finish composition of the metal and nanofluids. Contact angle nanofluid and solid of zircalloy 4 before and after oxidation in nanofluids media contain nanoparticle alumina is rarely studied. During LOCA accident the surface of zircaloy 4 is covered by oxide layer ZrO2 due to oxidation zircalloy 4 by atmosphere. The purpose of the present study is to examine the contact angle of aquades contain nanoparticle alumina on the metal surface of zircalloy 4 before and after oxidation at 700°C to form oxide layer ZrO2 with different spreading time will be studied. The present study is to evaluate wettability behavior of nanofluid contain 0,01 gpl nanoparticle alumina on surface of zircalloy 4 before and after oxidation by using a sessile drop methods.

IOP Conf. Series: Materials Science and Engineering **622** (2019) 012007 doi:10.1088/1757-899X/622/1/012007

2. Experimental

2.1. Material and nanofluid

The specimens used in this experiments were circular discs of zircalloy 4 having 1 Cm in diameter and 3 mm in thickness. Figure 1 showed specimens a) before oxidation and b) after oxidation at 700°C in air furnace. The specimen is zircalloy 4 with the major chemical composition (in wt%) 98%Zr, 1,4Sn, 0.24Fe 0.15Cr. The specimens were roughed by wet grinding with different emery paper 230,500, 800 and finished at1000 grit, degreased with detergent after that cleaned with ultrasonic cleaner in alcohol solution. In the present study, Al_2O_3 was used as nanoparticles. Al_2O_3 nanoparticle from Inframat with purity 99.99, surface area (BET) 201 m^2/g and the average primary particle size 20nm. Figure 2 showed nanoparticle of Al_2O_3 examination by TEM with magnificiant of 20 nm. It can be seen that the shape of the particles were approximately rounded. The base fluid of aquades was used for this experiment. Nanofluid is produced by mixing of aquades with 0.01 gpl Al_2O_3 nanoparticles. The nanofluids were prepared by incorporating 0.01 gram of Al_2O_3 nanoparticles mixed into 1 L aquades and mixed by magnetit stirrer for 30 minutes. After mixing, ultrasonic vibrator operating at a frequency of 44 kHz was used to stabilise nanofluid for 30 min in order to the Al_2O_3 particles were dispersed and stabilized in aquades [12].

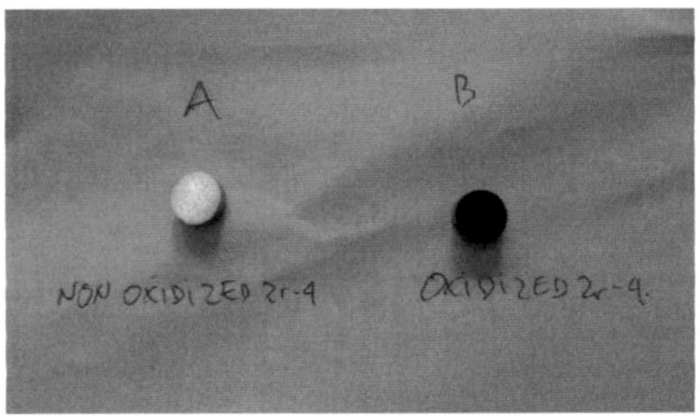

Figure 1. Specimens a) before and b) after oxidation

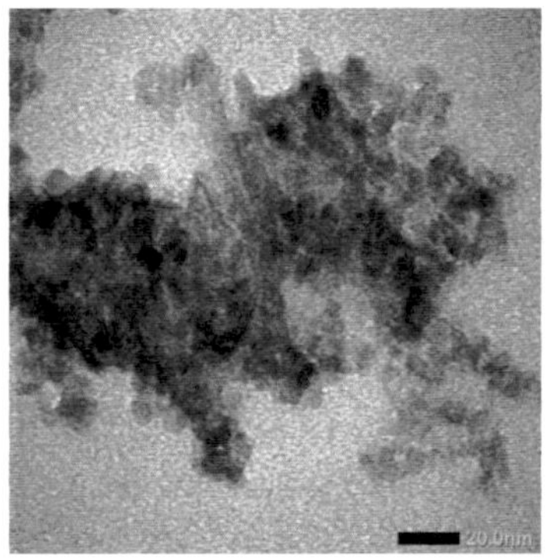

Figure 2. Nanoparticle Al_2O_3 examination by TEM with magnificiant of 20 nm

2.2. Sessile drop measurement

The wettability characteristics of nanofluid on the surface of the zircalloy 4 were carried out to evaluate contact angle measurement. The contact angle can be defined as the angle between the liquid phases formed on the surface of a stainless steel and the line tangent to the droplet radius from the point of contact with the surface of stainless steel. The Contact angle perform have been conducted by sessile drop method (13) equipment system from Kruss GmbH goniometer with a resolution of 0.1° in the measuring range of 1-180°. This equipment has been provided by software drop shape analysis (DSA 4) for controlling the experiment and analyzing the drop shape to calculate the contact angle. Droplets of known volume of 2 μL were dropped on a substrate of zircalloy 4 before after oxidation using an automatic dispenser and injected slowly onto the solid surface by a syringe. The experiments were conducted under similar air conditioned laboratory environments. The measured values of the contact angle, the images of the droplets were taken directly after 20 second deposition in order to eliminate the impact of droplet evaporation. The contact angle tests were conducted at an ambient temperature of 27 °C.

2.3. Sample characterization

Surface morphology of the zircalloy 4 before and after oxidation was examined by using an optical microscopy.

3. Results and discussion

3.1. Optical image and x ray diffraction of zircalloy 4 before and after oxidation

Figure 3(a) and (b) shows surface morphology of the zircalloy 4 before and after oxidation that was examined by using an optical microscope. Surfaces morphology of the sample relative the same roughness was produced by wet grinding of zircalloy 4. The color of oxidized zircalloy 4 relatively different with non-oxidized zircalloy 4 due to oxide scale was formed on the zircalloy 4 during oxidation of zircalloy at 700 °C. This is agreement with the Reif [14]. Oxidation of zircalloy at 600 and 700°C in air furnace resulted thin of oxide scale of zirconium oxide. Reaction zirconium with oxygen resulted ZrO2 [15].

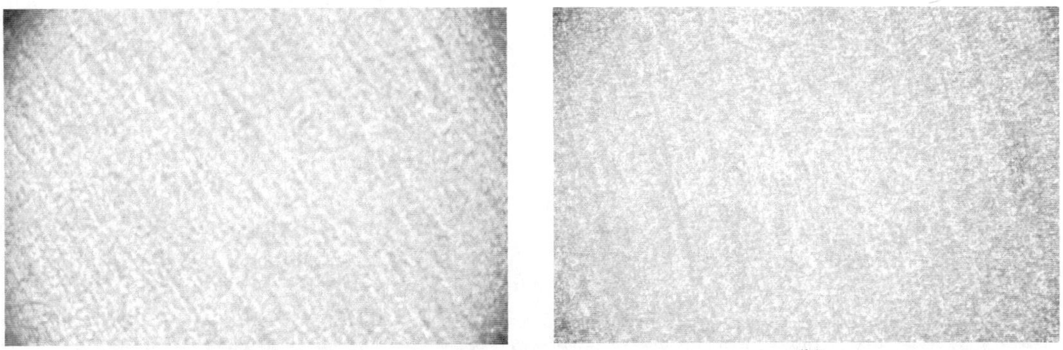

(a) (b)

Figure 3. Surface morphology of zircalloy4 a) before and b) after oxidation

3.2. Effect of spreading time on the wettability aquades and nanofluid on the surface of oxidized and non-oxidized Zircalloy 4

Dynamic wetting behavior was evaluated through measuring contact angles between aquadest and nanofluid epoxy and oxidized zirconium alloy and non-oxidized zirconium alloy as shown In Figure 4(a) and (b). From this figure depicted the contact angle of droplet aquadest and nanofluid on zircalloy 4 before and after oxidation as a function of spreading time multiple measurements. From Figure 4(a) and (b) , generally it can be noted that the contact angle value of aquadest and nanofluid on the surface of non-oxidized zircalloy 4 are higher compare with oxidized zircalloy 4 at different spreading time. From figure 4(a), it is shown that at early-stage spreading time, the contact angle of aquades were not different significantly on the surface of non-oxidized zircalloy 4 and oxidized zircalloy 4 . After 30 second elapsed time, the contact angle of aquadest on the surface of non-oxidized zircalloy 4 and oxidized zircalloy 4 was different significantly. The contact angle of aqueduct 66° on non-oxidized zircalloy 4 and the contact angle were decreased gradually with elapsed time. Finally, the contact angle was reach 42° and there was a decreased of 24° on the contact angles. Contact angle of aqueduct on surface of oxidized zircalloy 4 is described in figures 4(a). In the first spreading time, the contact angle of aquadest 64° on oxidized zircalloy 4 and the contact angle were decreased significantly during elapsed time. At the end elapsed time, the contact angle was reach 42° and there was a decreased of 24° on the contact angles. It can be concluded that contact angle of aquadest on the surface of oxidized zircalloy 4 lower compared with non-oxidized zircalloy 4.

Figure 4(b), depicted shown that at early stage spreading time, the contact angle of nanofluid were different significantly on the surface of non-oxidized zircalloy 4 and oxidized zircalloy 4. After 30 second elapsed time, the contact angle of nanofluid on the surface of non-oxidized zircalloy 4 and oxidized zircalloy 4 was big different. Initially, the contact angle of nanofluid 86° on non-oxidized zircalloy 4 and the contact angle were decreased gradually with elapsed time. Finally, the contact angle was reach 44° and there was a decreased of 42° on the contact angles. Contact angle of nanofluid on the surface of oxidized zircalloy 4 is described in figures 4(b). At the initial of spreading time, the contact angle of nanofluid 79° on oxidized zircalloy 4 and the contact angle were gradually decreased significantly during elapsed time. At the end elapsed time, the contact angle was reach 28° and there was a decreased of 31° on the contact angles. It suggested that the wettability of nanofluid on the oxidized zircalloy 4 became improved than that of non-oxidized zircalloy 4. The evident showed that the decrease of contact angle of oxidized zircalloy indicated dramatic changes the contact conditions due to chemistry behavior of surface zirconium alloy 4. According to Refs. [16, 17], there was a thin oxide film produced on the surface of sample that change behavior of wettability of nanofluid in the surface of oxidized zircalloy 4.

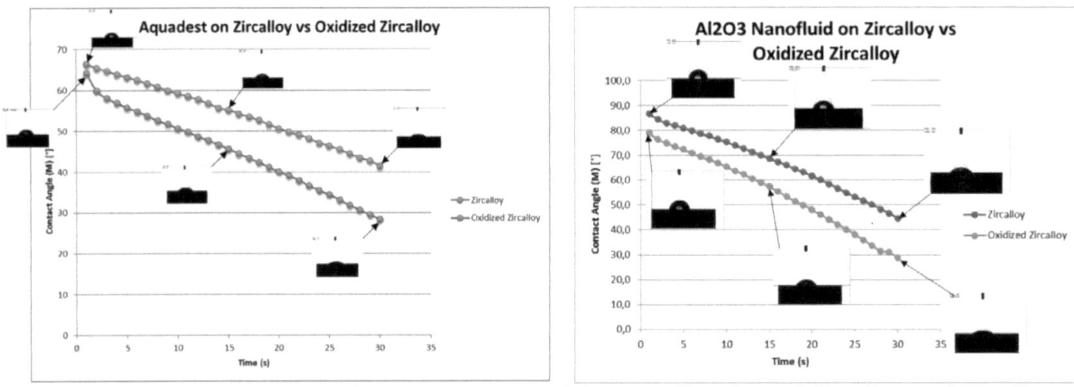

Figure 4. Measurements at ambient temperature of the spreading time of Zircalloy and Zircalloy oxidized on a) aquadest and b) nanofluid

The droplet image of aquadest and nanofluid on Zircalloy 4 before and after oxidation has been measured by using the sessile drop method is represented in Fig. 5 and 6 respectively. Generally, it is showed that the typical images of droplet at 0, 10,20 and 30s of aquadest and nanofluid on the surface of zirconium alloy 4 before and after oxidation are raise in diameter and contact angle decrease for longer spreading time. From figure 5, it is shown that at the early-stage spreading the liquid drop of low-viscosity aquades are different significantly on the surface of zircalloy 4 before and after oxidation. Fig.5(a) depicted the droplet of aquadest at 0,10,20, and 30s on the surface of zirconium alloy before oxidation. For long spreading time, the wetted area is found to grow and diameter of droplet increases while the contact angle is decrease significantly. The droplet of aquadest at 0, 10,20 and 30 minutes on zircalloy after oxidation showed in Figure 5(b). It can be seen that the wetting area of aquadest on the surface of zircalloy 4 after oxidation are more wide and contact angle relative small compare with zircalloy before oxidation. It can be concluded that wettability of aquadest on the surface of zirconium alloy after oxidation are considered good wettability's compare with zirconium alloy before oxidation.

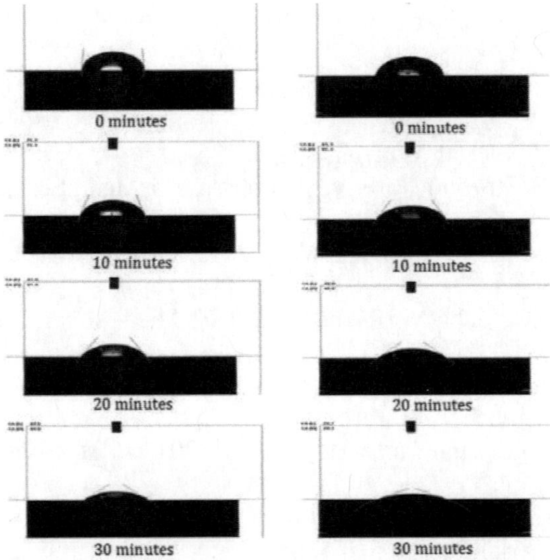

Figure 5. A droplet of aquadest on the surface of Zircalloy 4 a)before and b)after oxidation

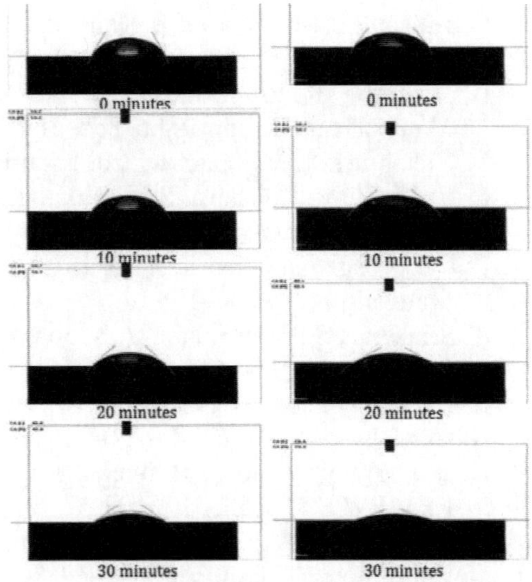

Figure 6. A droplet of nanofluid on the surface of Zircalloy 4 a) before and b)after oxidation

Figure 6 (a) and (b) depicted the different of the droplet image at 0, 10, 20 and 30 minutes of nanofluid on the surface of zircalloy 4 before and after oxidation, respectively. From figure 6, it is shown that at the early-stage spreading (0 minute) the liquid drop of nanofluid on the surface of zircalloy 4 before oxidation is different significantly with zircalloy 4 after oxidation. The droplet of nanofluid on the surface of zircalloy 4 after oxidation show semicircle with long diameter and small contact angle compared with zircalloy 4 before oxidation. A partially wetting on the substrate appeared on the surface of zircalloy 4 before oxidation. For long spreading time, the wetted area is developing and the diameter of droplet increase while contact angle decrease for wettability's of nanofluid on the surface of zircalloy 4 before and after oxidation. The droplet size of nanofluid on the surface of zircalloy 4 after oxidation shows wide diameter compared with zircalloy 4 before oxidation. For long spreading time, the contact angle of nanofluid on the surface of zircalloy 4 after oxidation is small compared with zircalloy 4 before oxidation. It can be concluded that wettability's of nanofluid show good performance on the surface of zircalloy 4 after oxidation compare with zircalloy 4 before oxidation. The perf romance of nanofluid on surface of zircalloy 4 after oxidation shows good wettability properties due to the function of oxide scale zirconium oxide on the surface of zircalloy 4 [18] .

4. Conclusion

The wettability behavior of aquadest and nanofluids contains 0.01 gpl nanoparticle alumina on the surface of zircalloy 4 before and after oxidation samples were studied using sessile drop methods. The results obtained can be summarized as follows:

(i) The values of contact angle decrease gradually with increasing spreading time.
(ii) The diameter of droplet of aquadest and nanofluid increase with increasing spreading time.
(iii) The wettability of the aquadest and nanofluid on surface of the oxidized zircalloy 4 are better compared with non-oxidized zircalloy 4.

Acknowledgement

The authors would like to extend thanks to National Nuclear Energy Agency for financial support, and the director of Research Centre for Nuclear Science and Technology.

References

[1] H.L. Yang , Y. Matsukawa , S. Kano, Z.G. Duan , K. Murakami, H. Abe, 2016, Journal of Nuclear Materials 481, p. 117-124
[2] H.L. Yang, S. Kano, Y.Matsukawa , Y.F. Li , J.J. Shen, F. Li, Z.S. Zhao, Y. Satoh, H. Abe,2016, Materials and Design 104,p. 355–364
[3] M. Steinbrück ⇑, M. Böttcher, ,2011,Journal of Nuclear Materials 414 p. 276–285
[4] É. V. Li, M. V. Koteneva, S. A. Nikulin, A. B. Rozhnov, and V. A. Belov,2015 Metal Science and Heat Treatment, 57, Nos. 3 – 4, p.215
[5] Li Nie, Yongzhong Zhan ⇑, Tong Hu, Xiaoxian Chen, Chenghui Wang 2013,Journal of Nuclear Materials 442, p.100–105
[6] K. Sridharan, S.P. Harrington, A.K. Johnson, J.R. Licht, M.H. Anderson, T.R. Allen, Mater. Des. 28 (2007) p.1177-1185
[7] A.S. Kuprin et all, 2015,Journal of Nuclear Materials,p 400-406
[8] Albadr J, Tayal S, Alasadi M, 2013, Thermal Engineering 1, p38–44
[9] Koen G. Winkels, Joost H. Weijs, Antonin Eddi, and Jacco H. Snoeijer, 2012,, PHYSICAL REVIEW E 85,,055301
[10] Belaud V, et al, 2015, Tribological International, 82, p343–349
[11] Vivian S. Tong, T. Ben Britton, 2017 Acta Materialia 129,p 510-520
[12] Djoko Hadi Prajitno and Dani Gustaman Syarif, 2015, Conference Series: Materials Science and Engineering 214 (1), 012015
[13] Prajitno D H and Syarif D G, 2016, Journal of Physics: Conference Series 739 012029
[14] M. Reif, F. Scherm, M. C. Galetz, U. Glatzel, 2014,Oxidation of Metals 82.,p. 99-112
[15] O. Coindreau, C. Duriez, S. Ederli, 2010, , Journal of Nuclear Materials 405, p.207–215
[16] Aiguo Feng a, Benjamin J. McCoy a, Zuhair A. Munir a,*, Domenick Cagliostro, 1998,Materials Science and Engineering A242 p 50
[17] Jiangyou Long, Minlin Zhonga, Peixun Fan, Dingwei Gong, and Hongjun Zhang, 2015,Journal of Laser Applications 27, S29107
[18] P. Tsai, R. C. A. van der Veen, M. van de Raa, and D. Lohse, 2010, Langmuir 26, p. 055301-1-055301-4

IOP Conf. Series: Materials Science and Engineering **622** (2019) 012008 doi:10.1088/1757-899X/622/1/012008

rGO based photo-anode in dye-sensitized solar cells (DSSC) and its photovoltaic characteristics

Waode Sukmawati Arsyad[1*], Aricia Pristianti[1], La Agusu[1], and Rahmat Hidayat[2]

[1] Physics Department, Faculty of Mathematics and Natural Sciences, Halu Oleo University, Jl. Prof. HEA. Mokodompit No. 1 Kendari 93231, South East Sulawesi, Indonesia

[2] Physics of Magnetism and Photonics, Physics Department, Faculty of Mathematics and Natural Sciences, Bandung Institute of Technology, Jl. Ganesha 10 Bandung 40132, West Java, Indonesia

*Corresponding author: wdsukmawati@gmail.com

Abstract. On the attempt to increase transport properties of the photo-anode in DSSC, we synthesized rGO powder from graphite bar (commercially available) using modified Hummer's method. The SEM-EDS results had confirmed the attachment of the rGO layer to the FTO substrate. For DSSC cells, we made two cells configuration, the first one was stacking layers of rGO and TiO_2 resulting configuration of TiO_2/rGO/ TiO_2 (A2) and rGO/TiO_2/rGO (A3), where rGO was deposited by spin coating and TiO_2 was deposited by a screen printing technique. The second one, the rGO powder was mixed with TiO_2 paste with several ratios in weight, namely TiO_2:rGO 40:1 (B1), 40:2 (B2), and 40:8 (B3) and then deposited on the FTO substrate by screen printing. The reference cell was assigned as A1 (TiO_2 only). From the conductivity measurement using the four-probe method, the utilization of rGO layer increased the conductivity of photo-anode layer, namely (1.37, 2.9 and 6.3)x10^{-2} $\Omega^{-1}cm^{-1}$ for A1 to A3 and (1.5, 2.5, and 3.7)x10^{-2} $\Omega^{-1}cm^{-1}$ for B1 to B3. From the photovoltaic measurement, we found that the efficiency of the DSSC cell firstly increased with the insertion of rGO layer, from 1.8% (A1) to 4.59% (A2), and decreased to 3.22%, as the conductivity increased in A3. While for the composite of TiO_2:rGO, the efficiency of the cell reduced with the increased amount of rGO, from 3.45% (B1) to 2.9% and 1.9% for B2 and B3. We found that the reduction of photovoltaic performance was affected by two main factors, specifically, direct contact between rGO and redox species in the electrolyte, which induced recombination process, and conductivity of the photo-anode layer. To fully achieve the advantage of rGO utilization in photo-anode, once must be considered was the use of protection layer on top of the rGO layer to avoid direct contact between the rGO/electrolyte interface.

Keywords: rGO, DSSC, photo-anode, conductivity, photovoltaic, efficiency

Content from this work may be used under the terms of the Creative Commons Attribution 3.0 licence. Any further distribution of this work must maintain attribution to the author(s) and the title of the work, journal citation and DOI.
Published under licence by IOP Publishing Ltd

1. Introduction

Since the first time reported [1] until now, there have been numerous effort to improve the performance of the dye-sensitized solar cell (DSSC) through optimization of all its constituent components [2, 3, 4, 5, 6]. Among other components such as a dye, redox electrolyte and counter electrode whose only has one function, photo-anode serve a dual function in DSSC, specifically as the dye adsorption site and medium for charge transfer and transport. Photo-anode of a DSSC consist of a dye-sensitized semiconducting layer on top of a conducting substrate. To be effective, photo-anode must fulfill some characteristics specifically, high surface area for dye loading, high transparency, high electron mobility, low recombination rate, and contains a hydroxyl group or defect states for dye attachment [7, 8, 9]. FTO is the most frequently used conducting substrate owing to its resistance that does not significantly change, even after heating up to 700^0C [10], while TiO_2 is the most utilized n-type metal oxide semiconductor in DSSC, due to its high chemical stability [11, 12], high dielectric constant and fewer defect states, which leads to less recombination [13, 14]. Despite TiO_2 nano-porous layer ability to provide large surface area for dye loading and light absorption [7, 15], but it also acts as the source of recombination site at TiO_2/electrolyte and FTO/electrolyte interfaces [7, 16]. The presence of the recombination process will induce photocurrent reduction and eventually affect the overall conversion efficiency of the DSSC cells. The major challenge for the DSSC performance improvement is to improve the charge transfer and transport in the TiO_2 layer since efficient charge transfer and transport will be able to compete with the recombination process. One way to accomplish efficient charge transfer and transport are by introducing charge carriers materials into photo-anode. This can be done through doping of semiconducting layer with Mg [17], Sb [18], Zr [18], N [19], S [20] and Cr [21], or through modification of photo-anode configuration, with the utilization of blocking layer [5, 22], scattering layer [23, 24] or incorporation efficient charge transport material such as carbon nanotube [25] and graphene [26, 27, 28].

Graphene has gained many attentions due to its excellent properties including optical transparency, high stability, low cost, and non-toxicity. Moreover, its electrical conductivities in two dimension can reach more than $2x10^5$ $cm^2V^{-1}s^{-1}$ [26, 29], and can easily anchor with TiO_2 through intermolecular forces [30] and thus provide a fast channel for electron transportation [31]. Graphene or its partly reduce part forming graphene-like sheets or reduced Graphene Oxide (rGO) can be used both in photo-anode and counter electrode part of DSSC. In photo-anode, rGO is used as blocking layer (BL) [32], mixed with TiO_2 particles to form composites or scaffold layer [33, 34] and stacked layer by layer with TiO_2 [35]. Although all the previously reported studies on the incorporation of graphene particularly rGO in DSSC stated that there are photovoltaic (PV) performance improvement with the utilization of rGO, some studies also reported the reduction of PV performance as the escalation amount of rGO content in the photo-anode [36]. The PV performance reduction, in this case, was attributed to the low light harvesting by the dye molecules due to the absorbing properties of rGO itself.

Herein, we investigated the possibility that there may be another mechanism responsible for PV performance reduction in DSSC utilized rGO. For that purpose, we fabricated two different structure of photo-anode. The first one is by stacking rGO and TiO_2 layer-by-layer forming configuration of TiO_2/rGO/TiO_2 and rGO/TiO_2/rGO, and the second is by mixing rGO and TiO_2 with different weight ratio to form composite or scaffold layer deposited on top FTO conductive substrate, where the photovoltaic parameters were analyzed from the current-voltage (I-V) measurement.

2. Experimental

2.1. Reduced Graphene oxide (rGO) preparation

2.1.1. Graphene oxide (GO) preparation. GO preparation was following the same method that already reported elsewhere based on modified hummers method re-described in [37, 38]. The graphite bar was crashed until smooth and strained using 250 mesh filter to form homogeneous graphite powder. 2 gr of graphite powder was mixed 98 ml of 98% Sulphuric Acid (H_2SO_4) and vigorously stirred for four hours in the ice bath environment at 0^0C temperature. During the stirring process, 8 gr of Potassium Permanganate ($KMnO_4$) and 4 gr of Sodium Nitrate ($NaNo_3$) was added very slowly into the mixture until the color of the solution turned into blackish green. The temperature then slowly raised to 35^0C for 24 hours. The color of the suspension slowly turned to light brown, and the suspension became more viscous. 200 ml of distilled water slowly added into the suspension while stirred for one hour, where the color changed into dark brown and became golden yellow with the addition of 15 ml of Hydrogen Peroxide (H_2O_2). The suspension was then precipitated for 24 hours to separate the liquid phase and solid phase. The solid phase was washed with 10 ml Hydrochloric Acid (HCl) 37% and distilled water several times until the pH neutralized, dried in the oven at 110^0C for 12 hours. 40 mg of the resulting graphite oxide was then diluted into 40 ml of distilled water, stirred overnight and sonicated with the frequency of 53 KHz for 120 minutes.

2.1.2. Reduction of Graphene Oxide using Zinc particles and thermal process. 10 ml HCl 37% added into the pre-prepared GO solution, to form an acidic environment. 1.6 gr of Zinc powder then added into the GO solution and let for 16 hours. The mixture then stirred for one hour and added 10 ml HCl 37% again to remove the Zn particles that did not react in the solution, washed with distilled water until the pH of the solution was neutralized and dried in the oven at 110^0C for 12 hours. The resulted rGO powder then subsequently dried in the microwave with 800 Watt power for two minutes with five seconds delay at every 30 seconds. The utilization of microwave in the drying process to produce homogeneous heat in relative very short time for speeding up the reduction time [38].

2.2. DSSC preparation

The materials and preparation methods for fabricating DSSCs here were basically the same as reported in [1], where all materials were used as received. The complete DSSC samples were made by the same fabrication method as reported in our previous work [39, 40] for preparing the A1 reference cell. First, the conductive substrate (FTO, surface resistivity = 15Ω/ per square) was cleaned with teepol for removing dirt, rinsed with ethanol (technical) and acetone (technical), and ultra-sonicated for 15 minutes in the mixture 1:1 (in volume) of Ethanol (Merck) and Isopropanol (Merck), and dried with air gun. One layer of transparent TiO_2 paste (18-nrt from Dye-Sol) was deposited by screen-printer (0.5 cm x 0.5 cm) followed by the heating process on the hot-plate for five minutes in 80^0C. The procedures were repeated two times for resulting thicker transparent TiO_2 layer. One layer of reflector TiO_2 paste (Wer-02 from Dye-Sol) was deposited using the same size of screen printer on top of former TiO_2 layers, followed with heating process on the hot-plate where the temperature gradually increased from 80^0C with the step of 20^0C / 6 minutes until it reached the final temperature of 500^0C, where it was maintained for 30 minutes. The resulting TiO_2 electrode then immersed in 0.7mM of cis-bis(isothio-cyanato)bis(2,2'-bipyridyl-4,4'-dicarboxylato) ruthenium(II)-bis-tetrabutylammonium (N719 dye, Solaronix) for 24 hours. The photo-anode and the counter electrode (1mm drilled FTO layered with Pt from Solaronix) were assembled and separated by a separator (Surlyn® of 50 μm thickness), the electrolyte solution (Mosalyte TDE-250 from solaronix) then injected through the hole of the counter electrode using a vacuum pump.

In this research, we made two cells configuration. In the first one, the rGO was made layer by layer with the TiO_2 layer resulting configuration of rGO/TiO_2/rGO (A2) and TiO_2/rGO/TiO_2 (A3). The rGO solution was dropped on top of FTO substrate, and let for five seconds before spin coated at 2000 rpm

IOP Conf. Series: Materials Science and Engineering **622** (2019) 012008 doi:10.1088/1757-899X/622/1/012008

for 30 seconds, and subsequently heated on the hot plate at 250^{0}C for two minutes. TiO_2 layer was deposited by a screen printing technique. In the second one, the rGO powder was mixed with TiO_2 paste with several ratios in weight, specifically TiO_2:rGO 40:1 (B1), 40:2 (B2), and 40:8 (B3) and then deposited on the FTO substrate by screen printing, following by annealing. In all samples, the procedures for preparing TiO_2 layers following the same steps for preparing the reference cell (A1).

2.3. Characterization

The resistance of the rGO layer was measured by using Four Probe Methode (FPM) at the bias voltage of 9 Volt. Morphology of rGO and stacking layer of rGO-TiO_2 was scanned using SEM. The fabricated DSSCs were characterized by the current density-voltage (J–V) measurements using a solar simulator (Newport, Model no. 11000) with the average light power density of 100mW/cm^2.

3. Result and Discussions

3.1. Morphology of rGO and stacked layer of rGO-TiO_2

Fig.1 shows the SEM results of one layer rGO, TiO_2/rGO/ TiO_2, and rGO/TiO_2/rGO. Fig.1(a) is the rGO layer that shows morphology like a stack of corn flakes. The distribution of the rGO flake is not homogeneously covered the substrate, which may cause a drawback when functioning as the blocking layer. Fig.1(b) shows the morphology of a stacked layer of TiO_2/rGO/ TiO_2, where TiO_2 was screen printed and the rGO was spin coated. The rGO is not evenly distributed on top of the TiO_2 layer and it seems that the spin coating process induces agglomeration of rGO particle forming bigger particles.

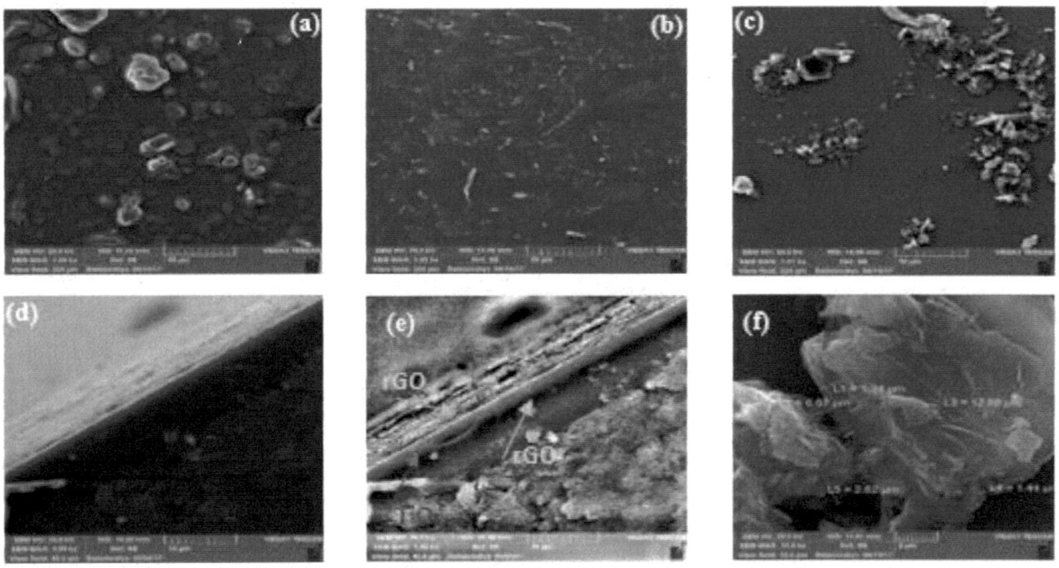

Figure 1. SEM results of one layer rGO (a), (b) TiO_2/rGO/TiO_2, (c) surface of rGO/TiO_2/rGO, cross section of rGO/TiO_2/rGO (d), enhanced version of cross section of rGO/TiO_2/rGO using ImageJ (e), and (f) thickness and width estimation of rGO flake.

Fig.1(c) shows the surface morphology of the rGO/TiO_2/rGO layer, the upper rGO layer agglomerates to form islands with a different size that scatter on top of the substrate. Fig.1(d) is the cross-section of the rGO/TiO_2/rGO layer. The interface between each layer can be seen, where the structure does form layer by layer, that was confirmed through the images contrast enhancement using ImageJ software in Fig.1(e). The SEM results confirmed that the size of the rGO particles that yielded is still too big, which may cause uneven surface, resulting in poor contact between the interfaces.

IOP Conf. Series: Materials Science and Engineering **622** (2019) 012008 doi:10.1088/1757-899X/622/1/012008

The thickness and width estimation of rGO flake is shown in Fig.1(f), where the thickness varies from (0.67 – 2.62) μm, while the widest rGO flake is 12.68 μm. This result is in good agreement with another study [41] who has also produced two-dimensional wrinkle sheets with the size of several micrometers. Fig.2 confirm the existence of rGO on the substrate through the EDS spectrum, with 60.98% of C atoms.

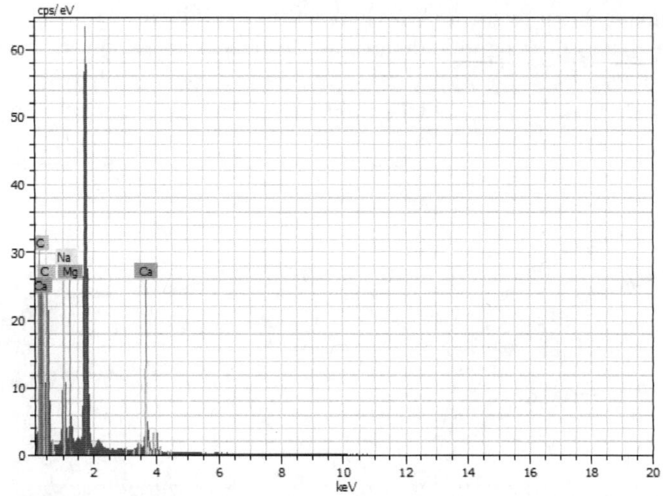

Figure 2. EDS result for one layer rGO on top of a glass substrate

3.2. Conductivity of rGO and rGO-TiO_2 composites layer

The conductivity of all the samples that were calculated from the V-I measurement using FPM is tabulated in Table 1. Overall, the utilization of rGO increases the conductivity of all samples compared to the reference cell (A1). The increasing of the conductivity in consecutive order are 111.6%, 359.8%, 9.5%, 82.5%, and 170.0% for A2, A3, B1, B2, and B3 compared to A1. For layer by layer structure (A2 and A3), A3 with TiO_2 in between two rGO layer shows 53.9% higher conductivity compared to A2 where rGO position is in between two TiO_2 layer. While for the composite of TiO_2-rGO (B1, B2, and B3), the conductivity increase with the amount of rGO added.

Table 1. Conductivity calculation of TiO_2, stacked layer of TiO_2 and rGO and composite of TiO_2 and rGO, measured with FPM at 9 Volt bias voltage.

Cell	V (Volt)	I (x 10^{-4} A)	Thickness (x10^{-3} cm)	ρ (Ωcm)	σ (x10^{-2} $Ω^{-1}$ cm^{-1})
A1	0.35	0.5	2.3	73.1	1.37
A2	0.36	1.1	2.3	34.14	2.9
A3	0.35	2.3	2.3	15.87	6.3
B1	0.37	0.6	2.3	64.66	1.5
B2	0.38	1.0	2.3	39.61	2.5
B3	0.38	1.5	2.3	25.53	3.7

Table 2. Photovoltaic parameters of DSSC incorporated rGO in the photo-anode

Cell	Configuration / composition	V_{oc} (Volt)	J_{sc} (mA/cm^2)	FF (%)	η (%)
A1	TiO_2 only (ref. cell)	0.67	4.62	59.0	1.80
A2	TiO_2/rGO/TiO_2	0.70	12.2	53.7	4.59
A3	rGO/TiO_2/rGO	0.70	10.0	46.7	3.22
B1	TiO_2:rGO 40:1	0.68	10.0	51.0	3.45
B2	TiO_2:rGO 40:2	0.70	7.68	55.0	2.90
B3	TiO_2:rGO 40:8	0.76	4.45	56.0	1.90

3.3. DSSCs photovoltaic properties

Fig.3 and Fig.4 show the influence of the exertion of rGO in photo-anode to the photovoltaic properties of the resulted DSSC cells, and the parameters including photocurrent (J_{sc}), open circuit voltage (V_{oc}), fill factor (FF) and overall conversion efficiency (η), tabulated in Table 2. The performance of the A1, A2 and A3 cells is depicted by its J-V curves under A.M 1.5 condition (Fig.3). As can be seen that rGO utilization has a pronounced impact on the photocurrent value, in which A2 configuration resulted in 164% photocurrent enhancement over the A1. This may be explained by the fact that graphene can act as an electron bridge in the photo-anode, shuttling electrons to the current collector and lowering recombination in the device [42]. Incorporation graphene materials in photo-anode also increase dye absorption capability [27, 32, 43], increased the light harvesting and eventually the photocurrent.

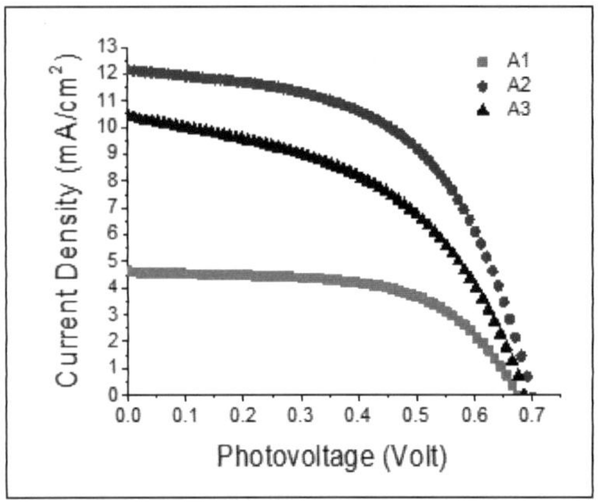

Figure 3. J-V curves of DSSC with the photo-anode configuration of TiO_2/rGO/TiO_2 (●), rGO/TiO_2/rGO (Δ), and the reference cell using TiO_2 only (▪)

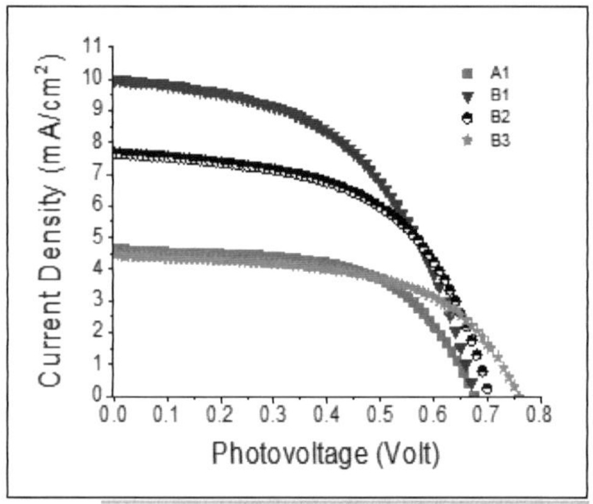

Figure 4. J-V curves of DSSC with photo-anode incorporate composite of TiO_2 and rGO with different weight ratio. The reference cell using TiO_2 only also shown.

In the A3 configuration, where TiO_2 is in between two rGO layer, the photocurrent decrease by 17.6%, although it has higher conductivity compared to A2. The decrease in the photocurrent may be due to the cell configuration, where the first rGO layer is in direct contact with the FTO and therefore act as a blocking layer (BL), providing a barrier between triiodide and FTO conductive substrate, but it would also provide an insulating barrier [7]. This insulating barrier will potentially retard electron diffusion from TiO_2 to FTO or lengthen electron path distance and opening up possibilities for recombination process to take place. Therefore, the rGO layer must be very thin so that the tunneling effect can be effective. Besides that, BL can absorb some light as much as 1.6% [7], that will reduce the light absorbed by the dye, which is in agreement with the result from Chen, et al [36], that reported the light utility of dyes to decrease by thick graphene coating on FTO. The existence of the second rGO layer, which is in direct contact with redox couple (I^-/I_3^-) also may contribute to the decrease in photocurrent. Graphene has excellent electrical conductivity, so there will be a large number of recombination between the interface of graphene and electrolyte that will reduce the photocurrent. Moreover, for a metallic or semi-metallic surface, the proximity of an ion induces a further effect associated with polarization of the surface that strongly affects the interfacial attraction/repulsion of the ions [44]. Graphene has an abundance of the aromatic ring with delocalized π-electrons [45], and from the computational simulation from Shi et.al, [46] revealing that there are strong anion-π interactions on the graphene flake that associated with effective donor-acceptor interactions between the halide anions and the graphene flake. This possible interaction between graphene and iodide ions in the electrolyte will reduce the number of iodide ions that supposed to regenerate the dye, and hence will diminish the amount of dye that can capture the incoming light, and eventually reduce the photocurrent (Fig. 3, Table 2).

In accordance with photocurrent value, the V_{oc} of the DSSC incorporated rGO layer also shows a 4.47% improvement for A2 and A3 compared to the reference cell (A1). The V_{oc} value strongly determines by the Fermi level between the TiO_2 conduction band edge and the potential energy of the redox couple. As the conductivity of A2 and A3 are higher than A1 (Table 1), the Fermi level then shifts upwards, inducing to an increase in Voc (Fig. 2, Table 1).

J-V curves for B1, B2, B3, and A1 as the reference cell and the tabulated parameters were shown in Fig. 3 and Table 2. The photocurrent shows an improvement with the incorporation of rGO in the TiO_2 composite or scaffold layer as much as 117.3% compared to the reference cell, due to increasing of light absorption at the N717 dye's absorption peak [34]. Further addition of rGO into the composite reducing the photocurrent by 23.5% and 55.5% for B2 and B3 respectively, compared to B1. The photocurrent reduction with the increasing content of rGO is inversely proportional with the conductivity value of the thin film (Table 1). This result indicated that there is a limit for graphene content in the composite of TiO_2:rGO to produce a maximum photocurrent. At the right amount, the inclusion of rGO could induce the formation of macropores due to the formation of voids formed from the further reduction process in the volume of graphene oxide during annealing process of the photo-anode layer which acted as the scattering layer [8, 31], and opened up a channel for electrolyte infiltration.

The Voc of the DSSC with a composite of TiO_2-rGO photo-anode also shows an improvement of 1.49% compared to A1. The increasing amount of rGO in the composite give rise the Voc by 2.94% and 11.76% for B2 and B3 compared to B1. The incorporation of rGO has the distinct effect to the photocurrent value for both photo-anode configuration (stacked layer and composite), and only slightly changed the Voc value (Fig. 2, Fig. 3 and Table 2).

4. Conclusion

The improvements in PV performance in DSSC with the utilization of rGO in photo-anode has been verified. Furthermore, increasing rGO content in the photo-anode will increase its conductivity, however as the conductivity increases the overall conversion efficiency of the DSSC cells were decreased which is mainly attributed to the reduction of the photocurrent and fill factor. Lower fill factor value implies that incorporation of rGO in photo-anode increases series resistance of the cell, due to imperfection interaction at the interface of rGO and TiO_2.

References

[1] O'Regan B and Gratzel M, *Nature,* vol. 353, no. 6346, pp. 737-739, 1991.

[2] Nazeeruddin MK, Kay A, Rodicio I, Humphry-Baker R, Miller E, Lisk P, Vlachopoulos N and Gratzel M, *Journal of the American Chemical Society,* vol. 115, no. 14, p. 6382, 1993.

[3] Wang M, Gratzel C, Zakeeruddin SM and Gratzel M, *Energy & Environmental Science,* vol. 5, no. 11, p. 9304, 2012.

[4] Kakiage K, Aoyama Y, Yano T, Otsuka T, Kyomen T, Unnoc M and Hanaya M, *Chemical Communications,* vol. 50, no. 48, p. 6379, 2014.

[5] Suresh S, Deepak TG, Ni C, Sreekala CNO, Satyanarayana M, Nair AS and Pillai VPM, *New Journal of Chemistry,* 2016.

[6] Dissanayake MAKL, Ekanayake EMBS, Bandara LRAK, Seneviratne VA, Thotawatthage CA, Jayaratne SL and Senadeera GKR, *J. Solid State Electrochem,* vol. 20, pp. 193-201, 2016.

[7] Yeoh ME and Chan KY, *Int. J. Energy Res,* vol. 41, no. 15, pp. 2446-2467, 2017.

[8] Sengupta D, Dasa P, Mondal B and Mukherjee K, *Renewable and Sustainable Energy Reviews,* vol. 60, pp. 335-376, 2016.

[9] Raj CC and Prasanth R, *Journal of Power Sources,* vol. 317, pp. 120-132, 2016.

[10] J. D. Roy-Mayhew and I. A. Aksay, *Chem. Rev.,* vol. 114, pp. 6323-6348, 2014.

[11] Gadisa A, Liu Y, Samulski ET and Lopez R, *Appl. Phys. Lett,* vol. 100, p. 253903, 2012.

[12] Tian H, Hu L, Zhang C, Chen S, Sheng J, Mo L, Liu W and Dai S, *J. Mater. Chem,* vol. 21, p. 863, 2011.

[13] Kim SK, Kim WD, Kim KM, Hwang CS and Jeong J, *Appl. Phys. Lett.,* vol. 85, p. 4112, 2004.

[14] Lee YJ, Davis RJ, Lloyd MT, Provencio PP, Prasankumar RP and Hsu JWP, *IEEE J. Sel. Top. Quantum Electron,* vol. 16, p. 1587, 2010.

[15] H. Yu, S. Zhang, H. Zhao, G. Will and P. Liu, *Electrochemica Acta,* vol. 54, no. 1, pp. 1319-1324, 2009.

[16] Noh SI, Bae KN, Ahn HJ and Seong TY, *Ceramics International,* vol. 39, no. 7, pp. 8097-8101, 2013.

[17] Kakiage K, Tokutome T, Iwamoto S, Kyomen T and Hanaya M, *Chemical Communication,* vol. 49, p. 179, 2013.

[18] Liu H, Tang J, Kramer IJ, Debnath R, Koleilat GI, Wang X, Fischer A, Li R, Brozowski L, Levina L and Sargent EH, *Advanced Matter,* vol. 23, p. 3832, 2011.

[19] Ma T, Akiyama M, Abe E and Imai I, *Nano Letters,* vol. 5, p. 2543, 2005.

[20] Sun Q, Zhang J, Wang P, Zheng J, Zhang X, Cui Y, Feng J and Zhu Y, *J. Renewable Sustainable Energy,* vol. 4, p. 023104, 2012.

[21] Kim C, Kim KS, Kim HY and Han YS, *J. Mater. Chem.,* vol. 18, p. 5809, 2008.

[22] Choi H, Nahm C, Kim J, Moon J, Nam S, Jung DR and Park B, *Current Applied Physics,* vol. 12, no. 3, pp. 737-741, 2012.

[23] Yu J, Li Q and Shu Z, *Electrochemica Acta,* vol. 56, no. 18, pp. 6293-6298, 2011.

[24] Song W, Gong Y, Tian J, Cao G, Zhao H and Sun C, *ACS Applied Materials & Interfaces,* vol. 8, no. 21, pp. 13418-13425, 2016.

[25] Brown P, Takechi K and Kamat PV, *J. Phys. Chem. C,* vol. 112, no. 12, pp. 4776-4782, 2008.

[26] Ding H, Zhang S, Chen JT, Hu XP, Du ZF, Qiu YX and Zhao DL, *Thin Solid Films,* 2015.

[27] Yang H, Guai GH, Guo C, Song Q, Jiang SP, Wang Y, Zhang W and Li CM, *J. Phys. Chem. C,* vol. 115, p. 12209, 2011.

[28] Yang NL, Zhai J, Wang D, Chen YS and Jiang L, *ACS Nano,* vol. 4, no. 2, pp. 887-894, 2010.

[29] Si Y and Samulski ET, *Nano Letters,* vol. 8, p. 1170, 2008.

[30] Williams G, Seger B and Kamat PV, *ACS Nano,* vol. 2, p. 1487, 2008.

[31] Fang X, Li M, Guo K, Zhu Y, Hu Z, Liu X, Chen B and Zhao X, *Electrochemica Acta,* vol. 65, p. 174, 2012.

[32] Kim SR, Parvez MK and Chowala M, *Chem. Phys. Lett,* vol. 483, p. 124, 2009.

[33] Peining Z, Nair AS, Shengjie P, Shengyuan Y and Ramakrishna A, *ACS Appl. Mater. Interfaces,* vol. 4, p. 581, 2012.

[34] Durantini J, Boix PP, Gervaldo M, Morales GM, Otero L, Bisquert J and Barea EM, *Electroanal. Chem.,* vol. 683, no. 43, 2012.

[35] Peiris DSU, Ekanayake P and Petra MI, *Organic Electronics,* vol. 59, pp. 399-405, 2018.

[36] Chen T, Hu W, Song J, Guai GH and Li CM, *Adv. Funct. Mater.*.

[37] Agusu L, Ahmad LO, Anggara D, Alimin, Mitsudo S, Fujii Y and Kikuchi H, *Journal of Physics Conf. Series,* vol. 1011, p. 012012, 2018.

[38] Agusu L, Ahmad LO, Alimin, Nurdin M, Herdianto, Mitsudo S and Kikuchi H, *IOP Conf. Series. Material Science & Engineering,* vol. 367, p. 012002, 2018.

[39] Arsyad WOS, Bahar H, Prijamboedi B and Hidayat R, *Ionics,* vol. 24, no. 3, p. 901, 2018.

[40] Arsyad WOS, Januari A, Jumalia, Ilmawati WS, Usman I and Hidayat R, *Journal of Nonlinear*

Optical Physics & Materials, vol. 27, no. 2, p. 1850019, 2018.

[41] U.-a. Kanta, V. Thongpool, W. Sangkhun, N. Wongyao and J. Wootthikanokkhan, *Journal of Nanomaterials,* vol. 2017, no. Article ID 2758294, p. 13 pages, 2017.

[42] Yang N, Zhai J, Wang D, Chen Y and Jiang L, *ACS Nano,* vol. 4, p. 887, 2010.

[43] Tsai TH, Chiou SC and Chen SM, *Int. J. Electrochem. Sci,* vol. 6, p. 3333, 2011.

[44] Sunner J, Nishizawa K and Kebarle P, *J. Phys. Chem,* vol. 85, pp. 1814-1820, 1981.

[45] Williams CD, Dix J, Troisi A and Carbone P, *J. Phys. Chem. Lett,* vol. 8, pp. 703-708, 2017.

[46] Guosheng S, Yihong D and Fang H, *Journal of Computational Chemistry,* vol. 33, pp. 1328-1337, 2012.

[47] Fang X, Li M, Guo K, Zhu Y, Hu Z, Liu X, Chen B and Zhao X, *Electrochemica Acta,* vol. 65, p. 174, 2012.

IOP Conf. Series: Materials Science and Engineering **622** (2019) 012009 doi:10.1088/1757-899X/622/1/012009

Effect of Sodium Dodecylbenzene Sulfonate as Anionic Surfactant on Water based Carbon Nanofluid Performance as Quench Medium in Heat Treatment

L Oktavio[1], C A Ramadhani[1], D Rakhman[1], S Harjanto[1], W N Putra[1]

[1]Department of Metallurgical and Materials Engineering, Faculty of Engineering, Universitas Indonesia, Depok, 16424, Indonesia

*Corresponding author: laurent.oktavio@ui.ac.id

Abstract. Quenching is performed as part of steel heat-treatment to enhance mechanical properties, by rapid cooling. Factors that affect the selection of quench medium are hardenability of material, geometry, and dimensions of the component. In recent developments, nanofluids are used to improve heat transfer capacity. In this research, nanofluids were synthesized using the two-step method. Milling of particles was done using a high energy ball mill for 15 hours at 500 rpm. Observation of particle size, material composition, and morphology of particle, and surface changes of the particle were measured by Field-Emission Scanning Electron Microscope (FE-SEM), and Energy Dispersive X-Ray Spectroscopy (EDX). Water-based nanofluids with a volume of 100ml were produced using the two-step method, with carbon concentrations of 0.1%, and 0.5% and Sodium Dodecylbenzene Sulfonate concentrations of 0%, 1%, 3%, and 5%. Samples of S45C steels were austenized at 1000°C for 60 minutes. Hardness testing results correspond to the severity of the quenching mediums, with peak hardness of 845 HV for 0.1% Carbon with 1% SDBS, and 878 HV for 0.5% carbon with 3% SDBS. Hardness testing results show a significant improvement over results without SDBS addition. Excess surfactant addition yields a lower hardness due to the re-agglomeration of particles..

1. Introduction

Quenching is the cooling process in heat treatment that hardens the treated component[1]. Hardening of steels done by quenching involves the transformation of the martensitic phase from the austenized steel component. The component is first heated above the austenization temperature (>723°C) which then it is put into contact with a heat transfer medium. Upon contact, the quenchant will facilitate heat transfer within the component-quenchant interface. Selection of heat transfer medium will affect the final physical properties, residual stresses and mechanical properties of the component after the heat treatment process. Nanofluids are enhanced heat transfer fluids that are commonly used to improve cooling performance in heat exchangers, solar collectors and, radiators. Nanofluids are obtained by adding nanoparticles into a base fluid. [2] Yang et.al.[3] reviewed that the addition of nanoparticles into a base fluid showed a significant increase in thermal conductivity. However, nanoparticle dispersion in the base fluid is showed to have an effect in the heat transfer capacity of the nanofluid. Azadeh et.al. [4] revealed that the stability of nanofluids is significantly improved with the addition of surfactants and ultrasonic processing. Sodium

Content from this work may be used under the terms of the Creative Commons Attribution 3.0 licence. Any further distribution of this work must maintain attribution to the author(s) and the title of the work, journal citation and DOI.
Published under licence by IOP Publishing Ltd

Dodecylbenzene Sulfonate (SDBS) is an anionic surfactant, commonly used in household products such as laundry detergent, dish soap, and floor cleaner [5]. The addition of the surfactant is done to reduce the surface tension between carbon nanoparticles and the base fluid, water. This study aims to observe the properties of quenched S45C steel using carbon nanoparticles suspended in water. Sodium Dodecylbenzene Sulfonate (SDBS) is added to assist the dispersion of the carbon nanoparticles. Ultrasonication is done to further stabilize the mixture. S45C steel samples were heated to 1000°C and held for 60 minutes followed by immersion quenching into the nanofluid.

2. Materials and Methods

Lab-grade carbon particles procured from Sigma-Aldrich were used to prepare nanofluids by using the standard two-step method. The two-step method was selected due to it being more cost and energy efficient. Nanofluids with concentrations of 0.1% and 0.5% carbon were prepared with varying SDBS concentrations of 0%, 1%, 3% and 5%. First, carbon particles were subject to High Energy Ball Mill (HEBM) at 500 rpm for 15 hours. Then, carbon particles were weighed accordingly and dispersed with SDBS to distilled water. Ultrasonication was done for 280 seconds to ensure the stability of the mixture and prevent agglomeration.

S45C steel rods were cut to a dimension of 15mm x 10mm x 10mm. The samples were then pre-heated in a furnace to 600°C and then left to austenize at 1000°C for an hour After exactly an hour, the S45C steel samples were immersed into the nanofluids. Microstructural observation was done using a Light Optical Microscope (LOM). The quenched S45C samples were then prepared for microstructural observation by using standard metallographic preparation procedures, which includes surface grinding of the S45C samples from 100 grit to 1500 grit sandpaper with 100 grit increments. Followed by polishing using velvet cloth and etching using 2% Nital.

Characterization of the carbon steel rod was done by Optical Emission Spectroscopy (OES) to determine its chemical composition. Field Emission Scanning Electron Microscope (FESEM) imaging was conducted to observe the average particle size of the lab-grade carbon powder. Microstructure analysis was carried out using a Light Optical Microscope (LOM) to observe phases formed after quenching. Preparation for metallographic analysis was done by using standard metallographic preparation procedure. The samples were etched using 3% nital solution which consists of 3 ml nitric acid and 97 ml ethyl alcohol on the polished surface of the sample. Etching of the samples was done by immersion into the etching solution for three seconds. Microstructure images were taken using an optical microscope at 100X.

Vickers hardness testing was done to observe the hardness value of each sample. Testing was done by applying a force of 300 gf using a diamond indenter on the samples' surface, and the load is given for 10 seconds. The resulting indent on the sample's surface is then examined using a microscope to be measured for its size and to be calculated for its hardness value. The testing was done on five different points on the samples' surface to obtain the mean hardness value of each point.

3. Results and discussion

3.1. Carbon powder characterization

Characterization of the carbon nanoparticles was done using a Field Emission Scanning Electron Microscope (FE-SEM) as observed in Figure 1. The FE-SEM imaging was conducted to observe the average particle size of the lab-grade carbon powder at magnifications of 500X and 1000X. From the result, it was determined that the average size of the carbon particle used in this study was 15.08 μm. Furthermore, the lab-grade carbon powder was also tested for impurities by using an Energy Dispersive X-Ray (EDX). The EDX result presented in Table 1 showed that the lab-grade carbon powder used in this research was free from any impurities.

Figure 1. FE-SEM image of lab-grade carbon powder at magnification of (a) 500X and (b) 1000X.

Table 1. The result of EDX for chemical composition (wt%) of laboratory-grade carbon powder

Laboratory-grade carbon powder	C	O
	94.63%	5.37%

It is seen in Table 1 that the EDS shows the carbon powder used in this study has a purity of 94.63%. It is observed that an oxygen content of 4.45% is present as an impurity. This confirms that the carbon used in this study consist of >90% carbon.

3.2. S45C steel characterization

Table 2. The chemical composition of S45C carbon steel (%) [6]

Fe	C	Si	Mn	P	S	Cr	Mo
98.3	0.47	0.287	0.718	0.0261	0.005	0.028	0.005
Ni	**Al**	**Co**	**Cu**	**Nb**	**Ti**	**V**	**W**
0.005	0.02	0.003	0.018	<0.002	0.035	<0.002	<0.01
Pb	**Sn**	**B**	**Ca**	**Zr**	**As**	**Bi**	
0.025	<0.002	0.0013	0.0005	<0.002	0.008	<0.03	

The OES results shown in Table 2 of the S45C sample show that the steel used is in accordance with the AISI 1045 steel standard. S45C steel was chosen because medium carbon steel is heat treatable and changes in properties by heat treatment process can be observed.

3.3. Microstructure analysis

It is observed in Figure 2, the microstructure of S45C after being quenched in water-based carbon nanofluid with varying surfactant concentration, the structure mainly consists of pearlite, which is represented in the brighter regions, while martensite structure appears in the darker regions. In figure 2a, for S45C quenched in pure distilled water, shows mostly pearlite regions, with few martensite regions. In figure 2b, it is observed that the contrast between the darker and the lighter regions are more apparent, this shows the addition of carbon particles increase the cooling rate of the S45C sample. In figure 2c, the martensite structure that occurs is quite apparent, as seen in the dark areas. As in figure 2d and 2e, martensite structures were not as heavily shown as previously. Figure 2f was quenched in 0.5% carbon without surfactant, very few martensite structures that appeared, this may happen due to the agglomeration of carbon particles in the fluid [7]. In figure 2g and 2h, quite significant numbers of martensite regions were observed, while figure 2i shows few martensite structures are observed

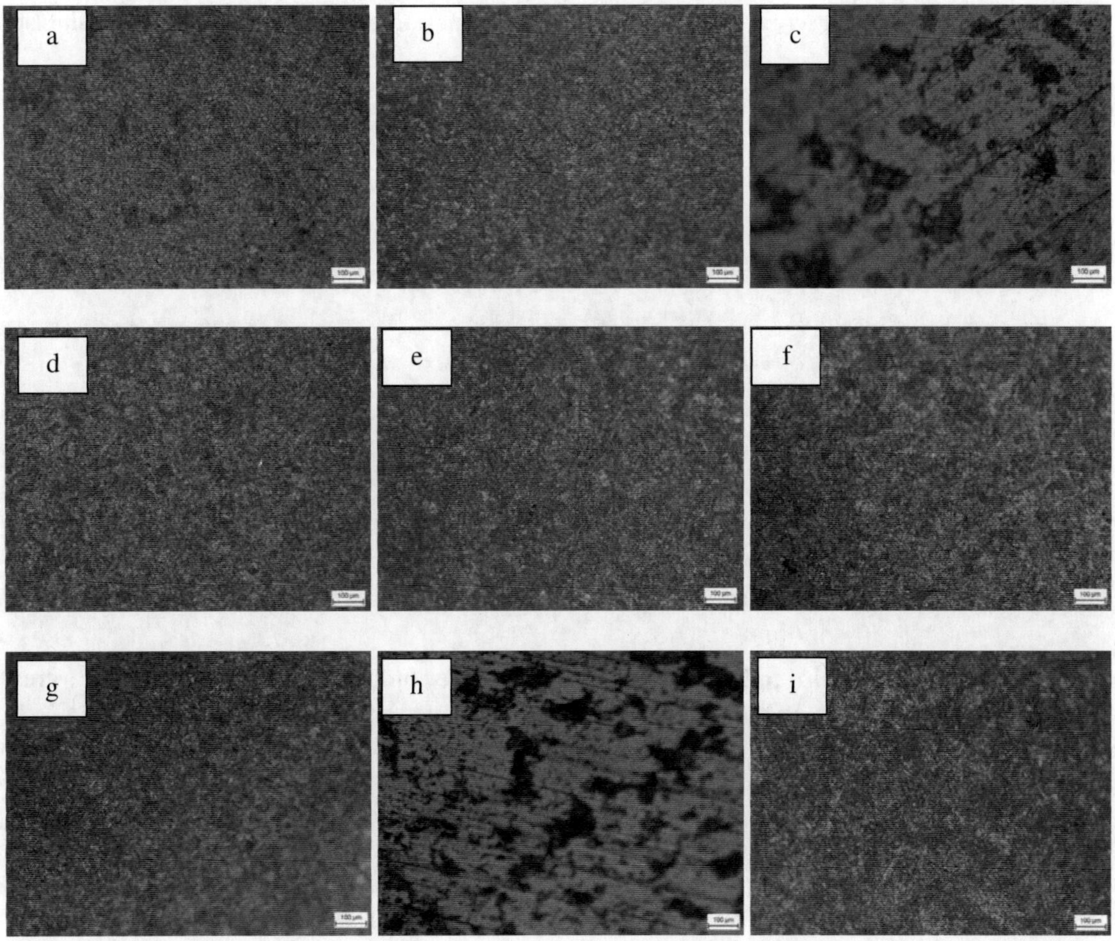

Figure 2. Microstructure of S45C after quenching in water based laboratory grade nanofluid (a) 0% vol. carbon, pure distilled water, (b) 0.1% vol. carbon, (c) 0.1% vol. carbon, 1% SDBS, (d) 0.1% vol. carbon, 3% SDBS, (e) 0.1% vol. carbon, 5% SDBS, (f) 0.5% vol. carbon, (g) 0.5% vol. carbon, 1% SDBS, (h) 0.5% vol. carbon, 3% SDBS, and (i) 0.5% vol. carbon, 5% SDBS. Magnification 100X.

Figure 3 compiled the hardness result for the S45C sample in all experiment variables. The sample quenched in pure distilled water has an average hardness of 667 HV. For a small increase in found in the sample quenched with 0.1% carbon nanofluid with no surfactant, which results in 712 HV. The hardness result for nanofluid without any added surfactant was similar to our previous result, whereas the 0.5% carbon was lower than 0.1% carbon nanofluid [8]. A significant increase in hardness found with the addition of 1% SDBS in the 0.1% carbon nanofluid, this results in a hardness value of 845 HV. However, after further addition of SDBS of 3% and 5% respectively, the hardness value was observed to be decreasing.

At 0.5% carbon concentration with no surfactant added, the hardness value was found to be quite low, which is due to the particle agglomeration [9]. However, with increasing surfactant addition at 1% and 3% SDBS addition, the hardness value increased and peak hardness value in this study was observed to be at 0.5% carbon with 3% SDBS. Further addition of 5% SDBS in 0.5% carbon nanofluid, caused the hardness value to decrease. The peak hardness for water-based nanofluid with SDBS addition was obtained at 0.5% vol. carbon nanofluid with 3% SDBS addition at 878 HV. The hardness value is in line with the microstructural observation, as the 0.5% carbon nanofluid with 5% SDBS shows a quite large martensite region.

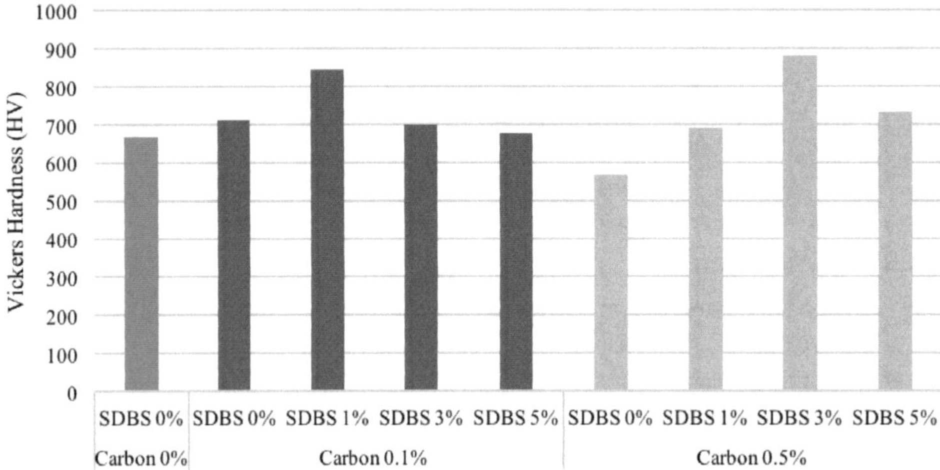

Figure 3. Graph displaying result of Vickers Hardness Testing vs. Concentration of respective variation

From the result, it is found that carbon nanofluid can be enhanced further by adding a surfactant to increase carbon stability in the fluid. An optimum amount of surfactant addition may enhance the quench severity of the nanofluid. The addition of surfactant helps the carbon reduce its surface tension with water, as carbon is hydrophobic in nature. The anionic surfactant, SDBS, physically adsorb to the carbon nanoparticle's surface, while its tail is in direct contact with water, causes the carbon to be able to be stably dispersed in water [10]. While the excess addition of the surfactant may cause the carbon to re-agglomerate, this is observed as the decreasing in hardness as excess surfactant concentration is added [11].

Conclusion

The effect of SDBS surfactant additions in various concentration for use as quenching medium in a carbon nanofluid is investigated in this study by using S45C medium carbon steel samples. The surfactant was added to carbon-distilled water nanofluid. Synthesis of the nanofluid was done by the two-step process. The variation of surfactant concentration is 0.1%, 0.3% and 0.5% volume and the variation of carbon concentration is 0%, 0.1% and 0.3% volume. The microstructure and hardness obtained in this study show that the concentration of surfactant added in the carbon nanofluids quite affected the quenching results. The quenched product by using a carbon nanofluid will result in a martensite structure, however, with the added surfactant an increase in the martensite structure is observed. At excess surfactant concentrations, the structure mainly consists of pearlite, with very few martensites. Hardness testing results also correspond to the severity of the quenching mediums, with peak hardness of 845 HV for 0.1% Carbon with 1% SDBS, and 878 HV for 0.5% carbon with 3% SDBS. The hardness value shows a significant improvement over hardness results without SDBS addition. Excess surfactant addition will also yield a lower hardness due to the re-agglomeration of particles. Overall, the hardness value was still improved compared to cooling without the addition of surfactant. This study concluded that surfactant addition into the nanofluid will enhance its heat transfer capability, which in turn will also improve the severity of the quench, at the appropriate concentration.

Acknowledgment

The authors would like to thank Ms. Ghiska Ramahdhita for the fruitful discussion, and gratefully acknowledged the financial support from Hibah PITTA 2018 funded by DRPM Universitas Indonesia No. 2555/UN2.R3.1/HKP.05.00/2018.

References

[1] Babu K and Prasanna Kumar T 2011 Effect of CNT concentration and agitation on surface heat flux during quenching in CNT nanofluids *International Journal of Heat and Mass Transfer* 54 106-117

[2] Vignesh Nayak U and Prabhu K 2015 Heat Transfer during Immersion Quenching in MWCNT Nanofluids *Materials Science Forum* 830-831 172-176

[3] Yang L, Xu J, Du K and Zhang X 2017 Recent developments on viscosity and thermal conductivity of nanofluids *Powder Technology* 317 348-369

[4] Azadeh G, Ibrahim H M 2013 The influence of surfactant and ultrasonic processsing on improvement of stability, thermal conductivity, and viscosity of titania nanofluid *Experimental Thermal and Fluid Science* 51 1-9

[5] Wusiman K, Jeong H, Tulugan K, Afrianto H and Chung H 2013 Thermal performance of multi-walled carbon nanotubes (MWCNTs) in aqueous suspensions with surfactants SDBS and SDS *International Communications in Heat and Mass Transfer* 41 28-33

[6] Kresnodrianto, Harjanto S, Putra W, Ramahdita G, Yahya S and Mahiswara E 2018 Characterization of water based nanofluid for quench medium *IOP Conference Series: Materials Science and Engineering* 348 012009

[7] Sadri R, Ahmadi G, Togun H, Dahari M, Kazi S, Sadeghinezhad E and Zubir N 2014 An experimental study on thermal conductivity and viscosity of nanofluids containing carbon nanotubes *Nanoscale Research Letters* 9

[8] Akhyar I and Sayuti M 2015 Effect of Heat Treatment on Hardness and Microstructures of AISI 1045 *Advanced Materials Research* 1119 575-579

[9] Yu W and Xie H 2012 A Review on Nanofluids: Preparation, Stability Mechanisms, and Applications *Journal of Nanomaterials* 2012 1-17

[10] Farbod M, Ahangarpour A and Etemad S 2015 Stability and thermal conductivity of water-based carbon nanotube nanofluids *Particuology* 22 59-65

[11] Yang K, Yi Z, Jing Q and Lin D 2014 Dispersion and aggregation of single-walled carbon nanotubes in aqueous solutions of anionic surfactants *Journal of Zhejiang University Science A* 15 624-633

Dried nata de coco for extend the shelf life of fruits

H. D. Rahmayanti, N. Amalia, E. Yuliza, E. Sustini and M. Abdullah*

Department of Physics, Faculty of Mathematics and Natural Sciences, Institut Teknologi Bandung, Jl. Ganesha no. 10, Bandung 40132, Indonesia

*Corresponding author: mikrajuddin@gmail.com

Abstract. We reported preliminary study about dried nata de coco for extend the shelf life of fruits We made nata de coco by mixing coconut water and adding Acetobacter xylinum. Nata de coco was dried and compacted with hot press. The design of nata de coco as an absorbent of water made similar to the design of silica gel. Moisture absorption test was carried out by storing dried nata de coco samples along with fruits (apple, mango, pear) in one container. Fourier Transform InfraRed Spectrometer (FTIR Alpha Platinum ATR A220) was used to check the water content in the dried nata de coco sample. The result showed that dried nata de coco succesfully to absorb water vapor.

Keywords: Nata de coco, water absorption, fruits

1. Introduction

Food is one of the primary needs of human. In this instant era, food preservatives are inseparable from food products. The use of food preservatives is intended to extend the shelf life of foods and to prevent microbial contamination that damages them. In the case of fruit storage, we can not store fruit for a long time because they will become runny and rotten. One way to keep the fruit is to maintain the humidity of the storage area. The material commonly used as a moisture absorbent on the storage of hygroscopic materials is silica gel [1]. Meanwhile, problem that arise if silica gel is accidentally consumed is a neuromuscular [2], nausea, vomiting [3], carcinogenic [4].

There are not many non-chemical food dryers on the market. Handayani et al. (2015) reported the results of research on the use of rice husk waste to made silica gel, but there was a weakness, namely that is the addition of chemical compounds such as NaOH, HCl and CH_3COOH, which made this material less safe [5]. Another organic materials commonly used as a moisture absorbers is bacterial cellulose [6]. Umartyotin et al (2016) reported that a mixture of bacterial cellulose and eggshells can absorb moisture in packaging products [7]. Bacterial cellulose has a high absorption capacity so that it can be used in chemical and physical modification processes [8]. Therefore, the study about the absorption properties of bacterial cellulose is an interesting subject to study. Nata de coco is one of the products of bacterial cellulose [9,10]. Researchers also investigated the application nata de coco as dye-absorbent in artificial wastewater [11], heavy metal absorbent for water treatment [12], ultrafiltration membranes [13], and materials for air mask [14]. Hence, in this study, the development of moisture absorbent material using nata de coco was carried out. Focus on this study for extend the shelf life of

Content from this work may be used under the terms of the Creative Commons Attribution 3.0 licence. Any further distribution of this work must maintain attribution to the author(s) and the title of the work, journal citation and DOI.
Published under licence by IOP Publishing Ltd

fruits. Perhaps, this is a simple research topic but could be an interesting research topic that has not been considered by many people as do many other researchers [15-17]

2. Experimental methods

2.1. Nata de coco preparation

Nata de coco was is made by mixing coconut water and adding *Acetobacter xylinum*, a bacteria that produces cellulose. Coconut water and *Acetobacter xylinum* were purchased from local market in Bandung, Indonesia. The coconut water was stored for more than 4 days to increase the acidity. Coconut water was mixed with 4.5 g of sugar and 4.5 g of urea. 4.5 ml of acetic acid was added to adjust pH into 3-4 and then boiled. After the mixture became cold, 9 ml of *Acetobacter xylinum* was also added [14]. It was then subsequently transferred into clean jars and covered with newspaper. After being storaged for 7 days, Nata de coco was ready to be harvested.

2.2. Sample preparation

Nata de coco was washed and soaked in water, the sugar content was assumed to disappear (nata de coco becomes tasteless). Next, the nata de coco was pressed with load until 5 ton at 120°C for 5-10 minutes. The dried nata de coco was cut into pieces of the same size.

2.3. Experiment

Moisture absorption test was carried out by storing dried nata de coco samples along with fruits (apple, mango, pear) in one container. Data collection was carried by directly weighing the mass of dried nata de coco before and after it was used to absorb. Fourier Transform InfraRed Spectrometer (FTIR Alpha Platinum ATR A220) was used to check the water content in the dried nata de coco sample.

3. Results and discussion

To prepare the dry material, nata de coco was pressed using a hot press while being heated to a temperature which allows the release of water in nata de coco easily but does not damage the physical properties of nata de coco so that dried nata de coco can be obtained (with a very low or near zero water content). The square specimen with dimensions 1 cm of length × 1 cm of width were cut from the sample (Figure 1(b)). Figure 1(a) is nata de coco before pressed.

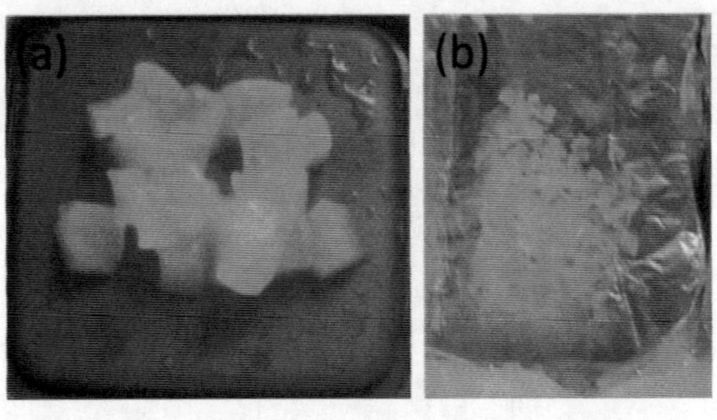

Figure 1. Image of nata de coco (a) before pressed (b) after pressed and was cut.

Figure 2. Dried nata de coco and apple were placed in the same closed container.

The smaller size may be preferred because it has a very large surface area, so it can absorb more water vapour. Surface area is defined as the total area of the material in contact with the air divided by the mass of the material and the unit is expressed in square meters per kilogram or equivalent units if expressed in a unit system different. Then, dried nata de coco were tested using FTIR to see the initial condition before the moisture absorption test was done.

The design of nata de coco as an absorbent of water made similar to the design of silica gel which is commonly used as a food dryer, where the material is wrapped in strong reticulation paper consisting of a waterproof layer, fiber layer, and polythene layer so it is safe to use [18].

Furthermore, dried nata de coco and fruits were placed in the same closed container. Figure 2 showed apple and dried nata de coco were placed in the same container. Storage was carried out for several days. Measurements were conducted by directly weighing the mass of all nata de coco. Figure 3 shows the weight of water absorbed by samples on different days. It seems that the rate of absorption initially increases rapidly with time and reaches a saturation value for a very long time. The increase in mass indicates the amount of water vapor absorbed. Its mean that dried nata de coco was able to absorb the moisture.

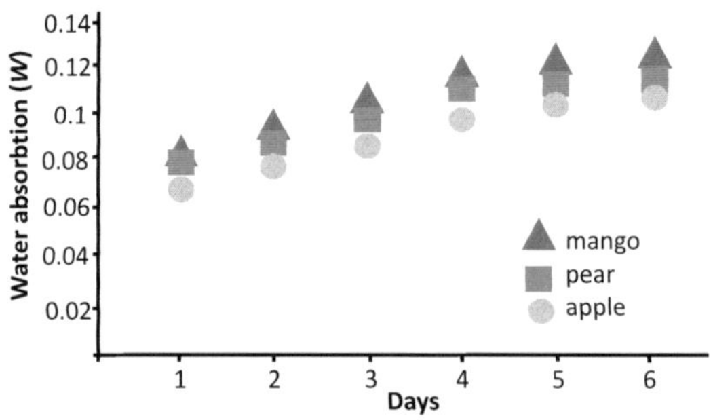

Figure 3. Image of nata de coco (a) before pressed (b) after pressed and was cut.

The highest water absorbtion was obtained from mango. Meanwhile, dried nata de coco absorbs less water vapor in apples. It also in line with the FTIR measurement graph which shows that the absorption area from mango is the largest (in the region of 3400-3500 cm-1) than another fruits, as shown in Figure 4. The amount of absorption area is calculated using an ImageJ TM software, a Java-based image processing program developed at the National Institutes of Health [xxx.] to predict the area of absorption curve. The result is shown in Table 1.

Table 1. The amount of absorption area.

Manggo	Pear	Apple
4725	4359	4143

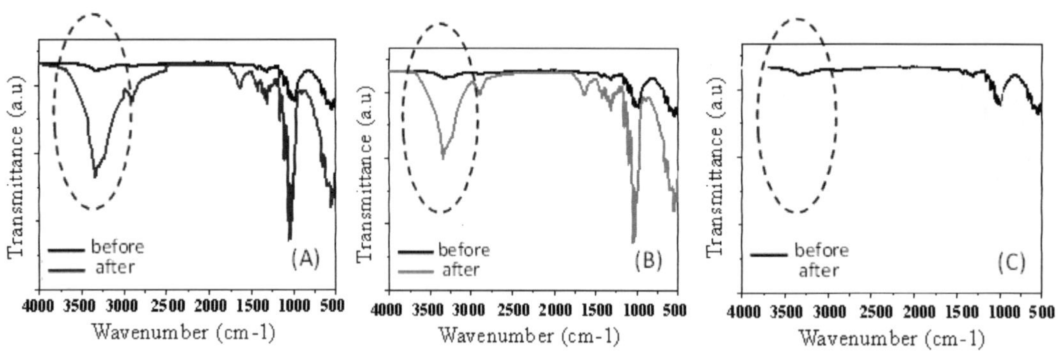

Figure 4. Graph of water absorption by the dried nata de coco for several days on fruit.

Figure 5 shows that the pear that stored in a sealed container without dried nata in days-6 showed juicier than pear that stored together with dried nata. This shows that dried nata de coco can absorb moisture around the storage container of pear so that pear not rotten. To prevent fruits do not rotten quickly, one is to maintain moisture levels. From the description above, we can conclude that dried nata de coco can extend the shelf life of fruits. However, further study must be performed in the future to clarify how long dried nata de coco can be used as a moisture absorber considering that nata de coco is an organic material that can rot.

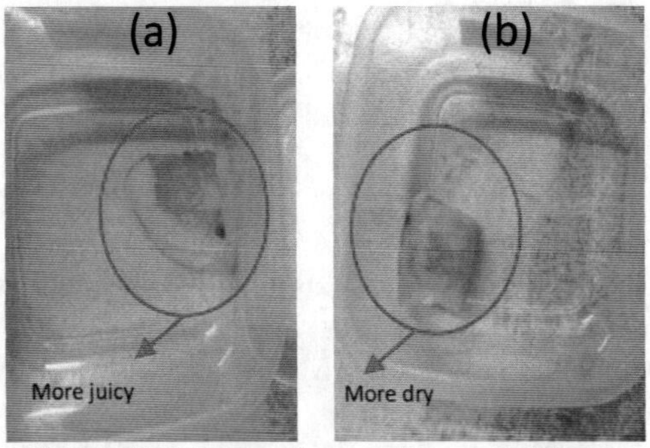

Figure 5. Figure of pear after stored for several days: (a) without dried nata (b) with dried nata.

4. Conclusion

We have succeeded in carry out a simple experiment to extend the shelf life of fruit using dried nata de coco. Based on the results of data analysis and discussion, the rate absorption dried nata de coco initially increases rapidly with time and reaches a saturation value for a very long time. The increase in mass indicates the amount of water vapor absorbed. Its mean that dried nata de coco was able to absorb the moisture.These results are consistent with results of characterization by FTIR.

Acknowledgement

We thankful for the PMDSU (Program Magister Doktor Sarjana Unggul) research grant from the Ministry of Research and Higher Education, Republic of Indonesia No. 535C/I1.C01/PL/2018.

References

[1] Li J, Kubota M, Watanabe W, Kobayashi N and Hasatani M 2004 Optimal design of a fintype silica gel tube module in the silica gel/water adsorption heat pump *J. Chem. Eng. Jpn.* **37**(4) 551–7

[2] Oremus, W 2011 What happen if I eat silica gel, *Slate article.* (https://slate.com/news-and-politics/2011/07/silica-gel-what-happens-if-i-eat-it.html, retrieved January 2018)

[3] POISONology 2016 What happen if I eat silica gel (University of Arizona: Arizona Poison and drug information center, retrieved January 2018)

[4] Schramm C, Kitzke A and Tessadri R 2017 Cobalt Chloride-based Humidity Attached to Sol-Gel Modified Cellulosic Material *Cellulose Chem. Technol.* **51**(3-4) 273-82

[5] Handayani P A, Nurjanah E and Rengga W D P 2015 Pemanfaatan Limbah Sekam Padi Menjadi Silika Gel *Jurnal Bahan Alam Terbarukan* **4**(2), 55-59

[6] Hidayat T 1996 Application of Bacterial Cellulose in Papermaking *Proceeding of the International Workshop Green Polymer* **1** 43-45

[7] Ummartyotin S, Pisitsak P and Pechyen C 2016 Eggshell and Bacterial Cellulose Composite Membrane as Absorbent Material in Active Packaging *J. Polym. Sci. A.* **1** 1-8

[8] Bielecki S, Krystynowicz A, Turkiewicz M and Kalinowska H 2005 Bacterial Cellulose (Germany: Wiley-VCH Verlag) 37-46

[9] Kongruang S 2008 Bacterial Cellulose Production by Acetobacter xylinum Strains from Agricultural Waste Product *Appl Biochem Biotechnol.* **148**, 245-56

[10] Byrom, D 1991 Microbial cellulose (New York: Stockton Press), 263-84

[11] Lindu, M, Puspitasari T and Ismi E 2008 Sintesis dan Uji Kemampuan Membran Selulosa Asetat dari Nata de coco sebagai Membran Ultrafiltrasi untuk Menyisihlan Zat Warna pada Air Limbah Artifisial *Jurnal Teknologi Lingkungan.* **4**(4),107-112

[12] Agustien R R, Indrayanti S D and Hastuti E 2014 Adsorbent Utilization Nata De Coco For Water Treatment Contaminated With Heavy Metal of Cu, Cd, And Cr Laboratory Scale *Jurnal Permukiman* **9**(3) 129-35

[13] Puspitasari T and Radiman C L 2006 Study of Graft Copolymerization of Acrylic Acid Onto Nata De Coco And Its Application As Microfiltration Membrane *Atom Indonesia* **32**(2) 119-28

[14] Rahmayanti H D, Amalia N, Dewi Y C, Sustini E and Abdullah M 2018 Development of Nata de Coco-based transparent air masks *Mater. Res. Express* **5**(5) 054004

[15] Rahmayanti H D, Munir R, Sustini E and Abdullah M 2019 Curling evolution of suspended threads replicates 2D self-avoiding walk phenomena and 1D crystallization process *Journal of Statistical Mechanics: Theory and Experiment* **1** 013401

[16] Rahmayanti H D, Utami F D and Abdullah M 2016 Physics model for wringing of wet cloth *European Journal of Physics* **37**(6) 065806

[17] Mayer H C and Krechetnikov R 2012 Walking with coffee: why does it spill? *Phys. Rev. E* **85** 046117

[18] Huang W, Xu J, Tang B, Wang H, Tan X and Aihua L 2017 Adsorption performance of hydrophobically modified silica gel for the vapors of n-hexane and water *Adsorption Science & Technology.* **0**(0) 1–16

IOP Conf. Series: Materials Science and Engineering **622** (2019) 012011 doi:10.1088/1757-899X/622/1/012011

Synthesis and Characterization of Gadolinium doped Cerium ($Ce_{0,9}Gd_{0,1}O_{1,95}$) by using Various Carbonates as a Precipitant

Ade Utami Hapsari [1], Jarot Raharjo[1], Hanif Yuliani [1], I.G. Agung Suradharmika[2], Damisih [1], Retna Deca Pravitasari[1], and Desiana P. Sari[3]

[1] Centre for Materials Technology, Agency for the Assessment and Application of Technology, Bldg. 224 PUSPIPTEK, Serpong, South Tangerang 15314, Indonesia.
[2] Laboratory For Ceramic Based Creative Industry, Agency for the Assessment and Application of Technology, Jl. By Pass Ngurah Rai, South Denpasar, Bali, Indonesia
[3] Department of Physic Engineering, Faculty of Engineering, Surya University, Tangerang, Indonesia

*Corresponding author: ade.utami@bppt.go.id

Abstract. Gadolinium-doped cerium ($Ce_{0.9}Gd_{0.1}O_{1.95}$, GDC) as an electrolyte in IT-SOFC has higher performance in nano-sized than the other sized. Carbonate was known as a precipitant agent that could affect the morphology and size of the particle. In this study, the effects of various carbonates as a precipitant agent on the synthesis of GDC by co-precipitation method were investigated. Cerium in the form of $Ce(NO_3)_3.6H_2O$ (cerium nitrate hexahydrate) used in the synthesis process. Three types of carbonates, i.e. ammonium carbonate, potassium carbonate, and sodium hydrogen carbonate, namely GDC1, GDC2, and GDC3, respectively, was used as a precipitating agent. The GDC powder resulted was then characterized by X-Ray Diffraction (XRD), Fourier Transform Infra Red (FTIR), Particle Size Analysis (PSA), and Scanning Electron Microscopy (SEM). The carbonates variation affected the physical properties and the formed particle of GDC. GDC2 showed the optimum carbonate that produces electrolyte GDC with better properties as an IT-SOFC electrolyte based on physical properties.

Keywords: co-precipitation, $Ce_{0.9}Gd_{0.1}O_{1.95}$, IT-SOFC, various carbonate

1. Introduction

Unlike most generating systems, solid oxide fuel cells (SOFC) do not generate vibration and produces little or no emissions of NO_x and SO_x [1]. So it can be said that SOFC is a technology with low pollution that can produce electricity electrochemically with high efficiency. SOFC currently has a high operating temperature of more than 800 °C [2]. This high SOFC operating temperature causes many problems to arise during SOFC operations, such as metal corrosion, cell component damage, long start-up times, high electrolytic sintering temperatures, and undesirable interfacial diffusion. High operating temperatures will also make the system require expensive high-temperature materials, resulting in high production costs [3].

Some of the problems arising from the high operating temperature of SOFC encourage the need for SOFC development that works at lower temperatures or so-called intermediate temperature solid oxide

Content from this work may be used under the terms of the Creative Commons Attribution 3.0 licence. Any further distribution of this work must maintain attribution to the author(s) and the title of the work, journal citation and DOI.
Published under licence by IOP Publishing Ltd

fuel cells (IT-SOFC). One is to develop an alternative electrolyte with high ionic conductivity at a temperature of 500-800 °C [4]. Gadolinium-doped cerium (GDC) is a potentially electrolytic material in IT-SOFC because of its transport and redox properties. Co-precipitation method is the most common method for GDC powder synthesis. This method is known for its ability to produce nano-sized GDC powders at low cost, easy procedure with short reaction time, and mass production capability [3]. Compared to other synthesis methods, the co-precipitation method uses a lower sintering temperature than the solid state reaction method that reaches 1700-1800℃. Currently, the development of nano-size in SOFC technology is the main target because nanostructures can improve performance on SOFC [5].

In the co-precipitation method, precipitant is an important factor affecting the characteristics of the final product [6]. Mostly, research to synthesize GDC powder by co-precipitation method using ammonium carbonate as its precipitation agent. However, the method of co-precipitation with ammonium carbonate is still not optimal because the size of the resulting particles is still large (~ 100 nm) [7] and agglomeration [8]. Besides ammonium carbonate, several other types of carbonates which can be used as precipitation agents in the synthesis of co-precipitation methods are potassium carbonate and sodium hydrogen carbonate. It was reported that the synthesis of CeO_2 nanoparticles using potassium carbonate and potassium hydroxide as its precipitating agent with a particle size of approximately 25 nm [9]. Sodium hydrogen carbonate is also one of the precipitant precursors ever used in the synthesis of $MgCO_3$ powder and successfully produces a powder with a high surface area [8]. Powder with small particle size has a high surface area which will affect the sintering process and increase ionic conductivity of the material. Powder with small particle makes it possible to produce thin electrolytes. Thin electrolytes have lower ohmic resistance, which allows the operation of SOFC at lower temperatures [10] and improves SOFC performance [11]. When the electrolyte thickness decreases, the maximum power density of the cell increases [10].

Therefore, this study aims to determine the optimal conditions for the use of carbonate compounds as precipitation agents to produce GDC powder with small particles and monodisperse.

2. Experiment

The materials used in this study were gadolinium (III) nitrate hexahydrate ($Gd(NO_3)_3.6H_2O$, Sigma Aldrich, 99.9%), cerium (III) nitrate hexahydrate ($Ce(NO_3)_3.6H_2O$, Sigma Aldrich, 99%), ammonium carbonate ($(NH_4)_2CO_3$, Sigma Aldrich), potassium carbonate (K_2CO_3, Sigma Aldrich), sodium hydrogen carbonate ($NaHCO_3$, Sigma Aldrich), and aquadest.

The synthesis of GDC10 powder uses 0.4514 gram $Gd(NO_3)_{3.6}H_2O$ and 3.9808 gram $Ce(NO_3)_{3.6}H_2O$ which is dissolved into 100 mL of distilled water (solution A), then some carbonate (ammonium carbonate, potassium carbonate, and sodium hydrogen carbonate, namely GDC1, GDC2, and GDC3) is dissolved in aquadest for mixing into solution A drop-by-drop while stirring with a magnetic stirrer at room temperature until precipitation occurs, resulting in a white solution. White solution then dried in an oven with a temperature of 110 °C for 24 hours. The dried sample was calcined at 900 °C for 90 minutes.

The resulting GDC powder is then characterized using x-ray diffraction (XRD), Fourier Transform Infra-Red (FTIR), particle size analysis (PSA), and scanning electron microscopy (SEM) to determine its physical properties

3. Results and discussion

3.1. Fourier Transform Infra-Red (FTIR)

Figure 1 shows spectral bands of different types of carbonate precipitates. This spectral band showed the same type. The existence of the spectral beam found at 700, 1127, 1300, 1472, 1630, 2360, 2979, 3300, and 3700 cm^{-1}.

In the FTIR graph, the spectrum below 700 cm^{-1} is the peak for organometallic compound GDC (finger print) products. In this research, there is a spectrum in the range of 700 cm^{-1}. The spectrum band in the region 850 - 400 cm^{-1} indicates the presence of a Ce-O bond [12]. This indicates Ce-O as an organometallic compound product and GDC is already formed.

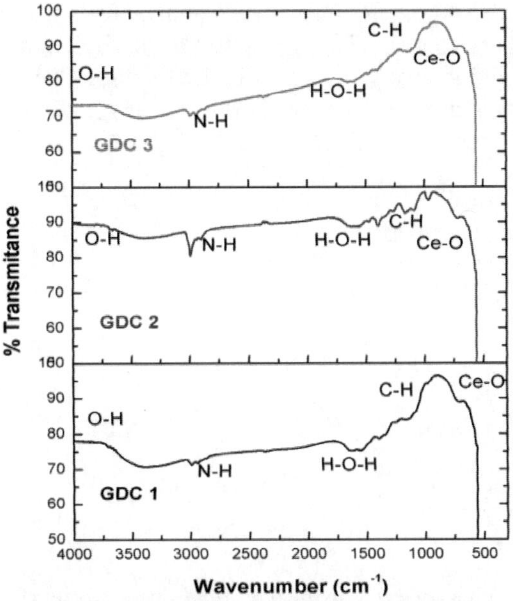

Figure 1. FTIR results of Powder GDC from different types of carbonates

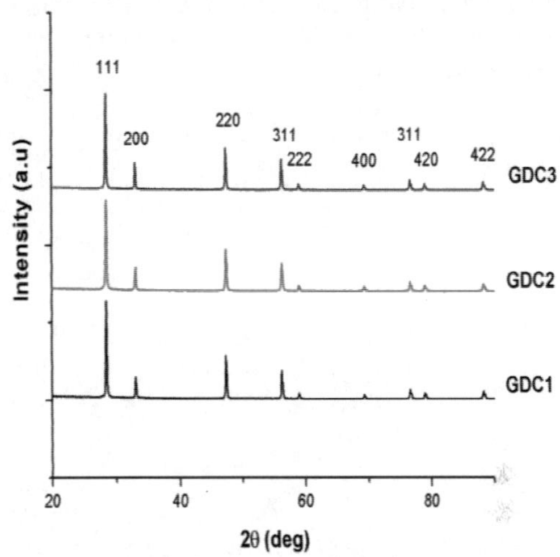

Figure 2. XRD patterns of Powder GDC from different types of carbonates

While the spectra above 700 cm-1 is a type of functional group absorption of organic compounds that indicate the presence of impuritis. Impuritis can be either synthesis or by-products. In this research, there are some spectra above 700 cm-1 that is in area 1127, 1300, 1472, 1630, 2360, 2979, 3300, and 3700 cm-1. This indicated that in the range 3000-3700 cm^{-1} and 1630 cm^{-1} are respectively associated with the hydroxyl group (-OH) [13] and the H-O-H bonds indicating the presence of water molecules in the sample [14]. This is commonly found when the material is hygroscopic.

Around 1270-1450 cm^{-1} are vibrations of the C-O, C-H [15] and O-H groups [9]. Sometimes there are also N-H bonds with typical characteristic graphic characteristics as shown in Figure 1. Overall, there is no impurity from the GDC synthesis reaction. The existence of the N-H bond formed is possible because there is still a nitrate bond with the rest of the carbonate molecule so as to produce a by-product of nitrate ($(NH_4)_2NO3$, K_2NO_3, $NaHNO_3$).

Sintering treatment at 900 °C temperature is expected to eliminate impurities and by products because the value of boiling point has been passed ($(NH_4)_2NO_3$ = 210 °C, K_2NO_3 = 400 °C, $NaHNO_3$ = 380 °C). Therefore, it is necessary to wash the process of synthesis before drying.

3.2. X-ray diffraction (XRD)

Figure 2 show the XRD graph of GDC in various carbonates. The diffraction patterns of all samples show that only one phase is formed with a cubic fluorite crystal structure. All the peaks corresponding to the composition $Ce_{0.90}Gd_{0.10}O_{1.95}$ (GDC10) indicate that Gd succeeded in doping into the Ce lattice [16]. There is no other detected crystal phase for all GDC samples indicates that only one phase (phase) of material and material has a high homogenity [17]. The sharper peak intensity indicates that the sample has a larger crystal size, while the wider peak (peak broaden) width indicates that the sample has a smaller crystal size [18].

All GDC samples have diffraction patterns that are very similar in both width and height of intensity, so it can be predicted that the crystallite size of the three GDC samples should not be much different.

The result of calculation the crystallite size of each GDC sample using the Scherrer equation [19] shown in Table 1.

Based on the calculation of crystallite size values, the best use of carbonate type as precipitant is potassium carbonate because it produces the smallest crystal size followed by ammonium carbonate and sodium hydrogen carbonate of 4.08 nm and 6.93 nm. The calculation results for all GDC samples correspond to readings of the width and height of the intensity of the diffraction pattern, having a diffraction pattern that is very similar in both width and height of intensity, thus having a crystalline size not much different.

Table 1. Crystallite size of GDC with different types of carbonates

Name Sample	Crystalite Size (nm)
GDC1	4.08
GDC2	3.47
GDC3	6.93

3.3. Particle Size Analysis (PSA)

The distribution and particle size of GDC materials are characterized using a particle size analyser (PSA) with the principle of dynamic light scattering (DLS). The surfactants used are triton x and acetyl acetone. Figure 3 showed the number distribution graph from PSA. Figure 3 shows the sample GDC1 having an average particle size of 22.0552 nm. Each particle size average for GDC 2 and GDC3 is 369.8 nm dan 27.524 nm. The average value of the smallest particle diameter for each sample was GDC potassium carbonate of 369.8 nm. This is proportional to the XRD results where the smallest crystallite size is obtained in potassium carbonate samples (3.47 nm).

There is still the possibility of agglomeration in the sample because of the nano-sized sample. This is seen from the presence of a number of nanoparticles measuring 161.5 nm in the ammonium carbonate sample. As we know that materials with nanoparticles tend to agglomerate. A fairly high average particle size in GDC samples is possible because high calcinations temperatures of 900 °C are used during the sample synthesis process. The particle size increases with the increase of the calcinations temperature [2]. The average particle size of GDC is much larger than the size of the crystallites. This indicates that the GDC powder formed has a polycrystalline structure wherein one particle there is more than one crystal inside [20].

The number distribution graphs of GDC1 show that the sample has a very wide particle size distribution meaning that GDC1 powder has a polydisperse system. This indicates that the average value of the particle size obtained does not represent the actual value. While in GDC2 and GDC3 powders, the graph of the distribution number appears to have a monodisperse system, which means that all particles have the same size.

3.4. Scanning Electron Microscopy (SEM)

Characterization of Scanning Electron Microscopy (SEM) aims to look at the morphology of GDC powders. The results of SEM characterization on GDC samples at 10,000x magnification with precipitant variations of ammonium carbonate, potassium carbonate, and sodium hydrogen carbonate are shown in Figure 4. Each sample indicates the occurrence of agglomeration. Synthesis of GDC using ammonium carbonate and sodium hydrogen carbonate results in the morphology of particles in the form of flakes, whereas when using potassium carbonate a more oval morphology is obtained [21].

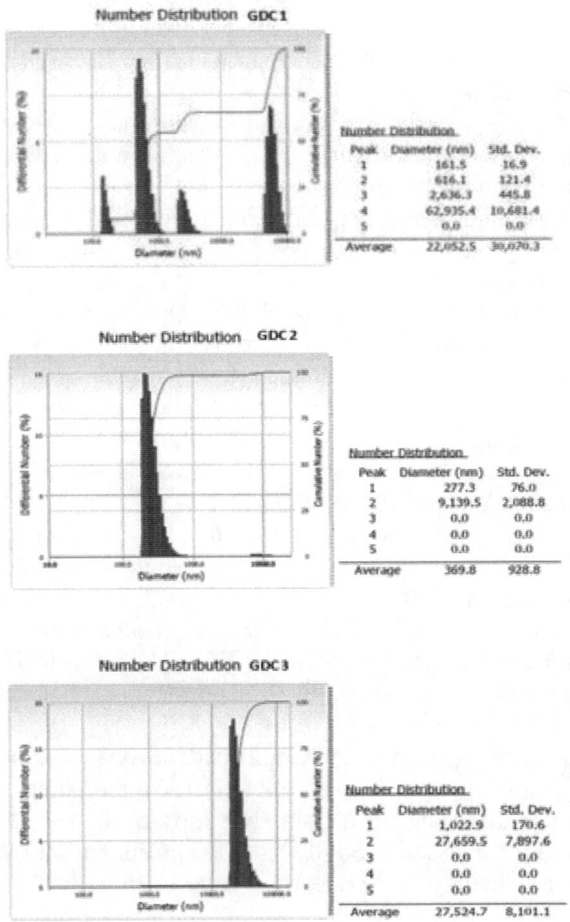

Figure 3. Number Distribution of graph in PSA of Powder GDC with different types of carbonates

Figure 4. SEM image of powder GDC of different types of carbonates

4. Conclusion

The Gadolinium doped cerium ($Ce_{0.9}Gd_{0.1}O_{1.95}$, GDC) powders was synthesized by co-precipitation method of raw material of cerium (III) nitrate hexahydrate (Ce $(NO_3)_3.6H_2O$ and various carbonates as precipitating agent. GDC phase formed is a cubic fluorite crystalline structure of the smallest crystallite size is obtained on GDC2 with potassium carbonate, which is 52.49 nm. All GDC has a polycrystalline structure and still agglomeration of each type of precipitant. Various carbonates as precipitate agent could affect the shape of the formed GDC 10 particles. Number distribution of PSA showed that GDC2 with potassium carbonates as a precipitate produces the smallest particle size 369.8 nm even though GDC1 with ammonium carbonates as a precipitate can produce a number of 161.5 nm sizes particles. This indicates that ammonium carbonate and potassium carbonate have the potential as a precipitate agent to produce GDC nano powder.

Acknowledgement

The authors gratefully acknowledge the financial support given by the Ministry of Research, Technology, and Higher Education, Republic of Indonesia via "INSINAS-2018", and the Centre for Materials Technology, Agency for the Assessment and Application of Technology (BPPT) through the development of rare-earth materials program.

References

[1] Singhal, S. C. 2013. Solid Oxide Fuel Cells: Facts and Figures, Past, Present and Future. In J. T. Irvine & P. Connor (Eds.), Solid Oxide Fuels Cells: Facts and Figures, Green Energy and Technology, pp. 1-25. London: Springer London.

[2] Raharjo, Jarot, et al. 2017. Synthesis and characterization of uniform-sized cubic ytterbium scandium co-doped zirconium oxide (1Yb10ScSZ) nanoparticles by using basic amino acid as organic precursor. International Journal of Hydrogen Energy 42, pp. 9274-9283.

[3] Joh, Dong Woo, et al. 2016. Sintering Behaviour and electrochemical performances of nano sized gadolinium-doped ceria via ammonium carbonate assisted co-precipitation for solid oxide fuel cells. Journal of Alloys and Compounds 682, pp. 188-195.

[4] Irvine, John T. S., Connor, Paul. 2013. Alternative Materials for SOFCs Opportunities and Limitations. In J. T. Irvine & P. Connor (Eds.), Solid Oxide Fuels Cells: Facts and Figures, Green Energy and Technology (pp. 163-). London: Springer London.

[5] Abdalla, Abdalla M., et al. 2018. Nanomaterials for solid oxide fuel cells: A review. Renewable and Sustainable Energy Reviews Volume 82, pp. 352-368.

[6] Huang, Y., Liu, T., Liu, M. 2013. The effect of precipitantson co-precipitation synthesis of MgO-ZrO2 powders. Advanced Materials Research Volume 788, pp. 132-135.

[7] Fan, L. Wang, C. Chen, M. Zhu, B. 2013. Recent development of ceria-based (nano) composite materials for low temperature ceramic fuel cells and electrolyte-free fuel cells. Journal of Power Sources Vol. 234, pp. 154–174.

[8] Zulfqar, Muzammal, et al. 2016. Investigations on Gadolinium doped ceria electrolyte prepared via sol-gel and co-precipitation routes for intermediate temperature solid oxide fuel cell (IT-SOFC) applications. Energy, Environment and Sustainable Development (EESD) 4th International Conference.

[9] Majid, F., Zarinkamar, M. 2015. Synthesis of nano-sized ceria (CeO2) particles via a cerium hydroxy carbonate precursor and the effect of reaction temperature on particle morphology. Journal of Ultrafine Grained and Nanostructured Materials Vol. 48, No. 1, pp. 5-10.

[10] Zuo, Chendong, Liu, Mingfei, Liu Meilin. 2012. Chapter 2 Solid Oxide Fuel Cells in Sol-Gel Processing for Conventional and Alternative Energy, by M. Aparicio, et al., 7-36. New York: Springer Science Business Media.

[11] Kim, Sun-Dong, Lee, Jong-Jin, Moon, Hwan, Hyun, Sang-Hoon, Moon, Jooho, Kim, Joosun, Lee, HaeWeon. 2007. Effects of anode and electrolyte microstructures on performance of solid oxide fuel cells. Journal of Power Sources. Volume169, Isusue 2, pp.265-270.

[12] Wang, T & Sun, DC. 2008 Prepara on and characteriza on of nanometerscalepowders ceria by electrochemical deposi on method. Mater Res Bull 43(7), pp 1754-1760.

[13] Accardo, et el. 2016. Electrical and microstructural characteriza on of ceramic gadolinium-doped ceria electrolytes for ITSOFCs by sol-gel route. J. Appl. Biomater Funct Mater 14 (1). Pp 35-41.

[14] Arabaci, A., Öksüzömerb, M. Faruk. 2012. Preparation and characterization of 10 mol% Gd doped CeO2 (GDC) electrolyte for SOFC applications. Ceramics International, Volume 38, Issue 8, pp. 6509-6515.

[15] Gondolini, et al. 2013. Effects of the microwave heating on the properties of gadolinium-doped cerium oxide prepared by polyol method. Journal of the European Ceramic Society 33, pp 67–77

[16] M.Jamshidijam, R.V.Mangalaraja, A.Akbari, Fakhrabadi, S.Ananthakumar, S.H.Chan. Effect of rare earth dopants on structural characteristics of nanoceria synthesized by combustion method. Powder Technology, Volume 253, February 2014, Pages 304-310

[17] Li,J-G.,Ikegami, T., Wang, Y.,Mori, T. 2002.a. Reactive Ceria Nanopowders via Carbonate Precipitation. Journal of The American Ceramic Society, 85 (9), 2376-2378.

[18] Theivasanthi, T., Alagar, M. 2012. Electroliytic Synthesis and Characterizations of Silver Nanopowder. Nano Biomedicine and Engineering, 4 (2), 58-65.

[19] Tok, A. I. Y, et al. 2004. Carbonate co-precipitation of Gd2O3-doped CeO2 solid solution nano-particles. Materials Science and Engineering.
[20] Mahato, et. Al. 2015. Progress in Material Selection for Solid Oxide Fuel Cell Technology: A Review, Progress in Materials Science.
[21] Spiridigliozzi, L., et al. 2018. Engineered co-precipitation chemistry with ammonium carbonate for scalable synthesis and sintering of improved Sm0.2Ce0.8O1.90 and Gd0.16Pr0.04Ce0.8O1.90 electrolytes for IT-SOFCs. Journal of Industrial and Engineering Chemistry 59, pp. 17-27.

Effect of Particle Size Distribution on the Preparation of Bonded NdFeB Permanent Magnet

C. Kurniawan[1*], S. Wahyuni[2], Ramlan[2], E. A. Setiadi[1] and P. Sebayang[1]

[1] Research Center for Physics, Indonesian Institute of Sciences (LIPI), Puspiptek Office Area, Tangerang Selatan, Indonesia 15314
[2] Department of Physics, Universitas Sriwijaya, Jl. Palembang – Prabumulih KM. 32, Indralaya Ogan Ilir, Sumatera Selatan, Indonesia 30662

*Corresponding author: cand002@lipi.go.id

Abstract. In this paper, the effect of particle size distribution on the preparation of bonded NdFeB isotropic permanent magnets was investigated. Different particle size distribution of NdFeB powders was prepared by sieving method using standard test sieves of #100, #200, and #325 meshes. Varied distribution of particle size was mixed with 3 wt% of epoxy binder and compacted by uniaxial hydraulic pressing machine to produce bonded isotropic green body. Bonded NdFeB green body was cured by vacuum drying machine under temperature 100 °C for 4 hours with vacuum pressure of 10 mbar. The characterization performed were bulk density measurement, particle size analysis, microhardness Vickers, SEM image and hysteresis curve. We find that the hardness of bonded NdFeB magnets decreased as the particle size decreases. Contrary to common belief that the finer powder with sharply particle size distribution not always give the highest product characteristics. The optimum characteristics of bonded NdFeB was obtained by the mixture of #100+#200 meshes powders with bulk density, hardness, magnetic remanence, coercivity, and energy product of 5.87 g/cm3, 88 HV, 6.5 kG, 11.15 kOe, and 9.4 MGOe, respectively. Therefore, the middle range of particle size distribution is effective to obtain high quality bonded NdFeB permanent magnets.

Keywords: Particle size distribution, bonded NdFeB, sieving, energy product

1. Introduction

The bonded NdFeB permanent magnet is one of the important part of functional materials which have been applied in various electromechanical application [1–3]. Bonded NdFeB permanent magnet typically uses polymeric matrix with magnetic NdFeB powder as a filler. There are two types of techniques to fabricate bonded NdFeB, such as compression moulding [4,5] and injection moulding [6,7].

The fabrication of bonded NdFeB was intended to make a high dense final product so that the larger energy product could be achieved. The starting particle size of NdFeB powder as a filler was kind of important factor to achieve good arrangement of isotropic bonded green body. In sintered type NdFeB, the fine particle with narrow distribution (~ 3 μm) could increase magnetic coercivity effectively, as studied by Namkung et al. [8]. the study of Li et al., showed that the coarser particle distribution by

Content from this work may be used under the terms of the Creative Commons Attribution 3.0 licence. Any further distribution of this work must maintain attribution to the author(s) and the title of the work, journal citation and DOI.
Published under licence by IOP Publishing Ltd

screening process yields a better magnetic energy product due to starting powder anisotropic effect on hot deformation process [9]. The element addition of NdFeB powder was also used to enhanced magnetic energy product by approaching exchange coupling interaction between hard-soft magnetic phases. Some kind elements such as Fe, Zr, Mn, Fe-Mn, and MnBi was introduced to NdFeB magnetic phases to obtained higher magnetic remanence and coercivity [10–14].

On the other hand, compaction and aging process were also important to make higher density and energy product of bonded NdFeB magnets. Various attempts have been made by some researchers to make denser green body.

Herchenroeder et al., use combustion driven compaction to achieved higher density bonded NdFeB up to 10% compared to the typical commercial product [15]. The uses of low vacuum curing process was also effective in producing better bonded NdFeB density and higher magnetic properties [16]. However, there is lack discussion of the particle size distribution in the making of bonded NdFeB.

In this paper, we have investigated the effect of particle size distribution on the preparation of bonded NdFeB isotropic permanent magnets by compression moulding followed by low vacuum curing process. Physical and mechanical properties of bonded NdFeB were also observed. It is found that the middle range of particle size distribution is effective to obtain high quality bonded NdFeB permanent magnets.

2. Experiment

The NdFeB powder from Magnequench Inc. type MQP-B+ was used in the experiment. Normal MQP-B+ powder was composed by mixture of Nd-Fe-Co-B alloy with theoretical density of 7.63 g/cm^3. The powder was isotropic type magnetic powder so no external magnetic field needed to form permanent magnet green body. The varied standard test sieves of #100, #200, and #325 meshes were utilized to separate particle size distribution of the powders. For optimizing particle distribution, the mixture of passed powder of #100 and #200 mesh with composition ratio of 1:1 was also made for bonded magnet.

Bonded magnet green bodies were made by uniaxially compacted the NdFeB powder after mixed with 3 wt.% of epoxy resin binder under pressure of 749 MPa. Based on optimizing parameters for curing process in our previous study, the green bodies were aged in the vacuum oven (≈10 mbar) in temperature of 100 °C for 4 hours [16].

The final bonded magnet samples were characterized by XRD (Rigaku Smartlab) for analyzing magnetic crystalline phase, PSA (Cilas) for particle size distribution measurement, Archimedes's method for bonded magnet density measurement, VSM (Dexing Magnet Ltd.) for analyzing magnetic hysteresis curve, SEM (Hitachi) for analyzing surface morphology and particle configuration, and Vickers hardness tester (LECO) for analyzing mechanical properties of bonded magnet samples.

3. Results and discussion

The XRD result as displayed in Figure 1 showed that the used starting powder of MQP-B+ is composed by Nd2Fe14B (2:14:1) magnetic phase compared to the data base number COD #96-100-8719. The composition of this 2:14:1 phase in the structure of bonded permanent magnet contributed to the high magnetic properties such as, magnetic flux density, remanence, and coercivity. Therefore, good crystallite structure is needed to produce high quality bonded permanent magnets [4].

The starting powder was measured its particle size distribution after sieved by multi-stage standard test sieved number 100, 200, and 325 meshes. The result of PSA analysis of the sieved powders is showed in Figure 2. It is showed in Figure 2(a), the starting powder has broad particle size distribution with 10% (Q1) and 90% (Q3) of cumulative values of 11.1 and 101.5 μm, respectively. On the other hand, median value (50% distribution or Q2) is around 42.4 μm, so that the starting powder has one peak Gaussian particle distribution. The sieved powder particle distribution was also measured by PSA and depicted in Figure 2(b). It is showed that the increased of sieved number (smaller opening size) results a smaller particle size distribution. However, there is an overlap distribution size from one sieve to another. This is affected by particle shape which is not perfectly rounded, so the larger size could be passed through rectangular lattice of the sieves.

IOP Conf. Series: Materials Science and Engineering **622** (2019) 012012 doi:10.1088/1757-899X/622/1/012012

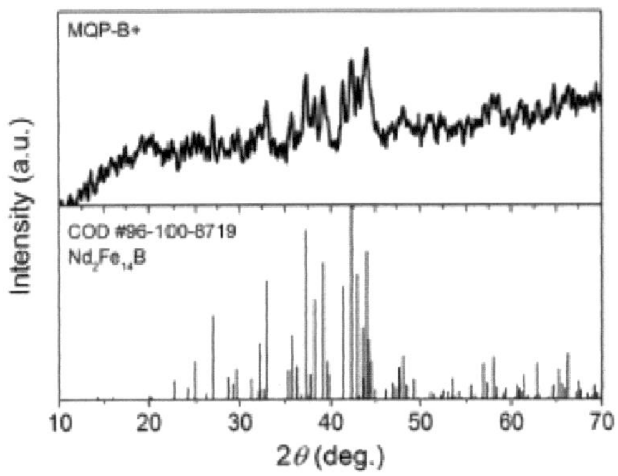

Figure 1. X-ray diffraction profile of NdFeB powder type MQP-B+

Each of the sieved powder distribution was compacted by compression molding method with 3 wt% of epoxy resin addition as a binder. After ageing process as mentioned in experimental detail section, the bonded magnet sample was measured its density by Archimedes method and the results were displayed in Figure 3. It is showed that the sample density is decreased as the particle size distribution decreased. This result was counter our intuitive that the smaller particle size should produce a higher density than the larger size. This is happened because of the contribution of narrow particle size and powder configuration in the bonded magnet products. However, the middle particle size distribution which composed by #100 and #200 meshes particle size with mix ratio 1:1 of mass produces highest density of 5.87 g/cm3. Compared to the starting powder of MQP-B+ particle size distribution, the sample density was relatively lower to the sieved powder samples with average bonded magnet density of 5.7 g/cm3. By analytic calculation of composite density of bonded magnets, the theoretical density of mixed Nd2Fe14B powder and 3 wt% of epoxy resin was around 6.71 g/cm3. So the highest density of mixed #100 and #200 meshes powder could improve the bonded magnet density by 1.5% from 86% to 87.5% compared to the original particle distribution of MQP-B+ powder. The higher sample density is correlated to the higher magnetic properties of bonded NdFeB products.

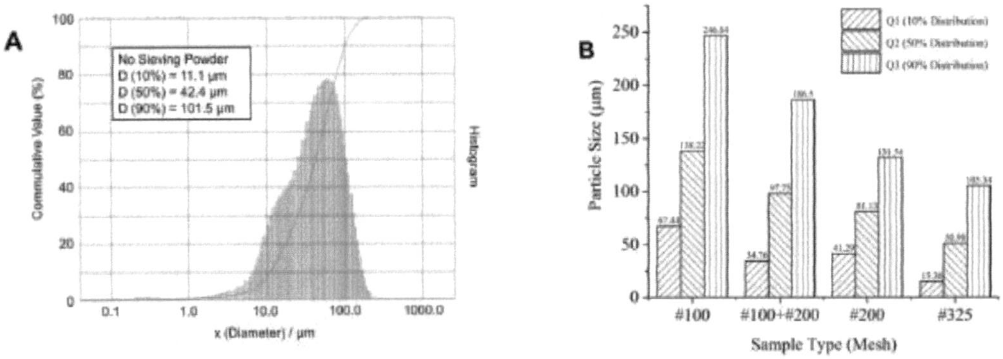

Figure 2. (A) Original particle size distribution histrogram of MQP-B+ measured by Particle Size Analyzer and (B) particle size distribution of 10%, 50%, and 90% from different sieved powder of #100, #100 + #200, #200, and #325 meshes.

The magnetic properties of bonded NdFeB magnet samples was measured by VSM as displayed in Figure 4 and the hysteresis curve was analyzed to achieved the intrinsic magnetic properties such as magnetic remanence, coercivity, and energy product (BHmax). It is showed from Figure 4(a) that the hard magnetic properties were dominant for all samples depicted by large hysteresis loops. From the Figure 4(b) as the enlargement of hysteresis loop, it is found that the remanence and coercivity are decreased as the particle size distribution decreased. This phenomenon is agreed with sample density measurement. The detail value of magnetic properties is showed in Table 1. It is showed that the highest magnetic properties was produced by the mixed powder of #100 and #200 meshes with highest magnetic saturation, remanence, coercivity, and energy product of 8.40 kG, 6.49 kG, 11.16 kOe, and 9.40 MGOe, respectively. So the middle range of particle size distribution is effective in the making of bonded NdFeB permanent magnets.

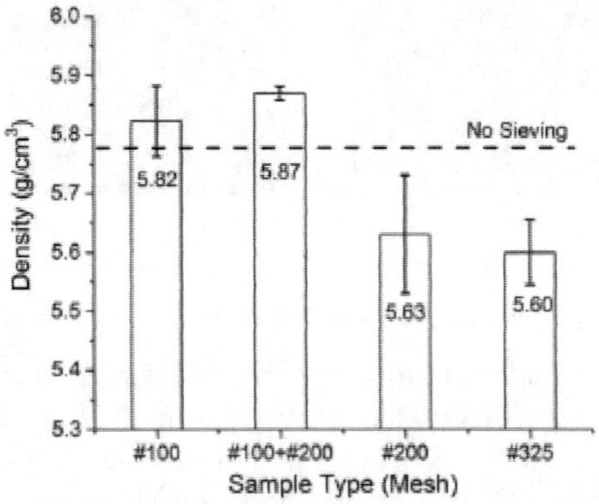

Figure 3. Sample density of bonded NdFeB made from different particle size distribution by sieving process

Table 1. Magnetic properties of bonded NdFeB magnet with different particle size distribution.

Description	$4\pi M_s$ (kG)	$4\pi M_r$ (kG)	H_{ci} (kOe)	BH_{max} (MGOe)
Original Powder	7.76	5.58	10.28	6.88
#100 mesh	8.19	6.08	11.06	8.13
#100+#200 mesh	8.40	6.49	11.16	9.40
#200 mesh	7.85	6.03	11.05	8.02
#325 mesh	7.60	5.66	10.77	7.09

To have a better analysis of powder configuration in bonded NdFeB system, the morphology of bonded magnet surfaces have been analyzed using SEM as showed in Figure 5. The different particle size distribution samples are used for comparison. It is showed that the smaller particle with narrow distribution produced a less density configuration in the epoxy matrix. The Figure 5(c,d) showed that the big empty filler of NdFeB powder was produce as shown as the dark spot in the SEM images. This emptiness reduced the density and magnetic properties of the bonded NdFeB samples. Compared to the larger size distribution in Figure 5(a,b), the filler configuration is relatively tight through the epoxy matrix. However, the narrower size from large particle only have made a wide range of filler emptiness as showed in Figure 5 (a). The SEM image also showed that the NdFeB powder has irregular shape. That is why the overlap intervals of sieved powder are existed.

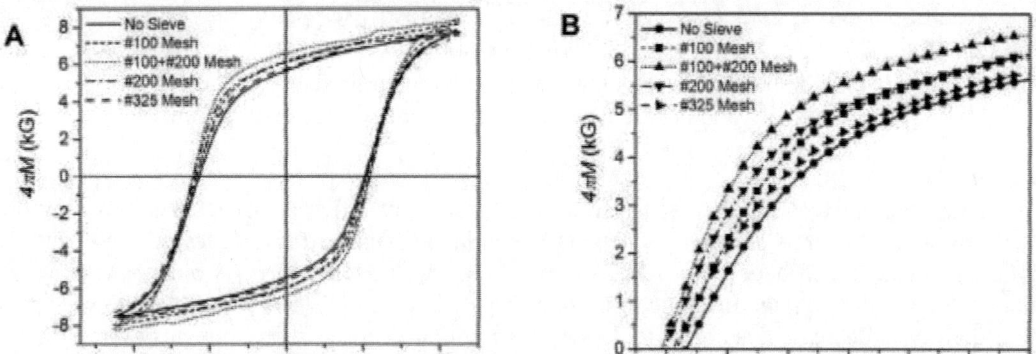

Figure 4. Hysteresis measurement of bonded NdFeB with different particle size distribution by Vibrating Sample Magnetometer with (A) full hysteresis loop and (B) magnetified second quadrant for enhanced view of remanences and coercivities.

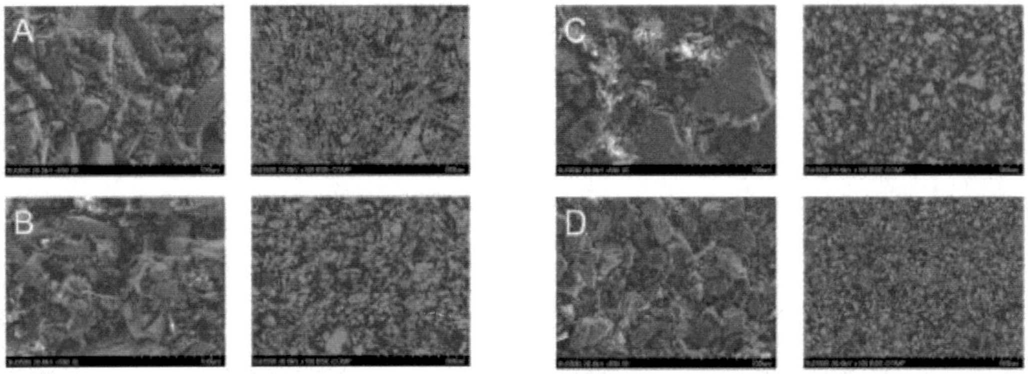

Figure 5. SEM analysis of bonded NdFeB with different particle size distribution of (A) #100 mesh, (B) #100 + #200 mesh, (C) #200 mesh, and (D) #325 mesh. The right images are from the lower magnification of SEM to show the particle and epoxy binder arrangement of bonded NdFeB.

The mechanical analysis of bonded NdFeB samples are important to achieves good performance of magnet products in real application. So the Vickers hardness analysis had been performed for all kind samples as showed in Figure 6 under load of 300 gram-force for 13 seconds. It is showed that the smaller particle size distribution produce lower value of hardness. Compared to the original powder, the hardness of the sample reaches the value of around 117 HV. The hardness values for #100, #100+#200, #200, and #325 meshes type powder distribution are 102 HV, 88.7 HV, 84 HV, and 80.4 HV, respectively. So the highest value of density and magnetic properties not always produce higher mechanical properties of bonded NdFeB due to the lower hardness value of the epoxy resin binder [17].

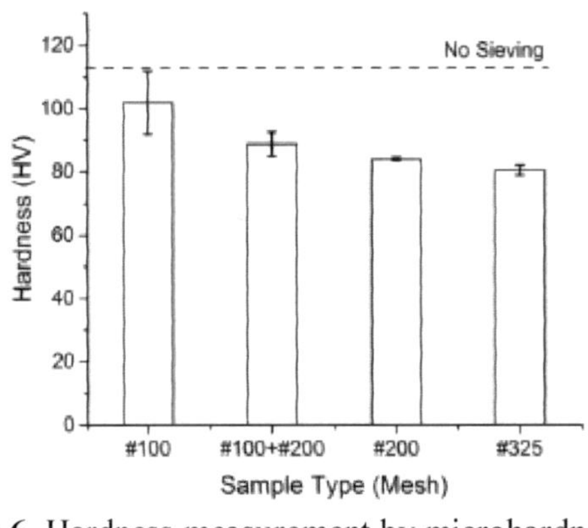

Figure 6. Hardness measurement by microhardness vickers tester of bonded NdFeB with different particle size distribution.

4. Conclusion

In this study, the bonded NdFeB magnets were produced from different particle size distribution of NdFeB powders which was prepared by sieving method using standard test sieves of #100, #200, and #325 meshes. Varied distribution of particle size was mixed with 3 wt% of epoxy binder and compacted by uniaxial hydraulic pressing machine to produce bonded isotropic green body. It is showed that the sample density is decreased as the particle size distribution decreased. The magnetic properties are decreased as the particle size distribution decreased, except for the bonded magnet produce from the mixture of #100 and #200 meshes powder. The optimum characteristics of bonded NdFeB was obtained by the mixture of #100+#200 meshes powders with bulk density, hardness, magnetic remanence, coercivity, and energy product of 5.87 g/cm3, 88.7 HV, 6.5 kG, 11.15 kOe, and 9.4 MGOe, respectively. So the middle range of particle size distribution is effective in the making of bonded NdFeB permanent magnets.

Acknowledgement
This study is supported by research program by DIPA of Research Center for Physics, Indonesian Institute of Sciences (LIPI) program year 2018 No: 64/IPT.1/A/2018.

References

[1] Sugimoto S 2011 Current status and recent topics of rare-earth permanent magnets *J. Phys. D. Appl. Phys.* **44**

[2] Lewis L H and Jiménez-Villacorta F 2013 Perspectives on Permanent Magnetic Materials for Energy Conversion and Power Generation *Metall. Mater. Trans. A* **44** 2–20

[3] McCallum R W, Lewis L, Skomski R, Kramer M J and Anderson I E 2014 Practical Aspects of Modern and Future Permanent Magnets *Annu. Rev. Mater. Res.* **44** 451–77

[4] Deng X, Liu Z, Yu H, Xiao Z and Zhang G 2015 Isotropic and anisotropic nanocrystalline NdFeB bulk magnets prepared by binder-free high-velocity compaction technique *J. Magn. Magn. Mater.* **390** 26–30

[5] Ma B, Sun A, Lu Z, Cheng C and Xu C 2016 Research on anisotropic bonded Nd-Fe-B magnets by 2-step compaction process *J. Magn. Magn. Mater.* **401** 802–5

[6] Nlebedim I C, Ucar H, Hatter C B, McCallum R W, McCall S K, Kramer M J and Paranthaman M P 2017 Studies on in situ magnetic alignment of bonded anisotropic Nd-Fe-B alloy powders *J. Magn. Magn. Mater.* **422** 168–73

[7] Li L, Tirado A, Nlebedim I C, Rios O, Post B, Kunc V, Lowden R R, Lara-Curzio E, Fredette R, Ormerod J, Lograsso T A and Paranthaman M P 2016 Big Area Additive Manufacturing of High Performance Bonded NdFeB Magnets *Sci. Rep.* **6** 1–7

[8] Namkung S, Kim D H and Jang T S 2011 Effect of particle size distribution on the microstructure and magnetic properties of sintered NdFeB magnets *Rev. Adv. Mater. Sci.* **28** 185–9

[9] Li Y, Kim Y B, Wang L, Suhr D S, Kim T K and Kim C O 2001 Influence of the powder particle size on the anisotropic properties of NdFeB magnets produced by single-stage hot deformation *J. Magn. Magn. Mater.* **223** 279–83

[10] Hou F, Li X, Zhang G, Hua Y, Li M, Lou L, Huang G, Li W and Zhang X 2018 Fabrication of bulk anisotropic Nd2Fe14B/α-Fe nanocomposite magnets with two-step high-pressure thermal compression *J. Alloys Compd.* **764** 718–23

[11] Han J, Liu S, Wang C, Chen H, Du H and Yang Y 2009 Effects of the conventional HDDR process and the additions of Co and Zr on anisotropy of HDDR Pr-Fe-B-type magnetic materials *J. Magn. Magn. Mater.* **321** 1331– 4

[12] Guozhi X, Yuping W, Xiaoyan L, Zehua W, Pinghua L, Benxi G and Youwei D 2006 Ferromagnetic/antiferromagnetic exchange coupling in melt-spun NdFeB nanocomposites *J. Non. Cryst. Solids* **352** 2137–42

[13] Kurniawan C, Purba A S, Setiadi E A, Simbolon S, Warman A and Sebayang P 2018 Effect of Fe-Mn addition on microstructure and magnetic properties of NdFeB magnetic powders Effect of Fe-Mn addition on microstructure and magnetic properties of NdFeB magnetic powders *IOP Conf. Ser. J. Phys. Conf. Ser.* **985** 012044

[14] Nguyen T X, Nguyen K Van and Nguyen V Van 2018 Enhancement of exchange coupling interaction of NdFeB/MnBi hybrid magnets *Phys. B Condens. Matter* **532** 130–4

[15] Herchenroeder J, Miller D, Sheth N K, Foo M C and Nagarathnam K 2011 High performance bonded neo magnets using high density compaction *J. Appl. Phys.* **109** 1–4

[16] Kurniawan C, Hutahaean R M and Muljadi 2015 The Effect of Low Vacuum Curing to Physical and Magnetic Properties of Bonded Magnet Pr-Fe-B *Adv. Mater. Res.* **1123** 84–7

[17] Zhang X H, Xiong W H, Li Y F and Song N 2009 Effect of process on the magnetic and mechanical properties of Nd- Fe-B bonded magnets *Mater. Des.* **30** 1386–90

IOP Conf. Series: Materials Science and Engineering **622** (2019) 012013 doi:10.1088/1757-899X/622/1/012013

Synthesis of Magnetite Nanoparticles via Co-precipitation Method

Ahmad Fadli[1], Komalasari[1], Arisman Adnan[2], Iwantono[3], Rahimah[1*], Adela Shofia Addabsi[1]

[1] Department of Chemical Engineering, Universitas Riau, Pekanbaru, Indonesia
[2] Department of Mathematics, Universitas Riau, Pekanbaru, Indonesia
[3] Department of Physics, Universitas Riau, Pekanbaru, Indonesia

[*]Corresponding author: rahimah29@yahoo.com

Abstract. Magnetite (Fe_3O_4) nanoparticles becomes a new innovation that gets attention of biomedicine scientists. Magnetite can be applied to cancer treatment as a drug carrier because it's good biocompatibility and very low toxicity. The purpose of this study was to determine the effect of temperature and stirring rate on the magnetite particle characteristics prepared by co-precipitation method. Firstly, $FeCl_3$ and $FeCl_2$ with 2:1 mole ratio were reacted with 10% NH_4OH at 40 - 80°C temperatures and stirring speed of 300 - 500 rpm in a beaker glass. Subsequently, the precipitate was separated using filter paper and it dried into air oven at 100°C. The properties of obtained magnetite powder were determined using XRD and SEM. From XRD pattern indicates that magnetite formed at all temperatures with crystallite diameter in the range of 10-12 nm. The SEM results indicate the agglomeration of the magnetite particles with size in the range of 2.7 to 3.3 µm. In the other hand, the higher of temperature and stirring rate of agitation will make the agglomeration of the particles become more uniform. The increasing of temperature and the stirring rate will increase the magnetite crystallinity level.

Keywords: drug delivery, co-precipitation, magnetite

1. Introduction

Nanoparticle technology become a new tren for biomedical science and used widely in various fields. One of the applications of nanoparticles in the medical field is drug delivery system that can be applied for cancer treatment. Nano drug delivery is a method of delivering drug compounds to the therapeutic target. The method coined by Paul Ehrlich, who was awarded a noble prize of medicine in 1908, principally conjugate drugs to nanoparticles as drug carriers which are injected into veins [1].

In the application, nano drug delivery is controlled with an external magnetic field to the target cell. Nano drug delivery also aims to reduce the potential for overdose, as well as to keep the drug concentration within the prescribed time frame [2].

Nanoparticles as drug delivery should have specific characteristics to achieve the desired target, in other words have a suitable combination of sizes, drug conjugation with nanoparticles, surface chemistry, hydrophobicity, surface functionalization, biodegradability, and response physical properties i.e temperature, pH, magnetism and so on. Among these factors, size is the most critical trait. Particular

Content from this work may be used under the terms of the Creative Commons Attribution 3.0 licence. Any further distribution of this work must maintain attribution to the author(s) and the title of the work, journal citation and DOI.
Published under licence by IOP Publishing Ltd

particle size can only access tissues or specific cells in the body and related to clearance in the blood circulation. Long-term circulation in the blood is important in drug delivery systems and drug release. In addition, the uniform distribution of particle size is also worth noting. Particles with a wide size distribution and a non-uniform shape can cause heterogeneous pharmacological effects [3]. Therefore, nanoparticles with controlled sizes and morphology are required. Because it's become a critical factor in drug delivery systems. The size and morphology of drug delivery particles influences its accessibility and residence time in blood circulation [4].

The material currently being developed as drug delivery is magnetite [5]. Magnetite nanoparticles (Fe_3O_4) are attractive materials and have a very wide application. In the spinel group, the magnetite material (Fe_3O_4) is a type of iron oxide mineral phase, in addition to maghemit (γ-Fe_2O_3) and hematite (αFe_2O_3). Magnetite nanoparticles (nanomagnetite) is one of the materials that gets a lot of scientific attention because of their advantages, such as very low toxicity and good biocompatibility [6], then low coercivity, high saturation magnetization, and low Curie temperature [7].

In biomedical fields, nanomagnetite was used in applications such as magnetic resonance imaging (MRI), cell separation, heat mediators for hyperthermia, biosensing, and drug delivery for cancer therapy [8].

The synthesis of nanomagnetite (Fe_3O_4) had been widely done using various methods, such as co-precipitation [9], hydrothermal [10], thermal decomposition [11] and the sol–gel technique [12] and in this research we will synthesis the magnetite nanoparticles via co-precipitation method. In the process of synthesis we have to know the effect of temperature and stirring rate of agitation on the magnetite particle characteristics, and the particle size of the magnetite.

2. Materials and methods

2.1. Magnetite Synthesis

Magnetite nanoparticles were synthesized via co-precipitation method. The synthesis process is carried out as follows: $FeCl_3.6H_2O$ (Merck, Germany), and $FeCl_2.4H_2O$ (Merck, Germany), with 2:1 mole ratio were each dissolved in 100 mL aquades, then the two solutions were mixed into a 500 mL beaker glass and then heated until various temperature were achieved for 10 min while stirring. After heating, 20 mL NH_4OH 10% (Merck, Germany) was gradually added while the solution was being vigorously stirred at 300, 400, and 500 rpm using a magnetic stirrer until the pH 10. Then continuously heating until 30 minutes at temperature of 40°C, 60°C, and 80°C. Further liquid and precipitate separated using filter paper and rinsed using aquades. The precipitate was dried into air oven at 100°C through 150 minutes.

2.2. Characterization of Magnetite

The crystallinity of the obtained magnetite was analyzed by X-ray Diffraction (XRD) X'PERT POWDER PW 30/40. Particles morphology was examined using SEM-EDX JEOL JSM-6510 LV analysis was performed to study the elemental analysis. The particle crystal size is determined by using the Scherrer equation.

$$\beta = \frac{K\lambda}{D.\cos\theta} = \frac{K\lambda}{D}.\frac{1}{\cos\theta}$$

3. Results and discussion

The magnetite crystals (Fe_3O_4) consist of Fe^{3+} and Fe^{2+} with a ratio 2:1. In this study $FeCl_3$ used as Fe^{3+} and $FeCl_2$ as Fe^{2+} source using co-precipitation method. NH_4OH 10% used as the settling agent. The formation of magnetite was characterized by thick black precipitate that can interact with external magnetic fields as seen in Figure 1.

Figure 1. (a) Magnetite dispersed (b) Magnetite attached to the external magnet.

Magnetite has a black pigment [13]. So, it called as black iron oxide. But iron oxides like hematite, maghemite and geotite have red, brown and yellow pigment. Based on the synthetic solution shown in Fig. 1, it shows the characteristic of magnetite nanoparticles (Fe_3O_4) that has black particles.

Magnetite powder that produced from the synthesis has magnetite characteristics, because when closer to the external magnet, magnetite attached to the external magnet. Figure 1.a is the condition of the magnetite solution before being brought close to the external magnet. And perfectly attracted by the magnet within 86 seconds as shown in Figure 1.b.

The magnetic properties of the precipitate appear when brought closer to the external magnet and then disappear when the magnetic field is kept away. It indicates that the magnetite sample has a paramagnetical property, i.e. a material containing a magnetic moment and can be pulled strongly by a magnet.

3.1. Crystalline and Size Analysis on Magnetite Synthesis Product

The analysis used to find the crystallinity level of this magnetite synthesis product is done by using XRD (X-Ray Diffraction) analysis. The XRD analysis was performed to analyze the composition or chemical compound contained in the sample and the characterization of the crystals. XRD analysis results are diffraction peak data showing the number of crystals and crystal size of the compounds contained in the sample. The higher of diffraction peak indicated the higher number of crystal compounds in the sample. While the width of the diffraction peak indicated the smaller size of the crystalline compounds in the sample.

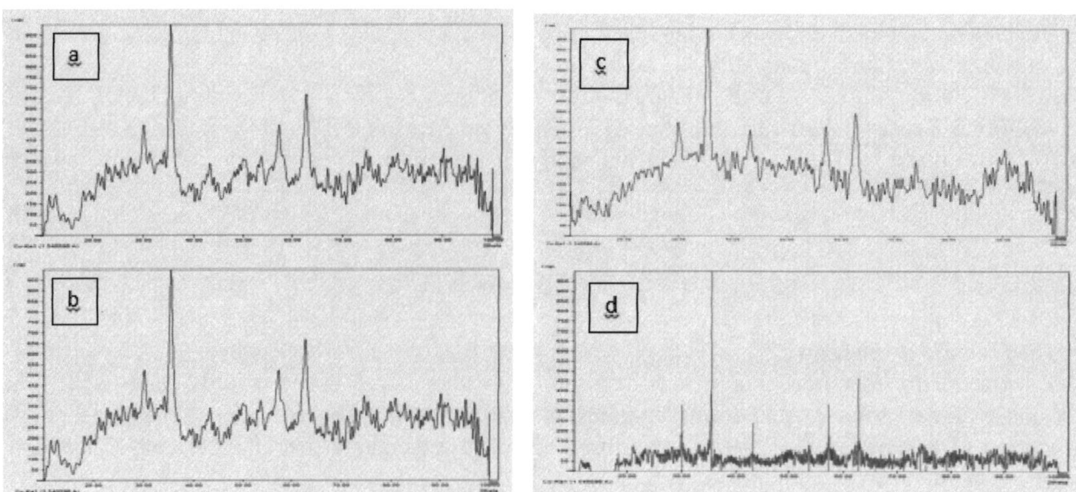

Figure 2. XRD pattern of as-syntesized sample at (a) 300 rpm 80°C (b) 500 rpm 80°C (c) 500 rpm 40°C (d) ICDD (01-075-0449).

Figure 2.d shows the standard peaks of magnetite. The standard peaks of International Center for Diffraction Data (ICDD) data are (220), (311), 400, 511, 440 with angle 2θ ie 30.28; 35,67; 43,36; 57,36; and 62.99. XRD analysis can determine the effect of temperature and stirring rate on chemical compounds contained in the resulting sample.

In Figure 2. it appears that there are five typical magnetite peaks corresponding between the 2θ angles from 30°-62°. These peaks correspond to the difractogram of the standard magnetite (01-075-0449). At 300 rpm and 80°C (Figure 2.a) has the highest intensity at 2θ 30.19°; 35.60° and 63.01°. 500 rpm 80°C (Figure 2.c) has an intensity at 2θ 30.27°; 35.26°; and 62.85°. But at the lower temperature (40°C) has a peak at 2θ angle which is not different at 2θ 30.17°; 35.62° and 62.85° (Fig. 2.b). This shows that magnetite crystals have been formed on the synthesis.

The magnetite particles produce at 500 rpm stirring rate with temperatures 40°C and 80°C werel1 nm and 10 nm with the cubic crystals that indicate characteristic of magnetite (Fe_3O_4) [14]. However, if the stirring rate is lowered at 300 rpm, the crystal size is to 12 nm. This shows that the stirring rate and temperature affects the size of the magnetite crystals. The size of the magnetite crystals obtained is suitable as drug delivery. The magnetite crystal size which can be used as drug delivery should be 10 - 200 nm [15].

3.2. *SEM Analysis*

The morphological analysis of the synthesized sample is shown in Figure 3 with 3000x magnification.

Figure 3. SEM Image of Magnetite syntesized sample (a) 80°C 300 rpm; (b) 40°C 500 rpm; (c) 80°C 500 rpm

The morphological analysis of the synthesized sample is shown in Figure 3 with 3000x magnification. The SEM results indicate there is an agglomeration the magnetite product and the uneven amount of particles. The effect of stirring rate on magnetic morphology is seen in Figure 3.a and Figure 3.c. From Figs 3.a and 3.c, the increase in stirring rate makes the clumps of magnetite particles smaller and irregularly shaped.

Increased temperatures also affect the morphology of magnetite nanoparticles. It can be seen in Figure 3.b and 3.c. Increased temperatures makes the magnetite particle size more uniform but there are still agglomerated. Measurement of particle size from SEM test image obtained magnetite particles size 2.7 to 3.3 μm. Agglomeration in magnetite synthesis is very difficult to avoid, because of its magnetism [16]. In addition, the agglomeration between particles can be caused by incomplete sample preparation. The higher temperature and speed rate also affect for magnetite crystallinity level. This is evident from the results of SEM-EDX test which can be seen in Table 1.

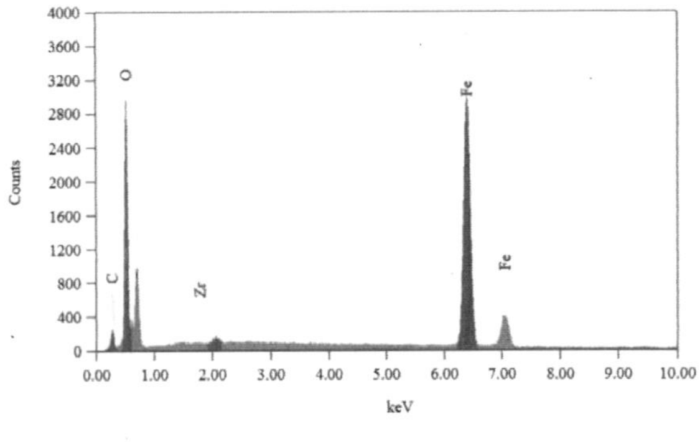

Figure 4. Spectrum EDX of magnetite prepared at 80°C and 500 rpm.

Table 1. Analysis of magnetite components

Variable	Component	Composition (%)
80°C; 300 rpm	Carbon, C	10,64
	Chloride, Cl	9,23
	Magnetite	69,75
40°C; 500 rpm	Carbon, C	7,68
	Chloride, Cl	0,37
	Magnetite	81,95
80°C; 500 rpm	Carbon, C	12,81
	Chloride, Cl	1,33
	Magnetite	85,87

EDX analysis at the point on the sample only give output in the form of a percentage of the elements. The EDX analysis performed to ensure that the compound formed in the sample was Fe_3O_4 based on the results of Fe and O elements in all samples. So it can be indicated that there are constituent elements of a magnetite. Figure 4 shows the EDX spectrum of the magnetite sample. The graph in the figure is the peak of 6.398 which is the Fe element, at 0.277 indicating the presence of element O and Cl element seen slightly at 2.042.

4. Conclusion

This research give the conclusion that magnetite can formed via co-precipitation method. All Various temperatures make crystallite diameter in the range of 10-12 nm. The SEM results indicate the agglomeration of the magnetite particles with size in the range of 2.7 to 3.3 μm. In the other hand, the higher of temperature and stirring rate of agitation will make the agglomeration of the particles become more uniform. The increasing of temperature and the stirring rate will increase the magnetite crystallinity level.

Acknowledgement

This work was supported by Ministry of Research, Technology and Higher Education of the Republic of Indonesia (Kemenristekdikti).

References

[1] Arruebo, M., Pacheco, F. R., Ibarra, R. M., and Santamaría, J. 2007. Nanotoday.2(3)22-32.
[2] Martien, R., Adhyatmika, A., Irianto, I.D., Farida, V. and Sari, D.P., 2012. Majalah Farmaseutik, 8(1)133-144.
[3] Arruebo, M., Pacheco, F. R., Ibarra, R. M., and Santamaría, J. 2007. Nanotoday.2(3)22-32.
[4] Bae, H. Y. and Park, K. 2011. Journal of Control Release.153198–205.
[5] Mahmoudi, A., Simchi, M., Imani, U, O., and Hafeli, J. 2009. Journal of Physical Chemistry C. 113 8124–8131.
[6] Okuda, M., Takeguchi, M., Ó Ruairc, Ó., Tagaya, M., Zhu, Y., Hashimoto, A., Hanagata, N., Schmitt, W. and Ikoma, T., 2009. Journal of electron microscopy, 59(2)173-179.
[7] Mornet, S., Vasseur, S., Grasset, F. and Duguet, E., 2004. Journal of Materials Chemistry, 14(14)2161-2175.
[8] Winarti, L. 2013. Stomatognatic. 10 75-81.
[9] Okuda, M., Takeguchi, M., Ó Ruairc, Ó., Tagaya, M., Zhu, Y., Hashimoto, A., Hanagata, N., Schmitt, W. and Ikoma, T., 2009. Journal of electron microscopy, 59(2)173-179.
[10] Liu, X., Kaminski, M. D., Guan, Y., Chen, H., Liu, H., & Rosengart, A. J. 2006. Journal of magnetism and magnetic materials, 306(2) 248-253.
[11] Sari, E. O., Fadli, A., & Amri, A. 2018. Jusami| Indonesian Journal of Materials Science, 19(1) 9-13.
[12] Asuha, S., Suyala, B., Siqintana, X., & Zhao, S. 2011. Journal of Alloys and Compounds, 509(6) 2870-2873.
[13] Unal, B., Durmus, Z., Kavas, H., Baykal, A., & Toprak, M. S. 2010. Materials Chemistry and Physics, 123(1) 184-190.
[14] Teja, A.S. and Koh, P.Y., 2009. Progress in crystal growth and characterization of materials, 55(1)22-45.
[15] Teja, A.S. and Koh, P.Y., 2009. Progress in crystal growth and characterization of materials, 55(1)22-45.
[16] Winarti, L. 2013. Stomatognatic. 10 75-81.
[17] Liang X., Wang, X., Zhuang, J., Chen, Y., Wang, D., dan Li, Y. 2006. Journal Advanced Functional Materials. 16 1805-1813.

IOP Conf. Series: Materials Science and Engineering **622** (2019) 012014 doi:10.1088/1757-899X/622/1/012014

The Role of Nano-SiC on Characteristics of Mg-Al-Sr/Nano-SiC Composites Produced by Stir Casting Route

Vioni Dwi Sartika[1] and Anne Zulfia Syahrial[1*]

Department of Metallurgy and Materials Engineering, Universitas Indonesia, Depok, Indonesia

*Corresponding author: anne@metal.ui.ac.id

Abstract. Magnesium composite with Mg-Al-Sr matrix reinforced by nano-SiC has successfully fabricated by stir casting process. Addition of nano-SiC into magnesium matrix varies by 0.05; 0.10; 0.15; 0.20; and 0.25 in percent of volume fraction (Vf-%). The optimum mechanical properties are found in composition of 0.15 Vf-% nano-SiC. The number of hardness, impact toughness, and wear rate of this composition are 68.4 BHN, 0.065 Joule/mm^2, 1.03310^{-5} mm^3/mm respectively. Addition of 0.15 Vf-% nano-SiC enhances the hardness by 26%, impact toughness by 23.57%, and wear resistance by 38.40% respectively. Furthermore, the existence of nano-SiC in Mg-Al-Sr matrix modify the microstructure of composite by dispersing the intermetallic compounds. However, it is observed that higher nano-SiC content tends to agglomerate thus the strengthen mechanism cannot effectively occur. Microstructure analysis using OM and SEM reveals that the addition of nano-SiC transforms the dendritic matrix to globular equiaxed. EDX result predicts the phases formed are α-Mg, Al_4Sr, $Mg_{17}Al_{12}$, MgAlSr, and XRD analysis finds the existence of SiC, SiO_2.

Keywords: magnesium composite, Mg-Al-Sr, nano-SiC, stir casting

1. Introduction

The increasing demand for economical use of energy resources and stricter control over emissions to lower environmental effect, the industries are always seeking for new materials as alternatives to conventional materials. This is one of the greatest challenge especially for automotive industry. In this context, the use of lightweight materials is an unavoidable solution to increase energy efficiency of vehicle. Magnesium is one such promising light metals for this problem [1].

Magnesium has the high specific strength with low density and good mechanical properties[2]. The strengthening process of magnesium is commonly by adding alloys or other elements[1] since this metal has hexagonal close-packed (HCP) crystal structure which make it difficult to do a cold working process[3]. The most common element used as a magnesium alloy is aluminum. The addition of strontium to Mg-Al alloys enhances the strength at elevated temperature [4]. Compared to monolithic alloys, the addition of various types of reinforcement to magnesium alloys increases their mechanical properties such as fatigue and wear resistance [5]. SiC particles as amplifiers can be uniformly distributed and have a good SiC-matrix interface [6]. Among the various types of metal matrix composite manufacturing processes, stir casting methods are commonly used. The stirring process has an advantage because it is simple and can be used for massive production [7].

Content from this work may be used under the terms of the Creative Commons Attribution 3.0 licence. Any further distribution of this work must maintain attribution to the author(s) and the title of the work, journal citation and DOI.
Published under licence by IOP Publishing Ltd

In this paper, Mg-Al-Sr/nano-SiC composite is fabricated by stir casting process. The use of Mg-Al-Sr matrix is based on Addition of nano-SiC into magnesium matrix varies by 0.05; 0.10; 0.15; 0.20; and 0.25 Vf-%. Nano-SiC used since it is particularly small, and it has large surface area which enhance the possibility of strengthening mechanism. The mechanical properties and microstructures are investigated.

2. Experimental Methods

2.1. Synthesis of Mg/SiC Composites

Magnesium composite was produced through stir casting process. Al-15Sr as master alloy was added as alloying elements. Ingot magnesium (99.9 wt%) was selected as the matrix and combined with 0.05; 0.10; 0.15; 0.20; and 0.25 Vf% of nano-SiC. In the making, magnesium was heated to its molten state in 860°C, then Al-15Sr in the form of a rod was added and stirred for 2 minutes. Before poured in the molten magnesium, nano-SiC was preheated in a muffle furnace for an hour in 900°C to remove moist and increase surface tension. The composite is poured into the mould which had been preheated in 900°C for 1 minutes and coated with zirconia to reduce thermal shock. The Mg, Al and Sr content of all casts were analyzed using Optical Emission Spectroscopy (OES) (the chemical compositions are shown in Table 1).

Table 1. Chemical composition of Mg-Al-Sr/Nano-SiC composite

Elm	Vf% Nano-SiC					
	0%	0.05%	0.10%	0.15%	0.20%	0.25%
Mg	96.3	96	96	96.5	95.8	95.3
Al	3.54	3.78	3.83	3.28	3.89	4.48
Be	<0.0001	0.0001	<0.0001	<0.0001	0.0002	0.0001
Cu	<0.0010	0.003	<0.0010	<0.0010	0.0168	0.0015
Mn	0.0645	0.0391	0.0477	0.0485	0.0524	0.0259
Zn	<0.0010	0.0343	0.0264	0.0339	0.0789	0.026
Ag	<0.0010	<0.0010	<0.0010	<0.0010	0.0047	<0.0010
Ca	0.0024	0.0031	0.0024	0.003	0.0031	0.0096
Cd	<0.0010	0.0011	<0.0010	<0.0010	0.0019	0.001
Sn	0.0327	0.0543	0.0291	0.0279	0.0834	0.0545
Sr	>0.0008	>0.0008	>0.0008	>0.0008	>0.0008	>0.0008

2.2. Mechanical Properties

All casts were prepared and tested to measure mechanical properties (hardness, wear resistance, impact strength). Some parts are cut into 2×3(cm) and ground to get a fine surface then tested by Brinell Hardness at room temperature. Wear test of the composite were conducted using the pin-on-disk machine by sliding a specimen under dry condition. Impact test was performed using a Charpy method at room temperature.

2.3. Metallography

All casts microstructurally characterized to observe the grain size of matrix alloy using an optical microscope (OM) and scanning electron microscope (SEM). They were cut, polished using alumina and ethanol, then etched using Nital reagent. EDX analysis is conducted for each sample and XRD analysis is conducted to confirm the phases formed in the cast.of FT 1000 Varian instrument at 450-4000 cm^{-1}. Vibrational mode of local structure collected with by Bruker senterra of Raman spectrometer at wavenumber range 30-1560 cm-1 and resolution ~3-5 cm-1 using green laser (532 nm) as the source of excitation. The reflectance properties were measured by UV-Vis Diffuse Reflectance spectroscopy (DRS) Shimadzu UV-2450 at wavelength range 100-900nm. Collected reflectance data (R) converted to Kubelka-Munk function ($F(R)=(1-R^2/2R)$). The band gap energy determined from threshold energy ($h\nu_0$) between $(F(R)h\nu)^{½}$ versus photon energy ($h\nu$) graph.

3. Results and discussion

3.1. Material and Microstructure

XRD Pattern of Indonesian Natural Zeolite has agreement with structural characteristic of mordenite topology (MOR Zeolite). Refinement process accomplished by using mordenite standard from IZA Structure. Refinement result revealed that mordenite's zeolite cystallized in orthorombic lattice with *Ccmc* space group (Figure 1). Unit cell parameter of modernite were a=18.1100 Å, b=20.5300 Å, c=7.5279 Å, α=90˚, asymmetric unit (Z)=1 and cell volume (V) = 2798.8979 Å^3. XRD pattern of $Ti_{0.997}V_{0.003}O_2$ on zeolite at various w/w ratio presented in Figure 2 along with undoped TiO_2 and vanadium doped TiO_2 ($Ti_{0.997}V_{0.003}O_2$). Undoped TiO_2 shown characteristic structure of anatase phase of TiO_2. Several peaks related to rutile phase of TiO_2 was detected while vanadium dopant introduced in TiO_2 lattice. The rutile phase formation may led by preparing condition including the nature of vanadium dopant, sonochemical method and temperature employed.

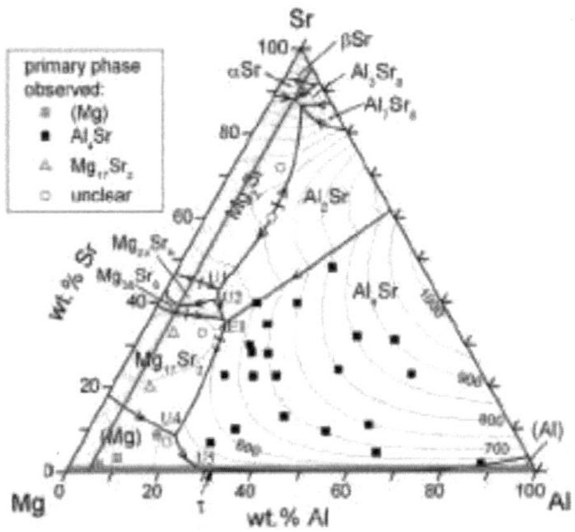

Figure 1. The liquidus projection on Mg-Al-Sr Mg-4.25Al-0.75Sr [8]

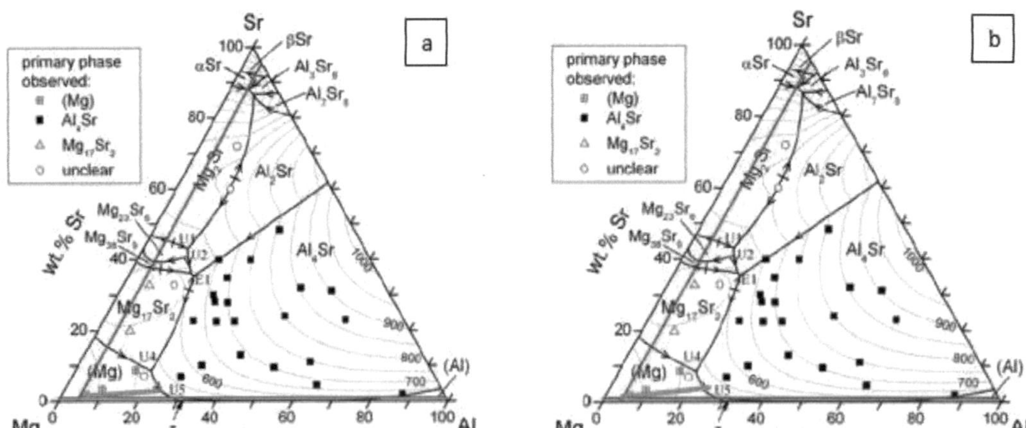

Figure 2. The liquidus projection on Mg-Al-Sr[8] (a)Mg-Al-Sr/0% nano-SiC; (b)Mg-Al-Sr/0.15% nano-SiC

Fig. 1 shows the phases formed based on liquidus projection in the ternary phase diagram of the Mg-4.25Al-0.75Sr alloy. Fig 2(a) and (b) show the liquidus projection of sample 0% and 0.15% nano-SiC which have the highest and lowest Sr content. The three diagrams show the intersecting line at U5-E4 (as the arrow showed in fig. 1b) with the transformation phase following the reactions (1) and (2)[8]. Based on the equation, the final phases formed are $Mg_{17}Al_{12}$, $Mg_{58}Al_{38}Sr_4$, α-Mg.

$$\text{U5} \quad : L+Al_4Sr \rightarrow Mg_{58}Al_{38}Sr_4 + \alpha\text{-}Mg \tag{1}$$
$$\text{E4} \quad : L \rightarrow Mg_{17}Al_{12} + Mg_{58}Al_{38}Sr_4 + \alpha\text{-}Mg \tag{2}$$

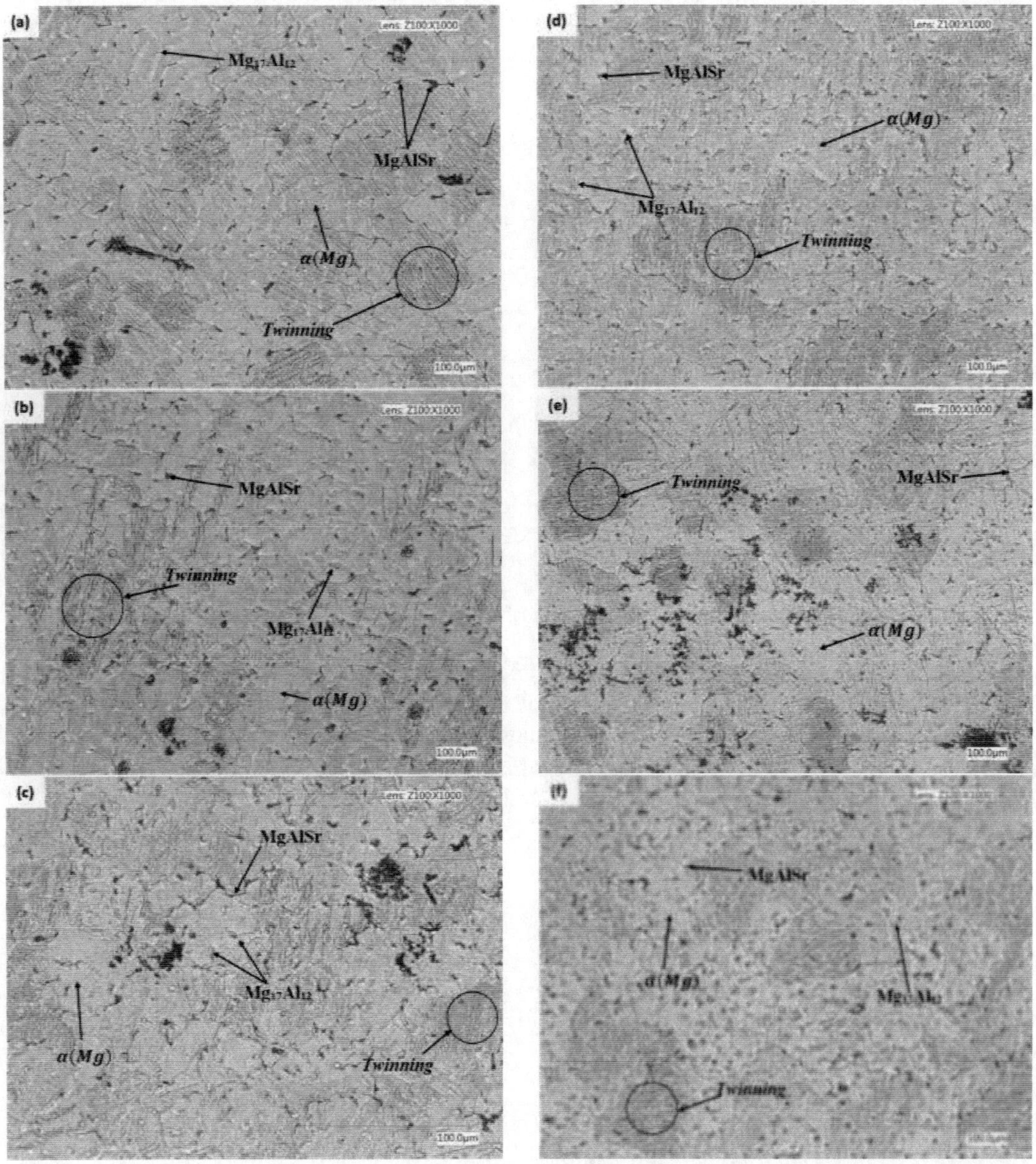

Figure 3. The microstructure of Mg-Al-Sr composites with Vf% of nano-SiC (a)0.00%; (b)0.05%; (c)0.10%; (d)0.15%; (e)0.20%; (f)0.25%

The microstructure of all composites in Fig. 3 is consists of α-Mg, $Mg_{17}Al_{12}$, MgAlSr. The $Mg_{17}Al_{12}$ and MgAlSr phases mainly distribute along a grain boundary of α-Mg with the morphology of discontinuous net. According to Zhang *et al.* [9], the two intermetallic compounds, $Mg_{17}Al_{12}$ and MgAlSr were found along the grain boundary. Intermetallic MgAlSr was observed in continuous network. This report is supported by Nayeri *et al.*[10] stated that $Mg_{17}Al_{12}$ in magnesium matrix found along dendritic regions in an irregular shape.

Fig. 3(a) observes in the dendrites form of the matrix. The addition of nano-SiC affects the transformation of the dendrites form into globular equiaxed as reported by Choi *et al.*[11]. These particles also modify the continuous intermetallic compounds into finer precipitates. Moreover, the twinning effect is found in all composites because of shear stress in the cutting process. Magnesium which has the HCP crystal structure is susceptible to twinning effect. Nano-SiC in the Mg matrix plays a role as a barrier to twining deformation at grain boundaries, so it needs more energy to move.

Visual observation does not show the significant change of grain size. The average grain size of all samples in fig.3(a)-3(f) is 59.74, 58.32, 60.70, 50.40, 52.27 μm respectively. The constant value of grain size is caused by the agglomeration of nano-SiC, and the grain strengthening is not effectively occurred.

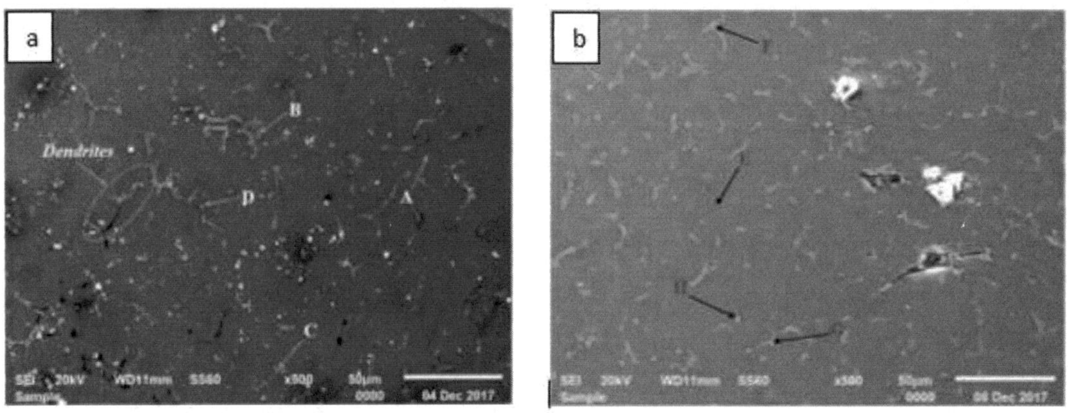

Figure 4. Microstructural analysis using SEM (a)0.00% nano-SiC; (b)0.15% nano-SiC

The microstructure analysis using SEM is restricted to composites with addition 0.00 Vf-% and 0.15 Vf-% nano-SiC. SEM analysis shows the dendrites formed in Mg-Al-Sr without the addition of nano-SiC in fig. 4(a). On the other hand, fig. 4(b) shows there is no dendrites phase and intermetallic compounds are finer than fig. 4(a). The average size of intermetallic compounds in composite 0.00% and 0.15% nano-SiC are 12.856 μm and 7.777 μm.

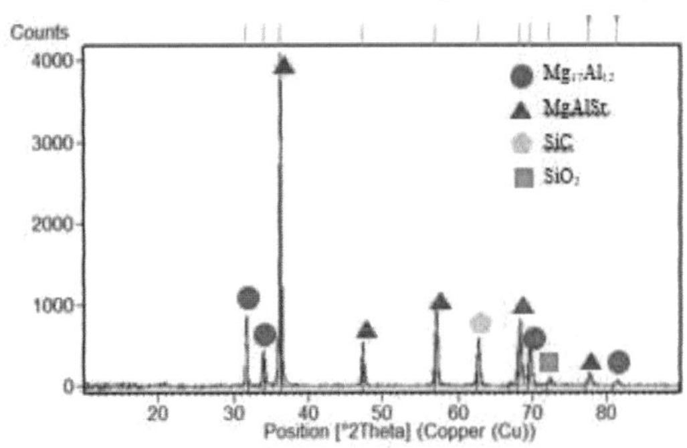

Figure 5. FTIR spectra of mordenite zeolite, TiO_2, $Ti_{0.997}V_{0.003}O_2$, and $Ti_{0.997}V_{0.003}O_2$/zeolite at various composition.

Table 2. EDX Result

Point	Wt (%)		
	Mg	Al	Sr
A	100.00	0.00	0.00
B	91.20	5.05	3.75
C	88.61	6.95	4.44
D	92.57	7.43	0.00
E	100.00	0.00	0.00
F	86.89	7.00	6.11
G	86.58	8.57	4.85
H	92.75	7.25	0.00

EDX analysis of point E, EDX result is presented in Table 2. predicts the compositions of point A, B, C, D are α-Mg, Al_4Sr, MgAlSr, and $Mg_{17}Al_{12}$ respectively. The same results found at point E, F, G, and H. The present of Al_4Sr phase is caused by the difficulties to control cooling rate, so it occurred at the non-equilibrium stage. The agglomeration of nano-SiC is still observed in Mg-Al-Sr/0.15% nano-SiC.

XRD analysis in fig. 5 is found $Mg_{17}Al_{12}$ and MgAlSr as predicted in SEM-EDX result. However, XRD data shows SiC and SiO_2 compounds at the position of 63 and 72.5 (2θ). SiO_2 compound is formed on the surface of SiC within the pre-heating process before casting. Li et al. [12] reported that pre-oxidation of nano-SiC at 973-1273K for 2 hours could produce the SiO_2 layer on the surface. This layer could enhance the adhesion force between the matrixes and reinforce.

3.2. Hardness and Wear Resistance

As nano-SiC content increased, the hardness of cast increase as well as seen in fig. 5. After certain addition, it gives no significant effect. The highest hardness value is 68.4 HB which observed in magnesium with addition 0.15% nano-SiC. The hardness of composites increased with higher nano SiC content approximately increased 26% compare to unreinforced Mg-Al-Sr alloys with possessed 50.79 HB in hardness. This hardness related to its wear resistance, as seen at 0.15% nano-SiC generated the highest hardness therefore they have good wear resistance which is indicated by the lowest wear rate. The wear rate of unreinforced is 1.67710^{-5} mm^3/mm, then decreased to 1.03310^{-5} mm^3/mm at composites with 0.15% nano-SiC.

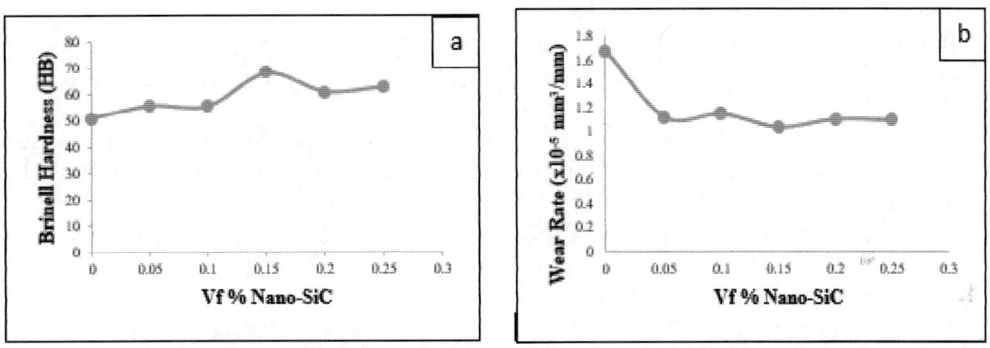

Figure 6. Hardness value (a) and wear rate (b) of composites

3.3. Impact Toughness

The impact toughness of composite increased with the addition of nano-SiC into a matrix as seen in fig. 6. The highest toughness is found at composite with 0.05% and 0.15% nano-SiC with the value of 0.067 and 0.065 Joule/mm^2. It has been reported by Cao et al. [13] that the addition of nano-SiC into Mg matrix increase the yield, tensile strength and ductility of the composite. The increase in strength and ductility indicate the enhancing of impact toughness which is in line with present work.

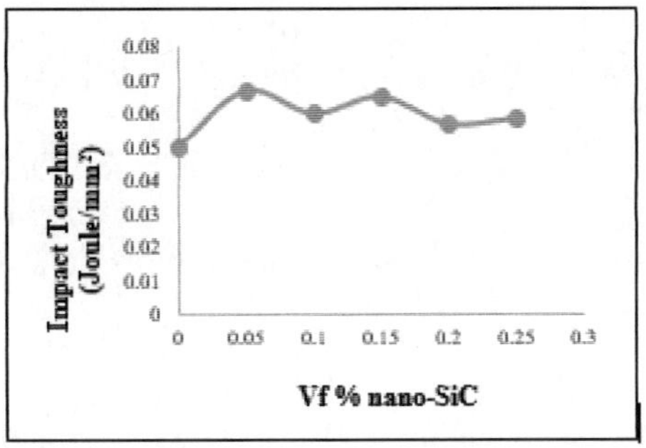

Figure 7. Impact toughness of composites

4. Conclusion

The effect of nano-SiC on characteristics of Mg-Al-Sr was investigated. From work discussed in this paper, conclusions can be drawn. Mg-Al-Sr/nano-SiC composite can be successfully fabricated through stir casting route. The optimum mechanical properties found in Mg-Al-Sr/0.15 Vf% nano-SiC. The number of hardness, impact toughness, and wear rate of this composition are 68.4 BHN, 0.065 Joule/mm^2, 1.03310^{-5} mm^3/mm respectively. Furthermore, microstructures observation showed the existence of some phases, α-Mg, $Mg_{17}Al_{12}$, MgAlSr, and Al_4Sr. Addition of nano-SiC modified the dendrite phases of α-Mg into equiaxed and refined the intermetallic compounds. XRD analysis found the existence of $Mg_{17}Al_{12}$, MgAlSr, SiC, SiO_2, SrO.

Acknowledgement

Authors would like to express their gratitude for financial support from the Directorate of Research and Community Services (DRPM), Universitas Indonesia, through International Publications Index for Final Year Students Projects Grant (PITTA) 2018 with contract number: 2378/UN2.R3.1/HKP.05.00/2018.

References

[1] M. Gupta and N. Sharon, Magnesium, Magnesium Alloys, and Magnesium Composites. Hoboken, N.J.: Wiley, 2013, pp. 1-2.
[2] C. Xu, J. Zhang, S. Liu, Y. Jing, Y. Jiao, L. Xu, L. Zhang, F. Jiang, M. Zhang and R. Wu, "Microstructure, Mechanical and Damping Properties of Mg–Er–Gd–Zn Alloy Reinforced with Stacking Faults", Materials & Design, vol. 79, pp. 53-59, 2015.
[3] M. Gupta and N. Sharon, Magnesium, Magnesium Alloys, and Magnesium Composites. Hoboken, N.J.: Wiley, 2013, pp. 118.
[4] F. Jinping, X. Bingshe, W. Shebin, L. Lu and F. Zhiyong, "Effect of Sr/Al Ratio on Microstructure and Properties of Mg-Al-Sr Alloy", Rare Metal Materials and Engineering, vol. 41, no. 10, pp. 1721-1724, 2012.
[5] K. Deng, K. Wu, Y. Wu, K. Nie and M. Zheng, "Effect of Submicron Size SiC Particulates on Microstructure and Mechanical Properties of AZ91 Magnesium Matrix Composites", Journal of Alloys and Compounds, vol. 504, no. 2, pp. 542-547, 2010.
[6] P. Poddar, V. Srivastava, P. De and K. Sahoo, "Processing and Mechanical Properties of SiC Reinforced Cast Magnesium Matrix Composites by Stir Casting Process", Materials Science and Engineering: A, vol. 460-461, pp. 357-364, 2007.
[7] J. Hashim, L. Looney and M. Hashmi, "Metal Matrix Composites: Production by The Stir Casting Method", Journal of Materials Processing Technology, vol. 92-93, pp. 1-7, 1999.
[8] A. Janz, J. Gröbner, D. Mirković, M. Medraj, J. Zhu, Y. Chang and R. Schmid-Fetzer, "Experimental Study and Thermodynamic Calculation of Al-Mg–Sr Phase Equilibria", Intermetallics, vol. 15, no. 4, pp. 506-519, 2007.
[9] Q. Zhang, S. Wang and H. Hu, "Effect of Strontium Content on Porosity and Tensile Properties of Squeeze Cast Mg-Al-Sr Alloy", Materials Technology, vol.24, no. 3, pp. 166-169, 2009.
[10] T. Nayeri, M. Yari and S. Sadreddini, " Effect of Sr on The Microstructure and Properties of Mg-6Al Alloy", Protection of Metals and Physical Chemistry of Surfaces, vol. 52, no. 2, pp. 273-278, 2016.
[11] H. Choi, H. Konishi and X. Li, "Effects of Silicon Carbide Nanoparticles on Mechanical Properties and Microstructure of As-Cast Mg-12wt.% Al-0.2wt.% Mn Nanocomposites", Magnesium Technology 2011, pp. 447-452, 2011.
[12] C. Li, Z. Wang, M. Zha, C. Wang, H. Yu, H. Wang and Q. Jiang, "Effect of Pre-Oxidation Treatment of Nano-SiC Particulates on Microstructure and Mechanical Properties of SiC/Mg-8Al-1Sn Composites Fabricated by Powder Metallurgy Combined with Hot Extrusion", Materials, vol. 9, no. 12, p. 964, 2016.
[13] G. Cao, J. Kobliska, H. Konishi and X. Li, "Tensile Properties and Microstructure of SiC Nanoparticle–Reinforced Mg-4Zn Alloy Fabricated by Ultrasonic Cavitation–Based Solidification Processing", Metallurgical and Materials Transactions A, vol. 39, no. 4, pp. 880-886, 2008.

Molecular dynamics simulation of platinum film growth based on thermal evaporation method

Rinaldo Marimpul, Toto Winata*, Fatimah Arofiati Noor, Ibnu Syuhada and Ahmad Rosikhin

Department of Physics, Physics of Electronic Materials Research Division, PECVD Lab, Institut Teknologi Bandung, Jl. Ganesha 10, Bandung, 40132, Indonesia

*Corresponding author: toto@fi.itb.ac.id

Abstract. Platinum film growth using thermal evaporation method was studied using molecular dynamics simulation. This platinum film was intended as catalyst film for graphene growth. Tersoff, Eam and Lennard-Jones potential were used to describe interaction of Si-Si, Pt-Pt and Pt-Si respectively. Deposition process was performed with low incident energy to represent thermal evaporation method. Our simulation found that heating temperature at 400 K produced platinum film with higher percentage of crystal structure than other heating condition 300K, 500K & 600K. We also found transition phase from fcc to bcc at 600K.

Keywords: Platinum, film, evaporation, molecular dynamics

1. Introduction

Platinum thin film had several application, such as resistive sensor and catalyst substrate on graphene growth. Platinum film was chosen because it had relatively low solubility [1]. It is one of important factor to grow graphene. It was reported that the primary growth mechanism of graphene on platinum film were segregation or surface deposition [2,3]. Another factor is crystal structure quality of substrate catalyst. Because it affected quality and grain area of the produced graphene [4]. Therefore, it was concern to control and to get ideal metal catalyst substrate.

There are several method for deposition thin film, such as sputtering method, thermal evaporation method, molecular beam epitaxy, chemical vapor deposition. Among of its method, thermal evaporation method is interesting method because it involved relatively small energy. So, it cannot damage the substrate or film. However, there are some lack of this method, is poor quality film (amorphous structure). One of solution of this problem is give heating temperature on growth process to get better structure [5].

In this paper, we study molecular dynamics simulation of platinum film growth on silicon substrate to know basic mechanism of its process. Effects of heating temperature on platinum film catalyst substrate were investigated. Based on our knowledge, our study has not investigated yet by any researcher, in issue of growth and annealing simulation study corresponding to thermal evaporation method with incident energy 0.35 eV.

Content from this work may be used under the terms of the Creative Commons Attribution 3.0 licence. Any further distribution of this work must maintain attribution to the author(s) and the title of the work, journal citation and DOI.
Published under licence by IOP Publishing Ltd

2. Simulations Method

Figure 1 shows silicon substrate system with dimension 4a x 4a x 3a (a = 5.43 Å, lattice constant of diamond structure) and consists of 416 atoms. Periodic boundary conditions were applied in the x and y directions.

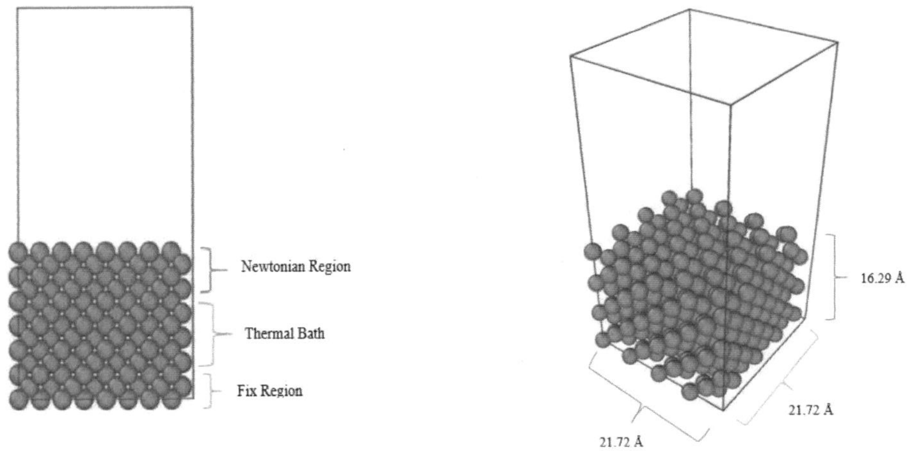

Figure 1. Dimension and region division of silicon substrate

There are three regions on silicon substrate, namely fixed region, thermal region and newtonian region. Fixed region consists of three layers of atom who prevent atom moving and fix in position when deposition process. Thermal region consists of six layers of atoms who control substrate temperature in constant condition. Newtonian region consists of four layers of atoms. The motion of atoms in thermal region and newtonian region is governed by Newton's equation. For solving the equation of motion, we used Verlet Algorithm with timestep Δt = 1 fs. We used Berendsen thermostat to control temperature with a damping parameter of 10 timesteps.

The simulation process was initiated by equilibration stage of silicon substrate at 300 K for 750 ps. Then the simulation was followed by deposition process of platinum atom on silicon substrate. The growth simulation was performed based on thermal evaporation method with incident energy 0.35 eV [6]. The deposition process was performed at 1 atom/ 3ps [7]. This proceeded until 700 atoms were deposited on the substrate. All of our molecular dynamics simulations were performed with using LAMMPS (Large-scale atomic / molecular massively parallel simulator) [8]. We used OVITO (Open Visualization Tool) to visualize the system from LAMMPS output data. By using Common Neighbour Analysis method [8], we identify the structure of Pt & Si atoms [9]. By using Visual Molecular Dynamics (VMD), we analyzed radial distribution function of platinum catalyst substrate [10].

In this study, we used very simple method for interaction between Si and Pt. Tersoff potential [11], Eam potential [12] and Lennard-Jones potential were applied to describe Si-Si, Pt-Pt and Si-Pt interaction respectively. Interaction between Pt-Si atoms was governed by equation:

$$\boldsymbol{E_{ij}} = 4\varepsilon\left[\left(\frac{\sigma}{r_{ij}}\right)^{12} - \left(\frac{\sigma}{r_{ij}}\right)^{6}\right] \tag{1}$$

where $\boldsymbol{E_{ij}}$ is potential energy between two atoms, ε is depth well potential, σ is the finite distance at which the inter-particle potential of Lennard Jones is zero and r_{ij} is distance between twoatoms. Interaction between Pt-Pt atoms was governed by equation:

$$E_i = F_\alpha\left(\sum_{j\neq i}\rho_\beta(r_{ij})\right)+\frac{1}{2}\sum_{j\neq i}\phi_{\alpha\beta(r_{ij})} \tag{2}$$

where F_α is the embedding energy which is a function of atomic electron density, ϕ is a pair potensial function, α and β are element types of atom i & j. The parameters of potential interaction function were given by table 1. Parameter R & D are cutoff parameters. Parameter A, B, $\lambda_1, \lambda_2, \lambda_3, \alpha, \beta, n, c, d, h$ are fitting parameters of Tersoff potential.

Table 1. Tersoff parameters for Si-Si interaction [11]

A = 3264.7 eV	B = 95.373 eV	
$\lambda_1 = 3.2394$ Å^{-1}	$\lambda_2 = 1.3258$ Å^{-1}	
$\alpha = 0$	$\beta = 0.33675$	$n = 22.956$
$c = 4.881$	$d = 2.0417$	$h = 0.0000$
$\lambda_3 = \lambda_2$	$R = 3.0$Å	$D = 0.2$Å

Table 2. Lennard Jones parameter for Si-Pt interaction

Parameter	$\varepsilon(eV)$	σ(Å)
Si-Pt	0.14523	3.1505

3. Results and discussion

In this simulation, we deposited platinum film on Silicon substrate. Deposited platinum film, namely platinum film catalyst substrate was produced by thermal evaporation method with incident energy 0.35 eV. Figure 2 shows percentage of each crystal structure on variation of substrate temperature.

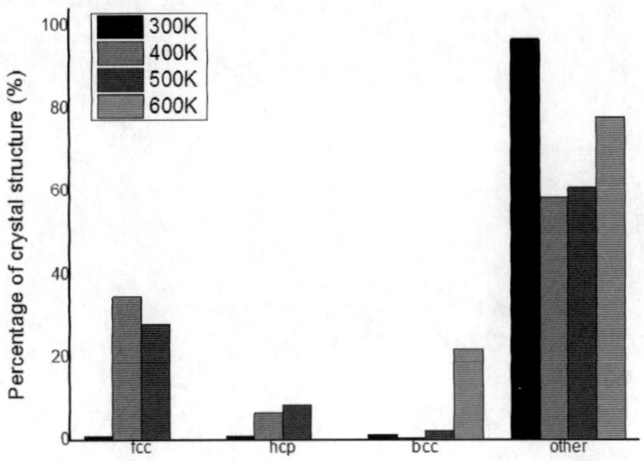

Figure 2. Percentage of crystal structure on variation of substrate temperature

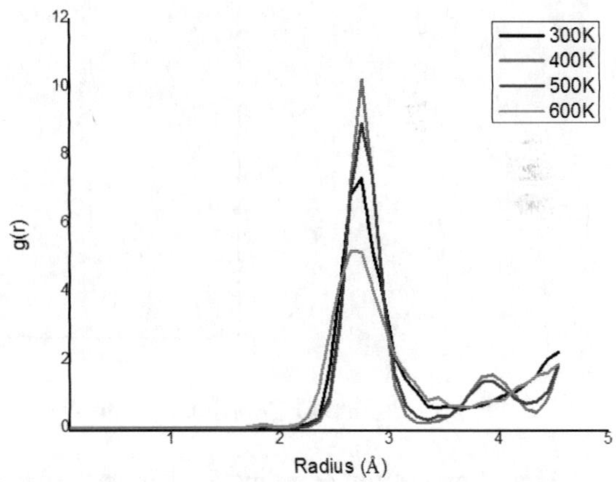

Figure 3. Radial distribution function of Pt-Pt bonds of platinum film catalyst substrate at melting point

From figure 2, it was shown that the percentage of crystal structure relatively increased as the increasing of substrate temperature and the percentage of other (unidentified structure) relatively decreased. With the total percentage of crystal structure at 400K, 500K and 600K are 41.5%, 38.9% and 22%. Meanwhile the percentage of other (unidentified structure) at 400K, 500K and 600K are 58.5%, 61.0% and 78.0%. During deposition process, we gave different treatment of heating temperature. The increase of temperature also affected velocity (or kinetic energy) was possessed by Pt atoms. Then Pt atoms can move from their lattice positions. When atoms had enough energy to move from their perfect lattice positions and then diffuse in surface. And it can affect Pt atoms in disorder structure.

We identified Pt atoms in different kind of environment: hcp (hexagonal close packed), bcc (base centred cubic), fcc (face centred cubic) and other (unidentified structure). We found that different heating on substrate affected the difference of dominant structure crystal in platinum film. The dominant structure at 400K, 500K and 600K are fcc, fcc and bcc respectively. As we know that each crystal structure need different energy required for the formation of the structures. It means difference in heating temperature also affected different energy having by each atom. From figure 2, we found Pt tend to form fcc at 400 K & 500 K. Ideally the structure of Pt atoms is fcc, however at 600 K, we found that Pt atoms tend to form bcc. Based on our knowledge, this phenomenon was occurred because energetically favourable structure by Pt atoms at 600 K.

We plot radial distribution function (rdf) graphic to know the effect of temperature on density of the atoms. As we know that the radial distribution function is a measure of the probability of finding atoms from another atom at specific distance. From this graphic, we want identify the structure of system. From figure 3, we can see at short distances, where distance is less than atomic diameter, $g(r)$ is zero because of strong repulsive forces between Pt atoms. We found that first peak formed at around r = 2.65-2.75 Å. This first peak corresponded to the nearest neighbour Pt atoms. We observe at first peak that shifting peak was occured as variation of temperature. From 300 K to 500 K, the intensity is relatively increase. It indicated atoms have moved from their lattice positions. However the intensity was drop at 600 K. Ideally, it was occurred because atoms have enough energy (or velocity) to move from their perfect lattice positions and then diffuse in surface. And it can affect Pt atoms in disorder structure. The second peak was formed at around r = 3.95 Å at 400 K & 500 K. This peak imply lattice constant of Pt atom.However the formed peak were still relatively wide that indicated the structure in short order range. As a validation, this rdf graphic had similarity with Akbarzadeh, et.al [13].

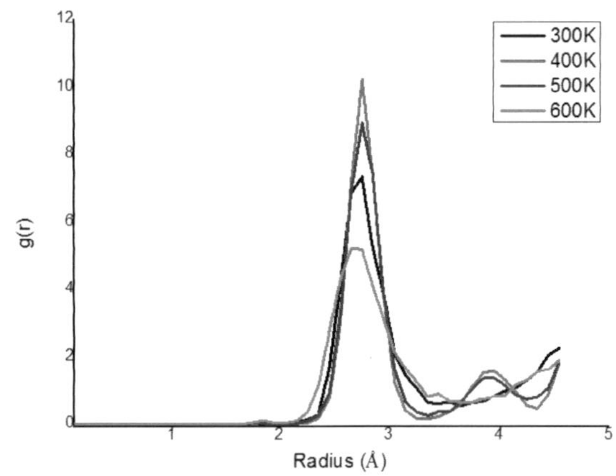

Figure 4. Coordination number curve of Pt after deposition process

Figure 4 shows the coordination number after deposition. The coordination number describes the number of neighbor atoms with respect to a central atom. We observed that relatively coordination number at 400 K & 500 K is higher than at 300 K. However at 600K, relatively had lowest coordination number at most interval radius. This is because Pt atoms get more energy with the temperature increase, then diffuse on surface substrate [14]. From figure 4, we observed that coordination number of Pt thin film at 400 K & 500 K is around 14. Theoretically, the coordination number is 12, because the ideal structure is fcc. The reason is Pt film still in different structure: fcp, bcc, hcp & other (unidentified structure). This phenomenon also occured at 600 K, because Pt film in bcc & other (unidentified) structure.

4. Conclusion
From our study we concluded that:
- Heating treatment on 400 K gave best result with highest percentage of crystal structure among of other heating treatment.
- The analysis crystalline structure of platinum film catalyst substrate with Common Neighbour Analysis (CNA) method indicated that the change of crystalline structure was occured from fcc to bcc at substrate temperature 600K.

References
[1] Chen X, Zhang L and Chen S 2015 Large area CVD growth of graphene *Synthetic Metals* **210** p 95-108.
[2] Kang B J, Mun J H, Hwang C Y and Cho B J 2009 Monolayer graphene growth on sputtered thin film platinum *Journal of Applied Physics* **106** p 104309-1 – 104309-6.
[3] Ma T, Ren W, Liu Z and et.al. 2014 Repeated growth-etching-regrowth for large-area defect-free single-crystal graphene by chemical vapor deposition *ACS Nano* **8** (**12**) p 12806-12813.
[4] Marinov G, Strijkova V, Vasileva M, Madjarova V, Malinowski N and Babeva T 2018 Effect of substrate temperature on the microstructural, morphological and optical properties of electrosprayed ZnO thin films *Advanced in Condensed Matter Physics* **2018** p 1-7.
[5] Kim K, Artyukhov V I, Regan W, Liu Y, Crommie M F, Yakobson B I and Zettl A 2012 Ripping graphene: preferred directions *Nano Letters* **12** p 293-297.
[6] Guo Z and Tan L 2009 *Fundamental and Applications of Nanomaterials* (London: Artech House).
[7] Hou Q, Hou M, Bardotti L, Prevel B, Melinon P and Perez A 2000 Deposition of Au N clusters on Au (111) surfaces *Physical Review B* **62** p 2825-2834.
[8] Plimpton S 1995 Fast parallel algorithms for short-range molecular dynamics *Journal of Computational Physics* **117** p 1-19.
[9] Stukowski A 2010 Visualization and analysis of atomistic simulation data with OVITO-the open visualization tool *Modelling and Simulation in Materials Science and Engineering* **18** p 015012.
[10] Humphrey W, Dalke A and Schulten K 1996 VMD: Visual molecular dynamics *Journal of Molecular Graphics* **14** p 33-38.
[11] Tersoff J 1988 New empirical approach for the structure and energy of covalent systems *Physical Review B* **37** p 6991-7000.
[12] Foiles S M, Baskes M I and Daw M S 1986 Embedded-atom-method functions for the fcc metals Cu, Ag, Au, Ni, Pd, Pt and their alloys *Physical Review B* **33** p 7983-7991
[13] Akbarzadeh H and Parsafar G A 2009 A molecular-dynamics study of thermal and physical properties nanoclusters *Fluid Phase Equilibria* **280** p 16-21.
[14] Wang S, Stamatakis M, Hansgen D A, Caratzoulas S and Vlachos D G 2010 Understanding mixing Ni and Pt in the Ni/Pt(111) bimetallic catalyst via molecular simulation and experiments *The Journal of Chemical Physics* **133** p 224503-1 – 224503-11.

Polyethylene Glycol Addition as Non-ionic Surfactant in Water-based Carbon Microfluid for Quench Medium in Heat Treatment Process

D. Rakhman, W. N. Putra*, C. A. Ramadhani, and S. Harjanto

Department of Metallurgy and Material, Faculty of Engineering, Universitas Indonesia, Depok, Indonesia

*Corresponding author: wahyuaji@metal.ui.ac.id

Abstract. The mechanical properties of a material depend on the quenching process. In this process, there is a rapid cooling from elevated to room temperature in a short time by using a quench medium. Therefore, the phase transformation from austenite to martensite occurs. The common medium used in the quenching process is water, oil, polymer, and gas. Nanofluids are started to be used as a quench medium because they offer better thermal conductivity compared with the conventional medium. Selection of carbon-based nanofluids as a quenching process medium aims to obtain high thermal conductivity values and controllable cooling rates. Thereby, the expected microstructure of the material could be relatively easier to form. In this paper, carbon particles were obtained using a top-down method with a planetary ball mill for 15 hours at 500 rpm. Based on the electron microscope and spectroscopy results, the particle dimension was average at 15 μm after milling, and the carbon purity of the powder used in this research was 99%. Carbon particles at 0.1%, 0.3%, and 0.5% with variation of non-ionic surfactant Polyethylene Glycol of 1%, 2%, 3%, 4% and 5% respectively was used in this research. AISI 1045 or JIS S45C carbon steel was used as a steel sample, and austenized at 1000°C for 1 hour and then quenched in the microfluid. The hardness obtained was up to 811 HV for the sample quenched in 0.5% carbon and 1% Polyethylene Glycol. The improvement was more than 100 HV, compared with the sample quenched in distilled water, which had a hardness only 666 HV.

Keywords: nanofluid, quenching, heat treatment.

1. Introduction

Heat treatment is a crucial process in the manufacturing industry, as it can modify the mechanical properties of a material. To increase the material hardness, for example, there is a rapid cooling from austenizing temperature to room temperature in a short time by using a quench medium, which is called quenching. Therefore, the phase transformation from austenite to martensite occurs, hence increasing the hardness. To obtain a martensite phase, the cooling rate should be equal to or higher than the critical cooling rate [1]. The standard medium used in the quenching process is water, oil, polymer, and gas.

Recently, nanofluids are started to be used as a quench medium because they offer a better thermal performance of and better thermal conductivity compared with the conventional medium [2]. Nanofluid

Content from this work may be used under the terms of the Creative Commons Attribution 3.0 licence. Any further distribution of this work must maintain attribution to the author(s) and the title of the work, journal citation and DOI.
Published under licence by IOP Publishing Ltd

is a mixture of fluids with nanoparticles (<100 nm). Many researchers have been done several research on the nanofluids performance as quench medium with different types of the particle [3-11].

However, several particles have a hydrophobic characteristic, which will reduce the performance of the nanofluid. Therefore, a surfactant is used to solve this problem. This study aims to show the effect of Polyethylene Glycol (PEG) addition as non-ionic surfactant and stabilizer for water-based quench medium with carbon microparticle. The effects of different concentrations of carbon and surfactants mixed in microfluids quench media were also discussed.

2. Materials and methods

The carbon particle used were purchased from Sigma-Aldrich with an average particle size of 21 μm. The particle size was reduced further by a top-down method with a planetary ball mill for 15 hours at 500 rpm. Observation of particle size, material morphology, and material composition were done using Field-Emission Scanning Electron Microscope (FE-SEM) and Energy Dispersive X-Ray Spectroscopy (EDS). To produce the microfluid, 100 ml water was used as the base, mixed with carbon particles at 0.1%, 0.3%, and 0.5% and variation of non-ionic surfactant Polyethylene Glycol of 1%, 2%, 3%, 4%, and 5% were also added. The micro-fluid was ultra-sonicated to increase its dispersion for 280 seconds.

JIS S45C carbon steel was used as a sample. Table 1 determine the chemical composition of the sample. This sample was the same used in the previous publication [12].

Table 1. The S45C carbon steel sample chemical composition (% weight) [12]

Fe	C	Si	Mn
98.3	0.47	0.287	0.718
P	**S**	**Cr**	**Mo**
0.0261	0.005	0.028	0.005
Ni	**Al**	**Co**	**Cu**
0.005	0.2	0.003	0.018
Nb	**Ti**	**V**	**W**
<0.002	0.035	<0.002	<0.01
Pb	**Sn**	**B**	**Ca**
0.025	<0.002	0.0013	0.0005
Zr	**As**	**Bi**	
<0.002	0.008	<0.03	

In this study, eight samples were used with dimension 15x10x10 mm. Samples were pre-heated at 600°C with heating rate 13.3°C/minute, and then austenized at 1000°C for 1 hour. Then the sample was quenched in the previously made micro-fluid.

To observe the microstructure of the quenched sample, standard metallography procedure was used. Samples were mounted with resin to help the sample handling. Then, samples were ground gradually from 80 grit to 1500 grit, and polished. The phase of the sample was made visible in the microscope by etching using nital (3% nitric acid, 97% alcohol) for 5 seconds.

To observe the mechanical properties of the quenched sample, vickers hardness testing was conducted. The applied force to press the indentor was 300 gf into the sample for 10 seconds. Three different locations were selected to measure the average hardness on each sample.

3. Results and discussion

Based on the FE-SEM results, the particle dimension of the laboratory-grade carbon powder was average at 15 μm after milling; therefore, it was still in micro range. Figure 1 showed the morphology of the carbon particle. EDS result in Table 2, showed that carbon purity of the powder used in this research was 99% [12].

IOP Conf. Series: Materials Science and Engineering **622** (2019) 012016 doi:10.1088/1757-899X/622/1/012016

Figure 2 showed the microstructure of non-heat treated sample, which ferrite and pearlite are the mainly phase. The bright area showed the ferrite structure, and the darker area showed the pearlite structure[12].

Table 2. The EDS result of the laboratory-grade carbon powder [12]

Laboratory-grade carbon powder	C	Cu
	99.9	0.1

Figure 1. FE-SEM image of laboratory-grade carbon powder with 1000x magnification.

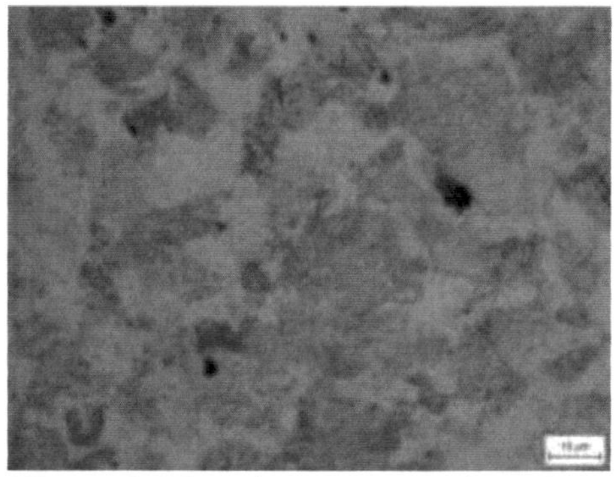

Figure 2. The microstructure of non-heat treated sample 1000x magnification

Figure 3 showed the microstructure of the sample after quenching. Martensite and bainite structure were observed in all of them but in different amount. However, it was difficult to justify the amount difference based on the microstructure only. Therefore, Vickers hardness testing was done to observe the hardness on each sample. Higher hardness could be related to a higher amount of martensite phase in the sample.

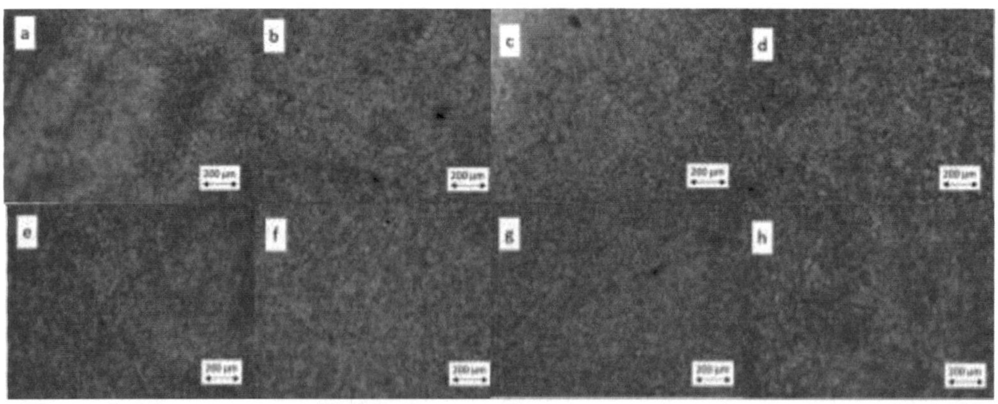

Figure 3. *Microstructure of S45C after quenching in water-based carbon micro-fluid 100x magnification* (a) 0.1% carbon, 0% PEG (b) 0.1% carbon, 1% PEG (c) 0.1% carbon, 3% PEG (d) 0.1% carbon, 5% PEG (e) 0.5% carbon, 0% PEG (f) 0.5% carbon, 1% PEG (g) 0.5% carbon, 3% PEG (h) 0.5% carbon, 5% PEG

The Vickers hardness testing of the samples was showed in Figure 4. The hardness of the sample that quenched without carbon and PEG was the lowest, i.e. 666 HV. The maximum hardness is 811 HV (0.5% Carbon, 0.1% PEG). While the maximum hardness for micro-fluid with 0.1% carbon is 769 HV (0.1% PEG).

This phenomenon showed clearly that the addition of carbon micro-particle improved the hardness of steel by increasing the cooling rate during quenching. The higher cooling rate would result in more martensite phase formed in steel, hence the higher hardness number.

On the other hand, the addition of PEG as a non-ionic surfactant and stabilizer in water-based carbon micro-fluid for quench medium with an appropriate concentration can enhance the quench severity, i.e. increasing the cooling rate. Carbon is hydrophobic in nature; therefore, it tends to agglomerate in the water. It can be seen from the result that higher carbon percentage in the quench medium lower the hardness.

To avoid this, a surfactant was added to reduce the surface tension of carbon particle in water. Without agglomeration, more surface area of carbon was available and improving the heat transfer in the quench medium. Excessive addition of surfactant, however, showed to decrease the effectivity of quench medium by introducing a re-agglomeration of carbon particle.

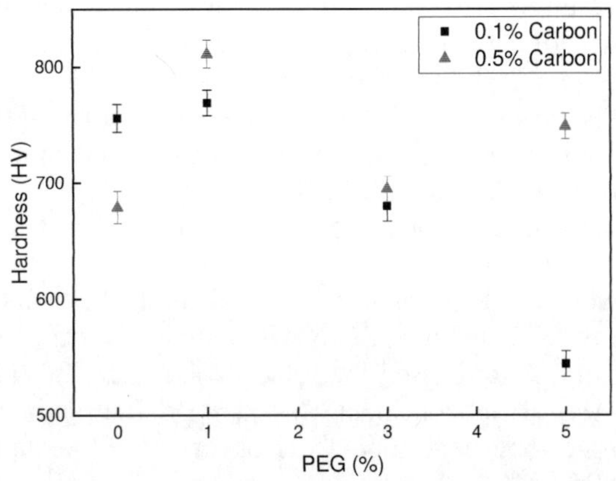

Figure 4. The Vickers hardness testing result.

4. Conclusion

The effect of PEG addition as non-ionic surfactant and stabilizer in water-based carbon micro-fluid for quench medium was experimentally investigated using JIS S45C carbon steel. The variation of carbon concentration is 0,1% and 0.5% with addition PEG 0%, 1%, 3%, and 5% each variation of carbon concentration. The result of metallographic observation and hardness testing showed that the concentration of carbon and PEG addition considerably affect the quenching process by increasing the cooling rate.

The maximum overall hardness was 811 HV (0.5% Carbon, 0.1% PEG). Meanwhile, the maximum hardness for micro-fluid with 0.1% carbon is 769 HV (0.1% PEG). In general, the hardness of steel quenched in normal water was lower than the sample quenched with carbon micro-fluid. PEG addition as a surfactant could increase the hardness in the appropriate concentration. Excessive addition of surfactant could reduce the quench medium effectivity.

From all the result, it can be concluded that addition polyethylene glicol as a non-ionic surfactant and stabilizer in water-based carbon micro-fluid with an appropriate concentration can enhance the heat transfer for quench medium in heat treatment process.

Acknowledgement

The authors would like to thank Ms. Ghiska Ramahdita for the fruitful discussion, and gratefully acknowledged the financial support from Hibah PITTA 2018 funded by DRPM Universitas Indonesia No. 2555/UN2.R3.1/HKP.05.00/2018.

IOP Conf. Series: Materials Science and Engineering **622** (2019) 012016 doi:10.1088/1757-899X/622/1/012016

References

[1] J. Grum, S. Bozic, and M. Zupancic, Influence of Quenching Process Parameters on Residual Stresses in Steel, J. Mater. Process. Technol., 2001, 114(1), p 57–70

[2] Dhinesh K D and Valan A A. 2016. A review on preparation, characterization, properties, and applications of nanofluids. (India: Elsevier) 21-40.

[3] Nayak, Vignesh U. Prabhu, Narayan. K. 2016. "Wetting Behavior and Heat Transfer of Aqueous Graphene Nanofluids.

[4] Xie H, Wang J, Xi T, Liu Y, Ai F and Wu Q, "Thermal conductivity enhancement of suspensions containing nanosized alumina particles", (J Appl Phys, 2002) vol. 91:4568-4572.

[5] S. Lee, S.U.S. Choi, S. Li, and J.A. Eastman, "Measuring thermal conductivity of fluids containing oxide nanoparticles", (Journal of Heat Transfer, 1999) vol. 121, pp. 280-289.

[6] X. Wang, X. Xu, and S.U.S. Choi, "Thermal conductivity of nanoparticle–fluid mixture," (Journal of Thermophysics and Heat Transfer, 1999) vol. 13, pp. 474-480.

[7] J.A. Eastman, S.U.S. Choi, S. Li,W. Yu, and L.J. Thompson, "Anomalously increased effective thermal conductivities of ethylene glycol-based nanofluids containing copper nanoparticles," (Applied Physics Letters, 2001) vol. 78, no. 6, pp. 718-720.

[8] Ramesh, Gopalan and Narayan Kotekar Prabhu, "Review of thermo-physical properties, wetting and heat transfer characteristics of nanofluids and their applicability in industrial quench heat treatment", (Nanoscale Research Letters, 2011) (15 pages)

[9] Xuan Y and Li Q, "Heat transfer enhancement of nanofluids", (Int J Heat Fluid Flow, 2000) vol. 21 pp 58-64.

[10] Kim, H., DeWitt, G., McKrell, T., Buongiorno, J., and Hu, L.W., "On the quenching of steel and zircaloy spheres in water-based nanofluids with alumina, silica and diamond nanoparticles", (International Journal of Multiphase Flow, 2009) vol. 35, pp 427–438.

[11] Khoshmehr, Habibi, Ahmad Saboonchi,and Mohammad Behshad Shafii, "The quenching of silver rod in boiling carbon nano tube-water nanofluid", (International Journal of Thermal Sciences, 2014) pp 95-104

[12] E. P. Mahiswara, *et.al.* 2018. *IOP Conf. Ser: Mater. Sci. Eng.* 299. 012068.

A Comparative Study on Commercial Grade and Laboratory Grade of TiO_2 particle in Nanofluid for Quench Medium in Rapid Quenching Process

C.A Ramadhani, W.N. Putra*, D. Rakhman, L. Oktavio, S. Harjanto

Department of Metallurgy & Materials, Faculty of Engineering, Universitas Indonesia, Depok, Indonesia

*Corresponding author: wahyuaji@metal.ui.ac.id

Abstract. Heat treatment of material particularly quenching requires a high thermal conductivity quench medium. Hardenability of material, dimension, and geometry of the component are considerate on choosing quench medium. The cooling rate of quenching affects the properties and microstructures by creating specific phase transformation to occur. Enhancing the quench medium by accelerating the cooling rate can be attained by the addition of nanoparticle which has higher thermal conductivity. This nanoparticle-added medium is commonly termed as nanofluid. Commercial and laboratory grade of TiO_2 was used as the nanoparticle to distilled water as the nanofluid base to acquired higher conductivity on the heat treatment process. In this experiment, a top-down method was done to obtain TiO_2 particles by grounding using a planetary ball mill for 15 hours at 500 rpm. Nanofluid quench medium was mixed with TiO_2 in various concentration of 1%, 5% and 10% with a volume of 100 ml each. Samples of AISI 1045 or JIS S45C carbon steel were used to obtain different cooling rate on a different type of TiO_2 particles. Samples were heat treated by austenizing at 1000°C for 1 hour, followed by rapid quenching in nanofluid quench medium with the addition of agitation as quenching variable. Observation of particle morphology and size, material composition, and the change of surface ere measured by Field-Emission Scanning Electron Microscope (FE-SEM), and Energy Dispersive X-Ray Spectroscopy (EDX). Initial characterization showed that the TiO_2 particle size was at 150 nm range, and roughly free from any impurities. Martensite microstructures have the most significant area and the amount at laboratory-grade TiO2 in 0.2 wt% composition, followed by commercial-grade at 0.3 wt% composition.

Keywords: Nanofluid, TiO_2, Quenching

1. Introduction

Heat treatment is one of the major processes used in engineering applications in order to achieve improved properties of the material. Desired and better microstructures in steel are obtained by quenching process that involves heating it to austenite temperature then holding followed by rapid cooling by immersion into fluids. Cooling rate and fluids affect significantly on the properties of the material as well as the microstructure[1]. The ability of heat transfer is necessary for the fluid or quenchants since it directly connected to the cooling rate. The outcome of a fast cooling rate is martensite

Content from this work may be used under the terms of the Creative Commons Attribution 3.0 licence. Any further distribution of this work must maintain attribution to the author(s) and the title of the work, journal citation and DOI.
Published under licence by IOP Publishing Ltd

structure with desirable high hardness value. Nanofluids have been common to be researched as quenchants as it owns the ability to increase heat transfer in quenching [2].

Titanium dioxide in the form of powder is extensively used in many industries, from food to the cosmetic industry. Visible and ultraviolet light cannot pass TiO_2 regarding its opacity [3]. In the food industry, it contributes to longer the lifetime of the food preventing early degradation. On the other hand in the cosmetic industry, it helps to absorb ultraviolet light in the sun thus becoming sunscreen [4]. Possessing a white and bright colour, TiO_2 benefit the use of coating and painting as pigments [5]. Not only the coating of the paint, but it also used on coating paper producing brighter and cleaner paper. In the previous paper, the beneficiation of TiO_2 is also applied to protect the metal from corrosion regarding its high specific surface area and suitable hydrophobicity property [6]. Having a high chemical and thermal stability and wideband energy TiO_2 is also used as light harvesting material in dye-sensitized solar cells [7].

TiO_2 particles in nanometer-sized can be dispersed in liquid suspensions in a base fluid resulting nanofluids. In nanofluid synthesis, a surfactant is added to avoid agglomeration of suspended particles because the more significant surface area that can worsen the heat transfer in nanofluid as quench medium [8-9].

The objective of this research is to know the characteristics of TiO_2 nanofluid as quench medium in quenching after heat treatment and study the difference of the use of laboratory and commercial grade nano-sized TiO_2 that obtained by planetary ball mill on microstructure, composition, and hardness.

2. Materials and methods

2.1. Preparation of TiO_2 Nanofluids

Laboratory and commercial grade TiO_2 powder were prepared by planetary ball mill with a top-down method in order to obtain nano-sized particles for 15 hours at 500 rpm.

The addition of 0.1%, 0.2%, 0.3%, 0.4% and 0.5% of TiO_2 were used to the quenchants of distilled water until 100ml. Nanofluid preparation then obtained by dispersing TiO_2 nano-sized particles in distilled water with ultrasonic agitation for 480 seconds.

2.2. Material for Quenching Experiments

Steel S45C was prepared by cutting the material in 15mm x 10mm x 10mm dimension manually by saw. Heat treatment was carried out in above austenizing temperature 600°C for 15 minutes for pre-heating, then heated at 1000°C for one hour followed by quenching in TiO_2 nanofluids.

2.3. Experimental

Microstructure figure was observed by metallography preparation followed by etching with Nital 3% then followed by microscope examination. Hardness measurement was obtained by a Vickers hardness test with a 300gf load. Field Emission Scanning Electron Microscope (FE-SEM) from Field Electron and Ion Company (FEI), model type F50 and Energy Dispersive X-Ray (EDX) from Energy Dispersive Analysis X-Ray Company (EDAX), model Apollo X examination were done at Centre for Material Processing and Failure Analysis (CMPFA) Universitas Indonesia.

3. Results and discussion

3.1. FE-SEM and EDX Examination

Figure 1 and 2 showed the result of FE-SEM examination of TiO_2 laboratory and commercial grade. Figure 1a, 2a showed the magnification of 25000X, and figure 1b,2b showed the magnification of 50000X. Figure 2 showed the result of FE-SEM examination of the TiO_2 commercial grade. Commercial grade TiO_2 had slightly bigger dimension at roughly 160 nm than laboratory grade TiO_2 which in range of 150 nm.

Figure 1. FE-SEM image of laboratory-grade TiO_2 with (a) 2500x magnification and (b) 50000x magnification.

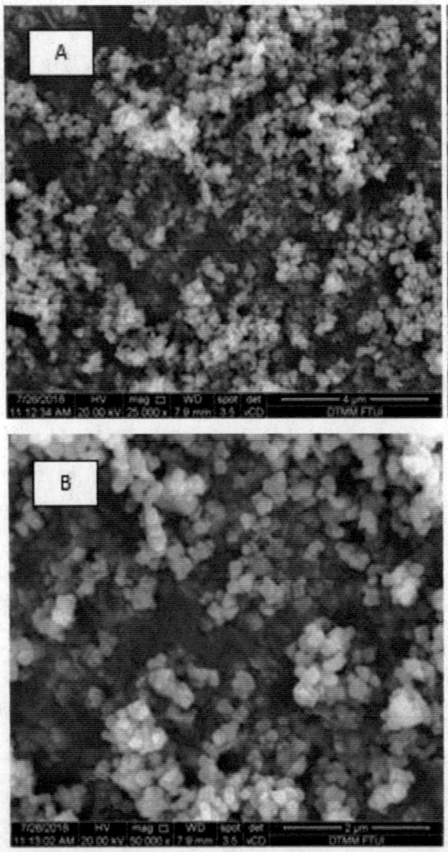

Figure 2. FE-SEM image of commercial-grade TiO_2 with (a) 25000x magnification and (b) 5000x magnification

Table 1. EDS Test result for chemical composition (wt %) of TiO_2 particles.

Sample	Ti	Si	P	S	Cl	K
Laboratory Grade Powder	99.06	-	0.43	-	0.51	-
Commercial Grade Powder	97.76	0.52	0.45	0.37	0.33	0.57

Table 1 showed the distribution of the elements in laboratory-grade and commercial-grade TiO_2. The element distribution shows that the purity of laboratory grade was higher than in the commercial grade TiO_2. On both powder, however, no significant metal contaminant was found.

3.2. Microstructure Observation

Figure 3 showed the microstructure of steel quenched in distilled water. The martensite structure was observed from the dark area. Figure 4 a-e showed the 0.1wt% - 0.5wt% composition of laboratory grade.Laboratory-grade TiO2 in 0.2 wt% composition showed a larger area of martensite structures than the rest of the composition, and it was shown on Figures 4. 0.3, 0.4 and 0.5 (wt.%) shows a smaller area of martensite structure than in 0.2 wt% composition.In microstructures, carbon has been spotted sputtering out marked by a black spot in microstructures. It could be caused by decarburization because S45C steel was medium carbon steel and react in austenizing temperature when heat and hold at 1000°C [10-11].

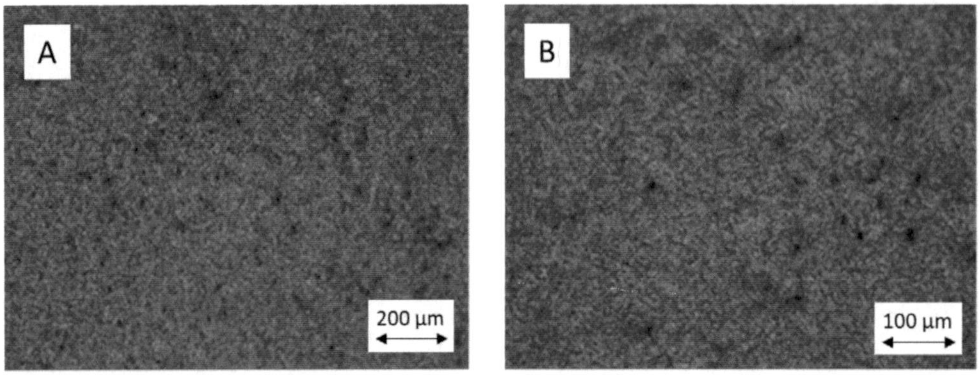

Figure 3. Microstructures of S45C as-cast in (a) 100x magnification and (b) 200x magnification.

The result of after-quenched steel immersed in commercial-grade TiO_2 can be seen in Figure 5. Carbon had been spotted sputtering out in figure 5a and figure 5d. A martensitic structure can be observed by the brown area. Figure 5c showed the largest martensitic structure area resulting in the highest hardness value obtained by commercial grade TiO_2. Figure 5d showed less amount of martensite structure known from the lesser area of a dark color[12-13].

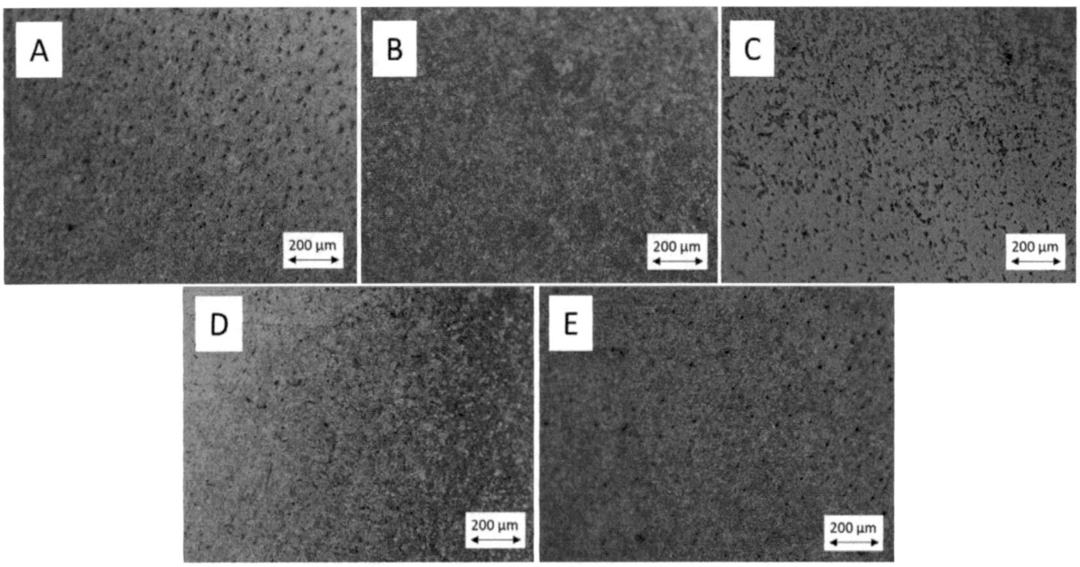

Figure 4. Microstructures of laboratory-grade TiO_2 as quench medium. (A) 0.1% vol (B) 0,2% vol. (C) 0.3% vol. (D) 0.4% vol. (E) 0.5% vol. taken at 100X magnification

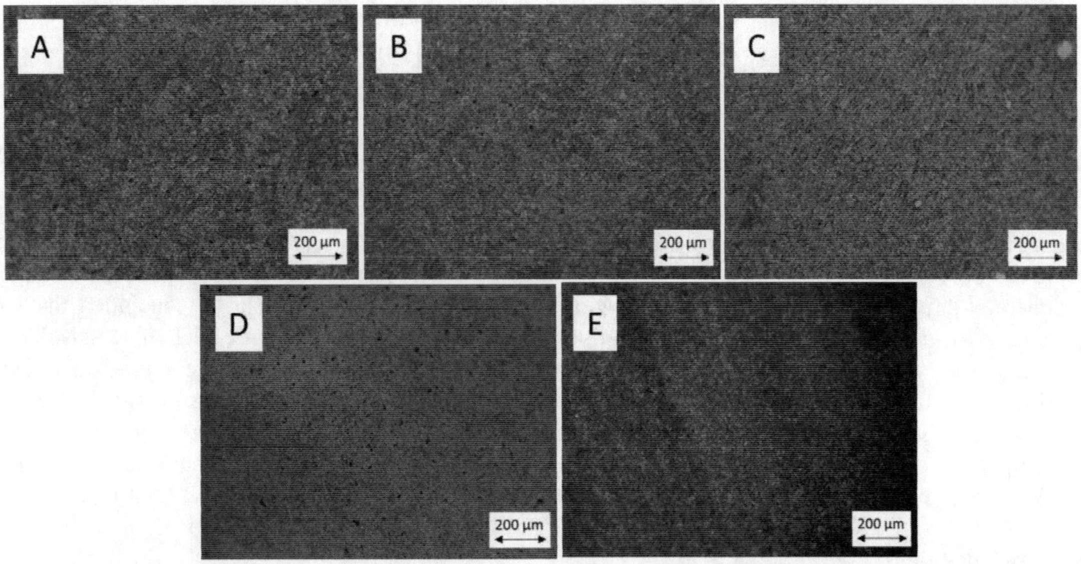

Figure 5. Microstructures of commercial-grade TiO_2 as quench medium. (A) 0.1% vol. (B) 0,2% vol. (C) 0.3% vol. (D) 0.4% vol. (E) 0.5% vol. taken at 100X magnification

3.3. Hardness Measurements

Hardness values were compared between steel quenched in laboratory-grade and commercial grade TiO_2 nanofluids. Table 2 showed the value of the hardness in HV. Addition of nano-sized particles in 0.1% has affected the hardness of the steel quite significantly than in 0% nanofluid or distilled water. Laboratory-grade TiO_2 0.1wt.% was harder than in commercial-grade TiO_2 nanofluids. Highest hardness values are owned by laboratory-grade TiO_2 nanofluid in 0.2 wit% composition, followed by commercial-grade TiO_2 nanofluid in 0.3wt% composition.

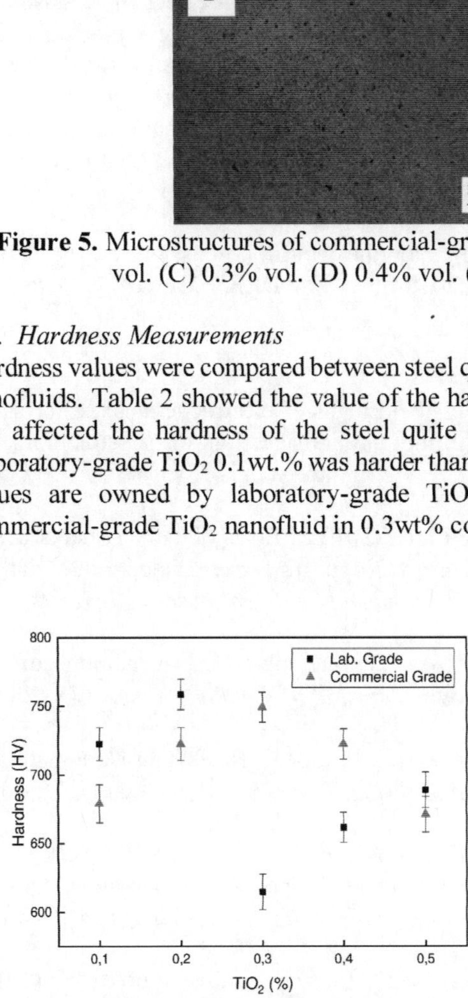

Figure 6. Hardness value comparison of laboratory grade and commercial grade TiO_2 in various composition.

Table 2. The hardness value of laboratory grade and commercial grade TiO_2 in various composition.

Composition		Hardness (HV)
0%		661.50
Laboratory Grade	0.1%	722.43
	0.2%	758.44
	0.3%	614.71
	0.4%	661.50
	0.5%	688.94
Commercial Grade	0.1%	678.94
	0.2%	722.43
	0.3%	749.19
	0.4%	722.43
	0.5%	671.11

Laboratory-grade TiO_2 peaked at 0.2wt% composition while commercial grade TiO_2 peaked at 0.3 wt% composition. Commercial grade TiO_2 showed a relatively steady chart while laboratory grade TiO_2 shows a fluctuating line with a decrease of hardness value in 0.3wt% composition shown in Figure 6. The fluctuating hardness might because of agglomeration occurred in this particular composition. In general, however, laboratory grade powder had better thermal conductivity compared with the commercial one. This result was different from our previous result where laboratory grade carbon had lower conductivity compared with commercial carbon powder [14].

4. Conclusion

Laboratory-grade and commercial-grade TiO_2 powder were grounded until reaching nano-sized particle then use as quenchants in nanofluids. The addition of 0.1%, 0.2%, 0.3%, 0.4% and 0.5% were added to distilled water as quench medium. Microstructures of the steel after a quench in nanofluid showed a greater amount of martensite structure than without the addition of TiO_2 powder. The highest hardness value of 758.44HV was obtained with laboratory-grade TiO_2 in 0.2wt% composition and followed by commercial-grade TiO_2 in 0.3wt% composition which was 749.19 HV. It can be concluded that, in general, laboratory-grade TiO_2 had bigger hardness value than commercial-grade TiO_2. The reason can be assumed because of the difference of Titanium element in laboratory and commercial grade that does not differ remotely. The largest area of martensite microstructures showed at laboratory-grade TiO_2 in 0.2wt% composition, followed by commercial-grade 0.3wt% composition. It can be concluded that addition of nanofluid can improve the hardness of the material and the most optimum composition is 0.2-0.3wt% composition.

Acknowledgement

The authors would like to thank Ms. Ghiska Ramahdita for the fruitful discussion, and gratefully acknowledged the financial support from Hibah PITTA 2018 funded by DRPM Universitas Indonesia No. 2537/UN2.R3.1/HKP.05.00/2018.

References

[1] S. Sun, X. Liao, G. Yin, Y. Yao, Z. Huang, X. Pu. (2016).Enhanced electrochemical performance of TiO_2 nanotube array electrodes by controlling the introduction of substoichiometric titanium oxides. *Journal of Alloys and Compounds*, 680, 538-543. https://doi.org/10.1016/j.jallcom.2016.04.17.

[2] Chen, L., Wang, Y., Huang, M., Li, X., Zhu, L., & Li, H. (2017). Effects of TiO_2crystal structure on the luminescence quenching of [Ru(bpy)2(dppz)]2 +-intercalated into DNA. *Spectrochimica Acta - Part A: Molecular and Biomolecular Spectroscopy*, *181*, 164–170. https://doi.org/10.1016/j.saa.2017.03.051.

[3] Mikhailov, M. M., & Yuryev, S. A. (2016). The temperature quenching of the cathodoluminescence bands in titanium dioxide. *Journal of Luminescence*, *175*, 28–34. https://doi.org/10.1016/j.jlumin.2016.02.012.

[4] Ullattil, S. G., Narendranath, S. B., Pillai, S. C., & Periyat, P. (2018). Black TiO_2Nanomaterials: A Review of Recent Advances. *Chemical Engineering Journal*, *343*(February), 708–736. https://doi.org/10.1016/j.cej.2018.01.069.

[5] Lata, S., Pandey, A., Labhansh, Sharma, A., Meena, K., Rana, R., & Lal, R. (2018). An Experimental Study and Analysis of the Mechanical Properties of Titanium Dioxide Reinforced Aluminum (AA 5051) Composite. *Materials Today: Proceedings*, *5*(2), 6090–6097. https://doi.org/10.1016/j.matpr.2017.12.214.

[6] Mengarelli, A., Verdini, F., Cardarelli, S., Di Nardo, F., Burattini, L., & Fioretti, S. (2018). Balance assessment during squatting exercise: A comparison between laboratory grade force plate and a commercial, low-cost device. *Journal of Biomechanics*, *71*, 264–270. https://doi.org/10.1016/j.jbiomech.2018.01.029.

[7] Liu, Y., Bian, D., Zhao, Y., & Wang, Y. (2018). Anti-corrosion performance of chemically bonded phosphate ceramic coatings reinforced by nano-TiO_2. *Journal of the Mechanical Behavior of Biomedical Materials*, *86*(June), 208–214. https://doi.org/10.1016/j.jmbbm.2018.06.045.

[8] Babu, K., Arularasan, R., & Srinath Ramkumar, S. (2017). Quenching performance of AISI 1010 in CNT nanofluids. *Materials Today: Proceedings*, *4*(10), 11044–11049.

https://doi.org/10.1016/j.matpr.2017.08.065.

[9] Sharma, A., & Das, S. (2009). Study of age hardening behavior of Al-4.5 wt%Cu/zircon sand composite in different quenching media - A comparative study. *Materials and Design*, *30*(9), 3900–3903. https://doi.org/10.1016/j.matdes.2009.03.022.

[10] Jing, Y. A., Yuan, Y. M., Yan, X. L., Zhang, L., & Sha, M. H. (2017). Decarburization mechanism during hydrogen reduction descaling of hot-rolled strip steel. *International Journal of Hydrogen Energy*, *42*(15), 10611–10621. https://doi.org/10.1016/j.ijhydene.2017.01.230.

[11] Burzic, D., Zamberger, J., & Kozeschnik, E. (2010). ARTICLE IN PRESS NDT & E International Non-destructive evaluation of decarburization of spring steel using electromagnetic measurement. *NDT and E International*, *43*(5), 446–450. https://doi.org/10.1016/j.ndteint.2010.04.006.

[12] Li, X., Shang, C., Ma, X., Subramanian, S. V., Misra, R. D. K., & Sun, J. (2018). Structure and crystallography of martensite–austenite constituent in the intercritically reheated coarse-grained heat affected zone of a high strength pipeline steel. *Materials Characterization*, *138*(January), 107–112. https://doi.org/10.1016/j.matchar.2018.01.042.

[13] Zhang, P., Chen, Y., Xiao, W., Ping, D., & Zhao, X. (2016). Twin structure of the lath martensite in low carbon steel. *Progress in Natural Science: Materials International*, *26*(2), 169–172. https://doi.org/10.1016/j.pnsc.2016.03.004.

[14] S. S. Yahya, S. Harjanto, W. N. Putra, G. Ramahdita, Kresnodrianto, and E. P. Mahiswara, *Characterization and observation of water-based nanofluids quench medium with carbon particle content variation*, AIP Conference Proceedings 1964, 020006 (2018); https://doi.org/10.1063/1.5038288.

10 Ghz Optical Modulator using CPS structure for communication and Sensing

P.Daud[1*], D.P. Kurniadi[1], D. Mahmudin[1], Y.N. Wijayanto[1], P.Putranto[1], E.J.Pristianto[1], A.Nurrahman[1],Andriana[2], Zulkarnaen[2],O.Vertus[2]

[1]ResearchCenter for Electronics and Telecommunication (PPET), Indonesian Institute of Sciences (LIPI), Bandung, Indonesia
[2]Faculty of Electro, Universitas Langlangbuana, Bandung, Indonesia

*Corresponding author: pamungkasdaud@lipi.go.id

Abstract. The combination of of wireless millimeter-wave (MMW) bands and fiber optic cables is able to compensate the propagation loss of millimeter waves in broadband communications and high-resolution imaging. This paper discusses the design and the realization of optical modulator using coplanar stripline (CPS) structure completed with the u-slot antenna patch for RF signal input. The substrate used for CPS structure is Lithium Niobat (LiNbO3) due to its large electro optical coefficient. A 10 Ghz radio wave frequency is inputted via microstrip u-slot antenna patch array 2 x 1 with input impedance 50 Ω, and acceptable return loss -10 dB. The light waves for the carrier are generated from the laser diode. Several measurements have been completely done and reported. The design is carried out by means of varying physical variable value which results in specification of the operating frequency 10 GHz. This is a pure research since the obtained results are not directly applied to the radio-over-fiber (RoF) technology.

Keywords: Coplanar Stripline(CPS), Radio over Fiber, Radio Wave (RF), light wave, optical modulator

1. Introduction

Wireless telecommunication system is increasingly growing [1]. This made radio frequency (RF) band is getting crowded so that the frequency usage is increasingly shifting to millimeter wave (MMW) / 30 GHz - 300 GHz wavelength and light waves (LW). One of the applications of RF–LW interaction whichis important in the telecommunications system is a device or circuit module called converter or better known as optical modulator. This converter has several advantages such as having low induction effect and low power consumption. This device usually be applied in telecommunication systems with wide bandwidth and in electromagnetic field measurement systems with low induction effects.

The use of a millimeter-wave (MMW) band optical modulator for wireless telecommunications systems combined with fiber optic cables can also compensate the propagation loss of high-frequency band (millimeters-wave band) which is used for broadband communications and high resolution imaging.MMW wireless technology is promising for fifth generation (5G) mobile communication systems due to its outstanding characteristics for broadband data transfer and massive connection capability. MMW is also important for high-resolution radar, remote sensing, and imaging systems.

Content from this work may be used under the terms of the Creative Commons Attribution 3.0 licence. Any further distribution of this work must maintain attribution to the author(s) and the title of the work, journal citation and DOI.
Published under licence by IOP Publishing Ltd

Therefore, studies on MMW components, devices, and sub-systems have recently been attracting a lot of researchers. However, MMW is not suitable for long distance (>10 km) free-space transmission since it has a large propagation loss in air [2]. Radio-over-fiber (RoF) technology is a solution to connect many cells over a small coverage area by the use of extremely low-loss silica fibers (~0.2 dB/km at λ = 1550 nm) in wireless MMW communication systems [3]. Moreover, RoF technology is useful for radar and sensing application systems. In MMW RoF systems, a signal conversion device from MMW to LW is significant. We have proposed several devices for converting MMW signals to LW signals using planar antennas and electro-optic (EO) crystals [4,5]. Although MMW has positive advantages for communication system, microwave (MW) is still better for long distance because of its small propagation loss.

In this paper, a new EO integrated modulator using 4 coplanar stripline array structure is proposed. With array structure, the gain of the electric field induced across the narrow gap for optical modulation are all increased. Therefore, the conversion efficiency from RF to LW is expected to be improved.

2. Theorem

2.1. Coplanar Stripline

There are several kind of exploitable structures to make an optical modulator. One among them is resonator electrode structure [6]. The main idea of this structure is that once the light wave enter the structure at passband frequency of the resonator, the electro-optic effect occurred and resulted in changes of refraction index. In the design, the utilized resonator electrode structure is Coplanar Stripline (CPS). CPS consists of substrate with 2 conductor plates arranged parallel and separated by a tiny space. The advantages of CPS are the ability to be put in series or shunt and that it is a balance transmission channel [3,7].

The characteristics of CPS are small losses, little distortion, insensitive to substrate thickness, and easy to build in either open-ended or short-ended [7,8]. To obtain the width of the conductor plate and the distance between both conductors, the equation below is given.To calculate the distance between conductor plates (*W*) is expressed as following equations, when the width of conductor plate (*S*).

$$\frac{W}{h} \leq \frac{10}{1+ln\,\varepsilon_r} \text{and} \frac{S}{h} \leq \frac{10}{3(1+ln\,\varepsilon_r)} \quad (1)$$

where, h is the substrate thickness (mm) and ε_r is the relative dielectric constant.
Hence,

$$S = W \times G(\varepsilon_r, h, Z_0, W) \quad (2)$$

With

$$G = \left[\frac{1}{8} exp\left(\frac{60\pi^2}{Z_0 \varepsilon_{eff}^{0.5}}\right) - \frac{1}{2}\right]^{-1} \text{ for } Z_0 \leq \frac{60\sqrt{2}\pi}{(\varepsilon_r+1)^{0.5}} \quad (3)$$

$$G = \frac{1}{4} \exp\left(\frac{Z_0 \varepsilon_{eff}^{0.5}}{120}\right) + \exp\left(-\frac{Z_0 \varepsilon_{eff}^{0.5}}{120}\right) \text{for } Z_0 > \frac{60\sqrt{2}\pi}{(\varepsilon_r+1)^{0.5}} \quad (4)$$

Where

$$\varepsilon_{eff} = \varepsilon_{eff}(\varepsilon_r, h, Z_0, W) = T_1 \times \left(1 + \frac{\varepsilon_r - 1}{Z_0(\varepsilon_r + 1)^{0.5}} T_2\right) \quad (5)$$

With

$$T_1 = 1 + tanh\left\{\frac{Z_0^2\varepsilon_r^7}{720\pi^3(\varepsilon_r+1)^8} \times exp\left[\left(1+0.0004\varepsilon_r Z_0 \frac{W}{h}\right) ln\left(\frac{W}{h}\right)\right]\right\} \quad (6)$$

And

$$T_2 = 84.85 ln\left(2\frac{1+g}{1-g}\right) \text{for} 0.841 \le g \le 1, \quad (7)$$

$$T_2 = \frac{837.5}{ln\left(2\frac{1+(1-g^4)^{0.25}}{1-(1-g^4)^{0.25}}\right)} \text{for} 0 < g \le 0.841, \quad (8)$$

$$g=\left\{\frac{\exp\left(\frac{\pi(1+Q)W}{2h}\right)-\exp\left(\frac{\pi W}{2h}\right)}{\exp\left(\frac{\pi(2+Q)W}{2h}\right)-1}\right\}^{0.5} \quad (9)$$

$$Q = G|_{\varepsilon_{eff}=\frac{\varepsilon_r+1}{2}} \quad (10)$$

where ε_{eff}= effective relative dielectric constant, Z_0 = line characteristic impedance (ohm). The CPS arm length can be obtained by using the equation (11).

$$L_{cps} = \frac{c}{2f\sqrt{\varepsilon_{reff}}} \quad (11)$$

2.2. Transmission Line Dimension

A microstrip-type transmission line consists of 2 conductors which are a plate with width W, and ground plane. Both of them are separated by substrate which has $\varepsilon_r$with thickness$h$.Because Z_0 is given, width of the microstrip line can be derived by the equation below.

$$\frac{W}{h} = \frac{8e^A}{e^{2A}-2} \text{for} \frac{W}{h} < 2 \quad (12)$$

$$\frac{W}{h} = \frac{2h}{\pi}\left\{B-1-\ln(2B-1)+\frac{\varepsilon_r-1}{2\varepsilon_r}\left[\ln(B-1)+0{,}39-\frac{0{,}61}{\varepsilon_r}\right]\right\} \text{for} \frac{W}{h} > 2 \quad (13)$$

$$A = \frac{Z_0}{60}\sqrt{\frac{\varepsilon_r+1}{2}} + \frac{\varepsilon_r-1}{\varepsilon_r+1}\left(0.23+\frac{0{,}11}{\varepsilon_r}\right) \quad (14)$$

$$B = \frac{377\pi}{2Z_0\sqrt{\varepsilon_r}} \quad (15)$$

The effective relative dielectric constant of microstrip lines can be determined by the equation below [2].

$$\varepsilon_{eff} = \frac{\varepsilon_r+1}{2} + \frac{\varepsilon_r-1}{2}\frac{1}{\sqrt{1+12h/W}} \quad (16)$$

2.3. Array Structure

Array structure is a structure that allows several components to be used simultaneously. In this design, T-Junction is used as divider circuit producing a parallel impedance distribution. Inputs are distributed with a pair of output lines so that the input impedance equivalent with the parallel circuit impedance in the output line [7].

2.4. Matching Imepdance Technique

Matching impedance technique is used to adjust the different impedances which are line characteristic impedance (Z_0) and load impedance (Z_l). In this design, transformer $\lambda/4$ is used as impedance matching technique in the transmission line. The length of transformer $\lambda/4$ line is gained by this equation [4,6].

$$l_f = \frac{\lambda_g}{4} \tag{17}$$

$$\lambda_g = \frac{\lambda_0}{\sqrt{\varepsilon_{reff}}} = \frac{c}{f\sqrt{\varepsilon_{reff}}} \tag{18}$$

3. Designing and Simulation

In this section, we will show the design, simulation, realization and measurement results, from prototypes of the antenna and optical modulator that the authors created.

3.1. Simulation of Antenna Patch U-Slot

In a microstrip slot the inductive coupling antenna is induced from the supply line to the slot. The greater coupling effect that occurs will reduce the circuit quality factor. The decreasing quality of the circuit quality factor will increase the bandwidth [9].

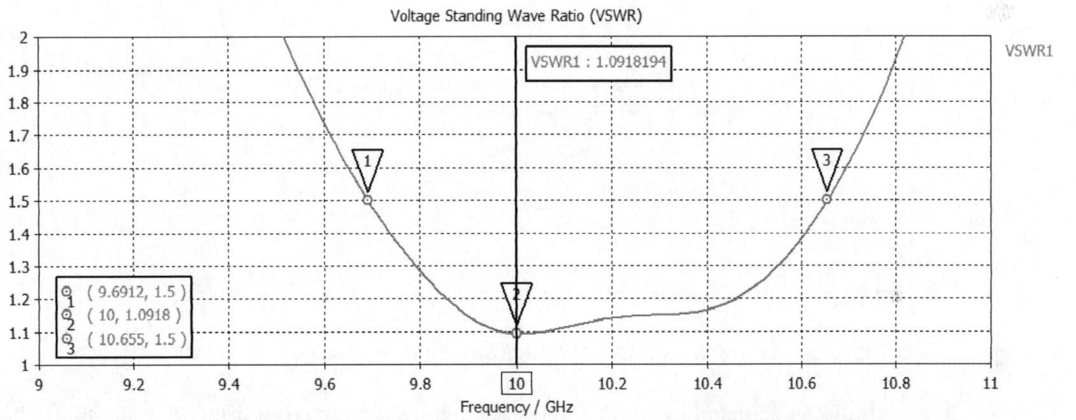

Figure 1. TheVSWR of antena Inset-Fed Arraywith U-Slot

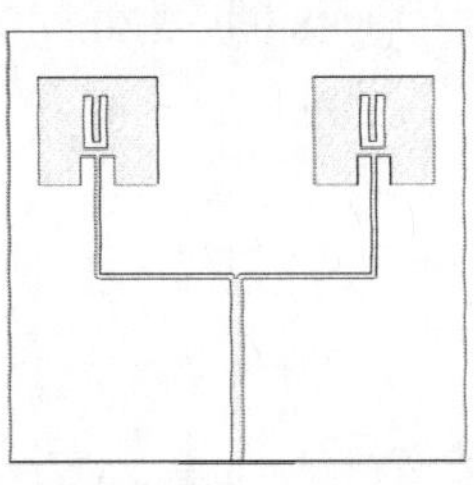

Figure 2. Design of Inset-Fed Antenna Arrays with U-SlotsAntenna

Figure 3. Realization ofInset-Fed Arrays with U-Slots

3.2. Simulation of Optical Modulator

The opticalmodulator consists of 4 CPS modules which are calculated to generate impedance values is shown in Fig. 4. The substrate used is Lithium Niobat (LiNbO3) which has a large electro optical coefficient;[[6,10,11]. Frequency of radio waves (RF) to be used for 10 Ghz which is planned to be fed

IOP Conf. Series: Materials Science and Engineering **622** (2019) 012018 doi:10.1088/1757-899X/622/1/012018

through Microstrip Patch-u slot 2 x 1 array antennas. Light wave used for carrier is generated from laser diode with wavelength 1550 nm.

The simulation of optical modulator using electromagnetic simulation software application [4] and the simulation results are shown in Fig. 5. It can be seen that frequency response shown by S_{11} parameter at 10 Ghz optical modulator works well with value of 90 dBm.

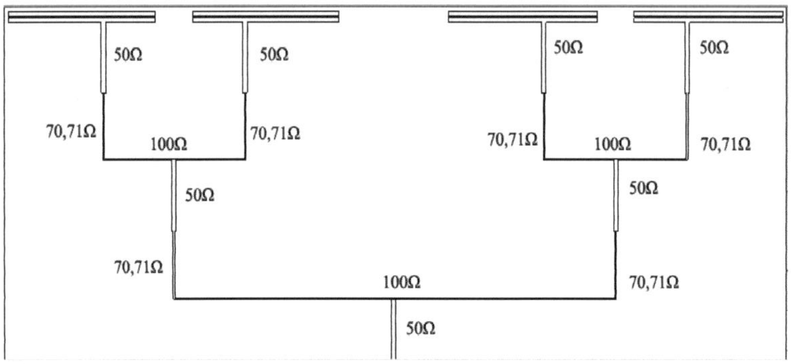

Figure 4. Designof Optical Modulator Structure 4x1 Array

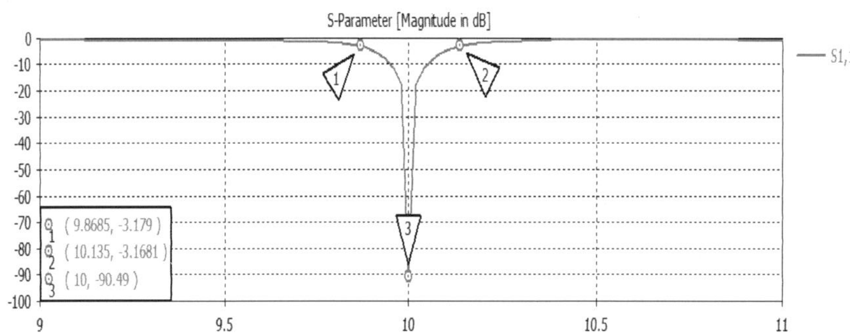

Figure 5. Graph S_{11} optimum optical modulator result of 4x1 array

The modulator design is depicted in Fig. 6. On the left hand side, there is a discrete modulator where the microstrip patch and the CPS are separated by a cable. On the right hand side, there is a compact structure of integrated modulator consists of CPS that is united with the patch, without using a cable.

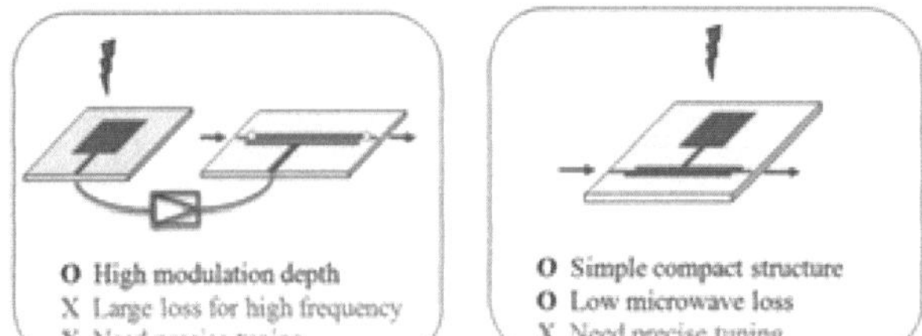

Figure 6. Block design of Discrete modulator and Integrated Modulator

4. Measurement and Analysis

The measurement detail of the fabricated antenna as shown in figure 3. isnot reported here , but the simulation result can be seen in figure 1.that the antenna is operated for 10GHz microwave bands and has VSWR 1.1. The measurement result of integratedmodulator using Vector Network Analyzer (VNA) can be seen infigure77. Based on this results, the both devices can be used and integrated between the patch antenna and optical modulator for receiving wireless 10GHz microwave bands and converting it to lightwave signals.

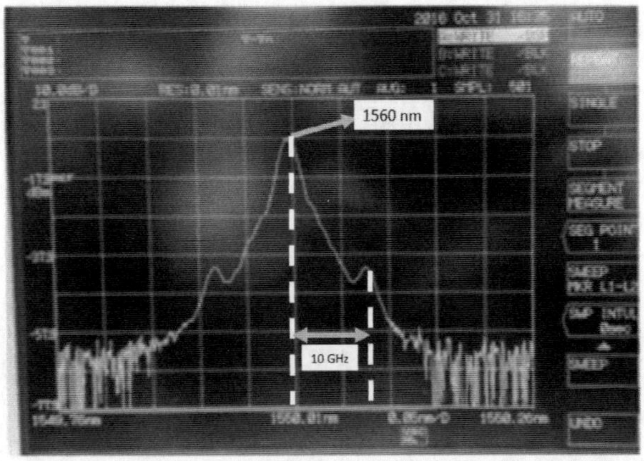

Figure 7. Frequency Response Measurement Results from Optical Modulator

5. Conclusion

The 10 GHz integrated optical modulator with CPS 2x1 array antenna has been made and evaluated. By analyzing the frequency response, it can be concluded that this system works well according to the simulation. The array structure can reduce the return loss, the more array structure used, the smaller of the return loss generated. With the decrease in return loss, the power received by the device approaches the power provided by the source. The CPS gap is related to capacitance so that the gap change from CPS can emit the change of the working frequency and impedance of the optical modulator. With the change of the impedance, the changes will affects to the return loss. In the future we will reported of the antenna measurement detail and combine both with the modulator measurement.

Acknowledgement

This research activity was conducted as part of the research project "Realization of light-based integrated sensor lighting sensors for heavy metal detection of industrial waste "through INSINAS research program 2018 from the Ministry of Research, Technology and Higher Education of the Republic of Indonesia.

References

[1] Suzhi Bi, Rui Zhang, Zhi Ding, and Shuguang Cui ,"Wireless communications in the era of big data", 2015 IEEE Communications Magazine Vol: 53, Issue: 10,Page(s): 190 – 199.

[2] P. Daud, J. A. Simalongo, D. Mahmudin, Y. N. Wijayanto, and P. Putranto, "Design of Vehicle Tracking System using MCU STM32F103RET6 and SIM908", 2017 Internetworking IndonesiaJournal, Vol. 9, No. 2 .

[3] P. Daud, D. Mahmudin, A.A. Fathnan, I.Syamsu, T.T.Estu, and Y.N.Wijayanto, "Inset-Fed U-Slotted Patch Antenna Array for 10GHz Radio-Over-Fiber Applications", IEEE-ICSE 2016 Proc. 2016, Kuala Lumpur, Malaysia.

[4] Y. N. Wijayanto, H. Murata, Y. Okamura ''High-Speed Guided-WaveElectro-Optic Modulators Using an Array of Gap-Embedded PatchAntennas with Phase Reversal'',2012 IEICE Technical Report MWP2011-72.

[5] N. Kohmu, H. Murata, Y. Okamura, "Electro-Optic Modulator UsingLiNbO3 Waveguide Suspended to Double-Antenna-Coupled Electrodeson Low-k Substrate for Millimeter-wave-Lightwave Signal Conversion",PA-6, APMP 2013, Gwangju, Korea.

[6] F. Darwis, A. Setiawan and P. Daud, " Performance of Narrowband Hairpin Bandpass Filter Square Resonator with Folded Coupled Line", 2015 International Conference on Radar, Antenna, Microwave, Electronics and Telecommunications (ICRAMET), Depok

[7] A.A. Fathnan, Taufiqqurrachman, Y.N. Wijayanto, D. Mahmudin, P. Daud , "Electromagnetic Coupling Mechanism in Vertical and Horizontal SRR-Loaded Coplanar Waveguide", 2016 Asia-Pasific International Symposium on Electromagnetic Compability & Signal Integrity and Technical Exhibition (APEMC) ,Shenzhen,China

[8] P. Daud, S. Hardiati, D. Mahmudin, T.T. Estu, A.A. Fathnan, Y.N. Wijayanto, N. Armi , "1x2 Array of U-Slotted Rectangular Patch Antennas for High-Speed LTE Mobile Networks", International Conference on Engineering and Technology for Sustainable Development (ICET4SD) – 2015, Jogyakarta

[9] Kai,Chang,.et.all.,"RF and Microwave Circuit and Component Design for Wireless System", New York : John Wiley and Son, 2002]

[10] Y. N. Wijayanto, A. Kanno, A.A. Fathnan, P. Daud, T. Kawanishi, D. Mahmudin, and N. Yamamoto,"Metamaterial Antenna Integrated to LiNbO3 Optical Modulator for Millimeter-Wave-Photonic Links", (ISAP) Tasmania-Australia 2015

[11] R. C. Alferness, Waveguide electrooptic modulators, IEEE Trans. Microwave Theory Tech. MTT-30 (1982) 1121–1137.] .

Polyetherimide nanofiltration membranes modified by interfacial polymerization for treatment of textile dyes wastewater

Gabriel Febrianto[1], Doni Karisma[1] and Dave Mangindaan[2*]

[1] Chemical Engineering Department, Parahyangan Catholic University, Bandung, West Java, Indonesia
[2] Professional Engineer Program, Faculty of Engineering, Bina Nusantara University, Jakarta, Indonesia 11480
[*]Corresponding author: dmangindaan@binus.edu

Abstract. Azo dyes are the major type of textile dye in the world, owing to their stability to light, microbial degradation, and physical degradation due to washing. However, these properties also lead to problematic removal or degradation of azo dyes that pollute the water body. In this research, nanofiltration (NF) thin film composite (TFC) membranes based on polyetherimide (PEI) polymer are utilized to remove an azo-based dye from a simulated textile wastewater, namely Reactive Black 5 (RB5). PEI is firstly dissolved by using N-methyl-2 pyrollidone (NMP) as solvent, combined with acetone as a non-solvent, and converted to be membranes via phase inversion method. The created membrane will be further modified by interfacial polymerization (IP) method using trimesoyl chloride (TMC) and m-phenylene diamine (MPD) as precursors of acyl chloride and amine, immersed in two immiscible liquids of hexane and water, respectively. This method fabricates a new selective layer composed of tightly-packed nylon-like polyamide layer that might improve the separation performance. Membranes from polymeric dope solution of PEI/acetone/NMP 15/65/20 (w/w) were employed due to acceptable flux and rejection, compared to other formulations. They were then modified by using IP method (0.05% TMC in hexane and 1.5% MPD in water) to create PEI-TFC membranes. The PEI-TFC membranes exhibited fluxes around 0.01 L m^{-2} h^{-1} psi^{-1}, with rejection of RB5 dyes up to 90%, which suggested the successful IP method on the PEI membranes. SEM and FTIR were carried out for comprehending the reasons behind the improved separation performance, and they revealed that the TFC nylon-like selective layer was successfully developed, from both physical and chemical perspectives, respectively. The fabrication of NF TFC membranes might open some new roads for environmental application of membranes in Indonesia.

Keywords: membranes, nanofiltration, dyes, polyetherimide

1. Introduction

The global scarcity of clean potable water is one of the major challenge faced by the humanity these days [1]. One of the largest industry sector that plays a pivotal role in the society's fundamental needs, and global economy as well as in the consumption of water is the textile industry [2]. This industry uses a huge amount of water for its dyeing process [3]. As an illustration, a kilogram of textile uses 60 to 100

Content from this work may be used under the terms of the Creative Commons Attribution 3.0 licence. Any further distribution of this work must maintain attribution to the author(s) and the title of the work, journal citation and DOI.
Published under licence by IOP Publishing Ltd

kg of water during dyeing and washing [4]. As a big issue of supply and demand of water, it is absolutely necessary to have the wastewater treated and recycled. However, the textile wastewater treatment is relatively not a straightforward process. The textile wastewater definitely must be decolorized, via some meticulous process of destruction of dyes molecules, or by separation of dyes from waters [5]. In order to broken down the dyes (which is generally made up of resilient azo structures), some processes namely chemical oxidation, photo-catalysis and biodegradation [6] could be taken. On the contrary, these methods could only be employed to some extent, as there are limitations on the intensive energy consumption, and also long treatment duration, due to the common azo molecular structure of dyes [7-9].

As an alternative resolution for the dyes destruction processes, the separation of dyes from wastewater (e.g. adsorption, coagulation and flocculation, and membrane-based treatment) are developed [5]. The adsorption process utilizing activated carbon-based materials is not uncommon; however it is limited by the removal capacity which requires vast amount of adsorbents, rendering uneconomical process especially for the application in the treatment of textile dyes wastewater. Moreover, after several time and regeneration processes, the adsorbents may suffer some significant reduction of adsorption performance [10-12].

Besides adsorption process, the coagulation and flocculation were also utilized in the treatment of wastewater due to their efficiency [12-14]. Coagulation and flocculation process agglomerates some insoluble particles of dyes, causing the dyes to clump in large quantity, thus could be effortlessly separated from the wastewater [15, 16]. However, they are not applicable in separating the soluble dyes [11].

A separation process that could remove both soluble and insoluble dyes is the membrane-based process that might be classified as ultrafiltration (UF), nanofiltration (NF) and reverse osmosis (RO) [7]. UF successfully separated insoluble dyes and also soluble dyes with large molecule structure, but ineffective for removing dissolved dyes having small molecular weight [7, 17]. The absolute separation of dyes could be obtained by application of RO, but limited with the high operation pressure (>50 bar). Therefore, it would be a good strategy to develop the membranes with properties between these two boundaries, which is the NF membranes. It demonstrates a good separation performance of dyes removal from textile wastewater [5], balanced with affordable operational cost [18]. This study targets to have a good performance in separating soluble textile dyes from wastewater, with relatively low operational pressure. This target is planned to be accomplished via further improvement of the previously developed polyetherimide (PEI) membranes, by preparing thin film composite (TFC) NF membranes [19].

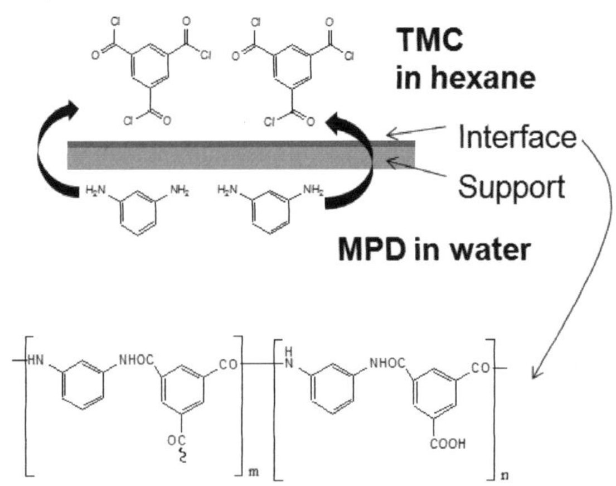

Figure 1. An illustration of interfacial polymerization involving TMC and MPD resulting in polyamide TFC membranes

To prepare TFC membranes, an unconventional polymerization processed called interfacial polymerization process was utilized. In brief, it is basically a polymerization reaction between two reactive precursors reacting at the border or interface of two immiscible solvents. Generally, the system comprises of amine dissolved in water, and acyl chloride in hexane, both in very low concentration to avoid rapid build of reaction temperature that may lead to explosion. The product of the interfacial reaction of the reactive precursors is a thin Nylon-like polyamide layer. In this study, m-phenylenediamine (MPD) and trimesoyl chloride (TMC) were used as the precursors [20, 21]. The illustration of interfacial polymerization to form TFC is shown in Figure 1.

2. Experimental methods

In this study, we prepared the membranes based on PEI materials (polyetherimide, Sigma-Aldrich, CAS 61128-46-9), using the previously developed formulation of polymeric dope solution [19]. The dope consists of using PEI polymer, acetone as non-solvent, and N-methyl pyrollidone (NMP) solvent (PEI/acetone/NMP mixture), having the dope compositions of 15/20/65 and 16/20/64 w/w. Acetone and NMP (CAS 872-50-4) were obtained from Merck. After the dope solutions were successfully prepared, membranes were casted via flat sheet casting method. The details of flat sheet membrane casting process could be found in the previous work [19, 22].

Briefly, the polymeric dope solutions were mixed until visually homogeneous, and then went into stationary condition for about two days to ensure removal of micro gas bubbles from the dope solution. The dope solution was poured onto A4 glass plates and then casted with thickness of 250 μm at room temperature. The casted membranes were removed from solvent residues via copious immersion in methanol and hexane for 3×30 mins each.

The casted membranes were then proceed to the stage of interfacial polymerization, by being immersed in aqueous MPD solution for 2 mins, subsequently followed with immersion in TMC for also 2 mins to form PEI-TFC membranes. Aqueous MPD solution is dissolved in deionized water at concentration of 1.5 wt%, while the TMC is dissolved in hexane at 0.05 wt%. with the molecular form listed in Figure 1 [23]. After the interfacial polymerization processes, the membranes were dried 70°C for 30 min in an oven. MPD (CAS 108-45-2) and TMC (CAS 4422-95-1) were obtained from Sigma-Aldrich.

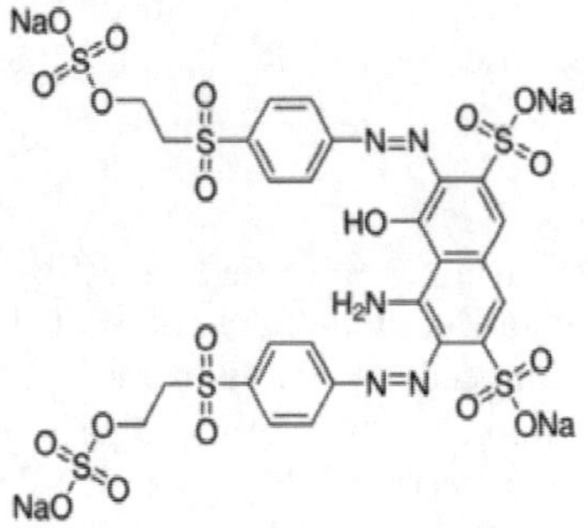

Figure 2. Molecular structure of RB5

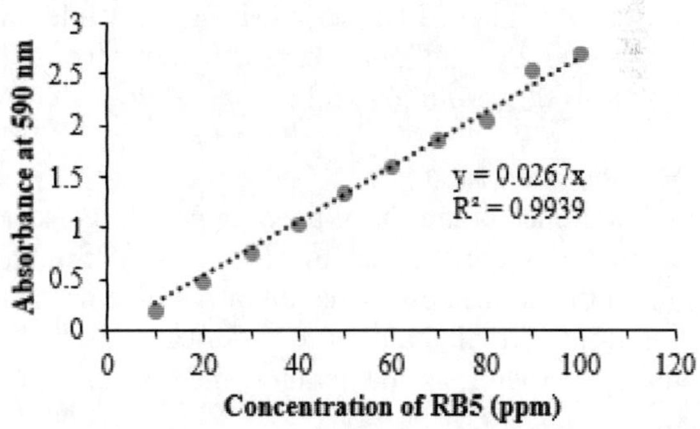

Figure 3. Standard curve of RB5 dyes at various concentration, with absorbance measured at λ_{max} 590 nm

The fabricated TFC membranes were objected to a test for separation performance, to assess both the quantity and quality of the separation performances. For the quantity aspect, the test was carried out by using a permeation cell combined with pressurized N_2 gas (pressure 40-80 psi). The separated object is Reactive Black 5 textile dye (RB5, CAS number 17095-24-8, Sigma-Aldrich, Figure 2) dissolved in water as a simulated wastewater, with concentration of 100 ppm and 150 ppm. The detailed procedure and illustration of the equipment for testing of the TFC membranes are shown elsewhere [19]. The assessment for the quality of the separation is obtained by using UV-Vis spectrophotometer. The spectrophotometer was utilized to quantify the colour intensity of the permeate, at a certain wavelength, in this case for that of RB5 dyes is at the wavelength that provides maximum absorbance $\lambda_{max} = 590$ nm, and shown in Figure 3.

The quantity of the separation is demonstrated by the parameter of permeate flux, while for that of quality, the parameter of rejection is utilized. The permeate flux flow rate is written in Equation 1.

$$J = \frac{V}{A\,t\,p} \tag{1}$$

with J = flux (L m^{-2} s^{-1} psi^{-1}), V= volume of the permeate (L), A= membrane surface area (m^2), t= time required to contain the permeate (s), and P = the pressure required for the feed to pass through the membrane (psi).

The quality of the membrane separation is exhibited by using the parameter of color rejection, as shown in Equation 2, as follows:

$$\%R = \left(1 - \frac{C_p}{C_f}\right) \times 100\% \tag{2}$$

with $\%R$ = Rejection, C_p=Concentration of the permeate (ppm), and C_f= Concentration of feed solution (ppm). Moreover, the pore radius of the fabricated membrane is estimated by using the Laplace equation, as follows [24]

$$r_{pore} = \frac{2\gamma}{\Delta P}\cos\theta \tag{3}$$

Pore radius r_{pore} of a capillary shaped pore completely wetted (contact angle θ= 0°) by water (surface tension γ= 72.3 mN/m), with the highest operating pressure of ~80 psi (~5.44 bar= 545000 Pa) is around 267 nm. Therefore, the pore of the fabricated NF membrane is confirmed to be in the nanometer range. The physical morphology (cross-section) of the membranes were analyzed by using scanning electron microscopy (SEM, JEOL JSM-5600LV), while the chemical properties of the membranes were characterized by employing Fourier transform infrared spectroscopy (FTIR, Perkin-Elmer FTIR Spectrum 2000, resolution of 2 cm^{-1}) [25, 26].

3. Results and discussion

As the PEI membranes have been prepared, the separation performances were then investigated from the perspectives of the quantity and quality. Similar to that of in the previous work [19] we firstly explored the permeate flux (quantitative parameter. Before any molecular separation were performed, we subjected the membranes to pure water first in order to comprehend the ability to pass the water molecules through the membranes, mentioned as PWP (pure water permeability). The PWP result is shown in Figure 4.

From Figure 4, it could be observed that the membrane casted from the PEI16 (from dope solution with concentration of PEI/acetone/NMP 16/20/64 w/w) delivered much slower PWP compared to that of PEI15 (15/20/65 w/w), i.e. only 0.25 vs up to 1.50 L m^{-2} h^{-1} psi^{-1}, respectively. The operating pressure requirement of PEI16 is also higher than that of PEI15, i.e. 50-80 psi vs. 40-70 psi, respectively. Based on this result we directed the PEI15 to be further processed to the interfacial polymerization stage, for additional attachment of polyamide selective layer on top of PEI15, thus creating the TFC membranes.

The separation performance of the PEI15-TFC membranes in separating RB5 dyes (100 ppm) is shown in Figure 5. It could be observed that there the flux is decreased significantly after the membranes were modified by TFC, which is from around 0.25-0.40 to be about 0.05 L m^{-2} h^{-1}, respectively. It is quite common to obtain the reduction of flux for the membranes coated with additional layer, creating TFC membrane [21] or for membranes subjected to crosslinking processes [25, 26]. The reduction is likely due to the addition of tighter layers or caused by polymer chain repacking in microscopic level [25]. The quality of the separation was improved slightly, from between 20-50% to 50%-67% (with varied operating pressure from 40-70 psi), for PEI15 and PEI15-TFC, respectively.

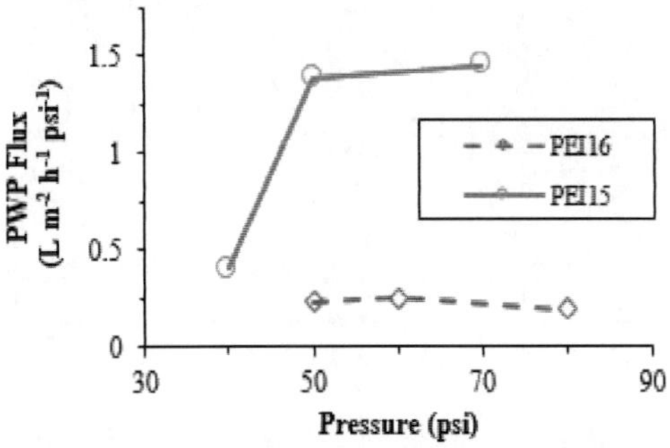

Figure 4. PWP of PEI15 (PEI/acetone/NMP of 15/20/65 w/w) and PEI16 (16/20/64 w/w) membranes as function of operating pressure

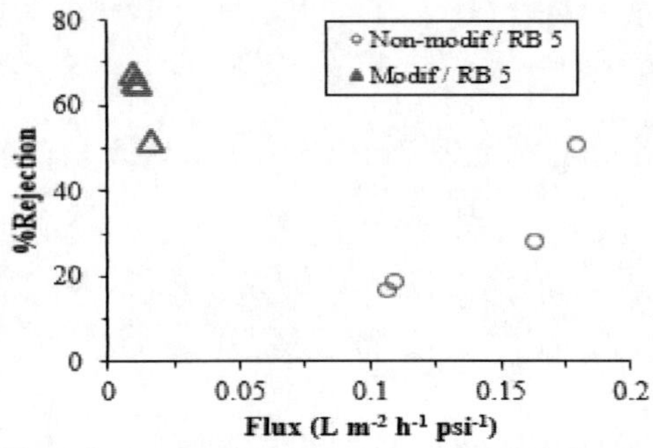

Figure 5. Separation performance of unmodified (PEI15) and modified (PEI15-TFC) membranes for separating RB5 dyes (100 ppm) at operating pressure of 40-70 psi

The low quality of rejection of RB5 dyes shown in Figure 5 might be a result of low concentration of dyes in the simulated wastewater containing RB5 dyes. Therefore, to corroborate this suggestion, addition experiment for testing the effect of the concentration of dyes were conducted. The nanofiltration process of RB5 dyes with initial dyes concentrations of 50, 100, and 150 ppm were conducted at uniform operating pressure of 40 psi, and the result is reported in Figure 6. It could be observed that the higher the concentration of RB5, then the higher the rejection, where it could achieve 92.2% rejection for initial dyes concentration of 150 ppm.

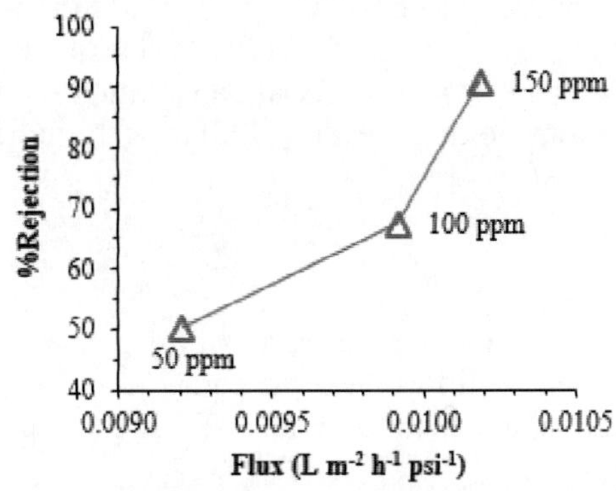

Figure 6. Separation performance of PEI15-TFC for separating RB5 dyes at various concentration at 40 psi

In order to understand the structure-performance relationship of the membranes, or how the membrane modification influenced the separation performance, the cross-sectional view of the membranes as captured via SEM is shown in Figure 7.

It could be observed that the top surface of PEI15 (Figure 7a) is still thin and has numerous voids in the structure. However, the PEI15-TFC (Figure 7b) exhibited a denser top surface than that of PEI15, having less voids and giving an impression of being stiffer than that of PEI15. This is the structure that is desired for a top selective layer.

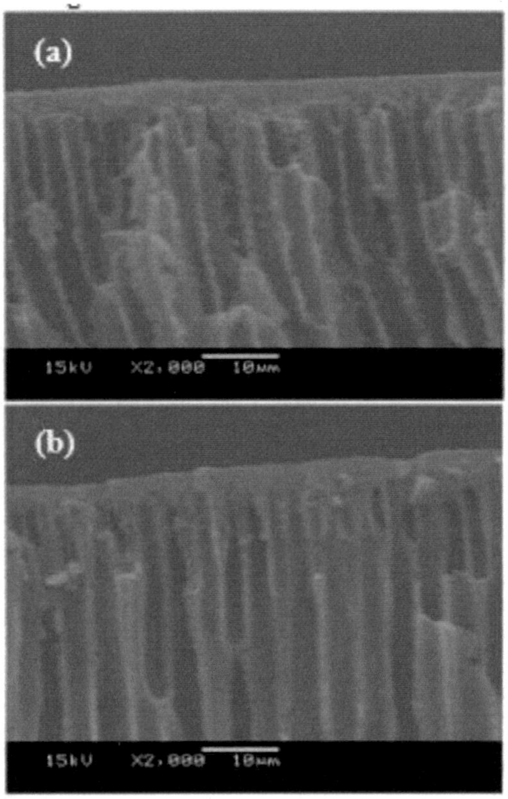

Figure 7. Cross-sectional view of membranes, (a) PEI15, and (b) PEI15-TFC

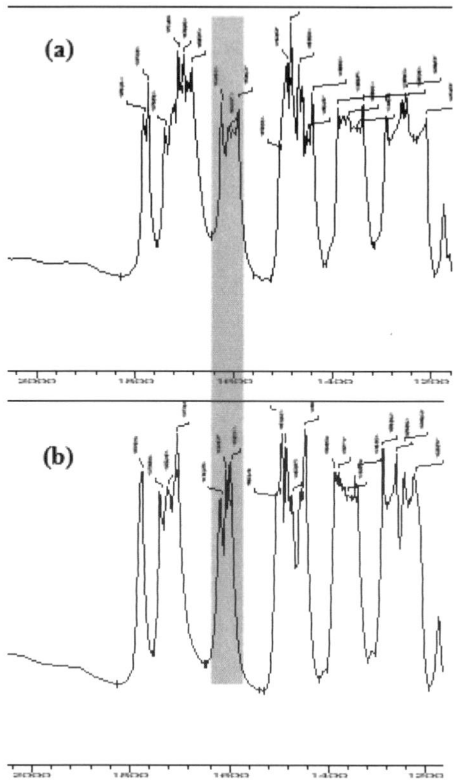

Figure 8. FTIR spectra of (a) PEI15, and (b) PEI15-TFC membranes

The chemistry behind the fabrication of PEI-TFC membranes is discussed in Figure 8, based on the results from the FTIR characterizations, for PEI15 and PEI15-TFC, in Figure 8a, and 8b, respectively. The FTIR spectra of PEI15-TFC showed N-H deformation vibration and C=C stretching vibration [27] at around 1600-1640 cm^{-1}, and aromatic ring breathing [28] at as the indicator of polyimide groups.

4. Conclusion

It could be concluded that PEI-TFC membrane was successfully prepared via interfacial polymerization between MPD and TMC. The fabricated PEI-TFC membrane could separate textile dyes wastewater (RB5, 150 ppm) with 92% rejection, due to the additional TFC selective layer which was confirmed by SEM and FTIR characterizations.

Acknowledgement
The authors would like to thanks Indonesia Toray Science Foundation who has awarded the research grant at 21st Science and Technology Research Grant to the Principal Investigator (Ir. Dave Mangindaan, PhD, AMRSC, AMIChemE, IPM) for conducted this study.

References

[1] Han G., Zhang S., Li X., Widjojo N. and Chung T.-S. 2012 *Chemical Engineering Science* **80**.
[2] Dasgupta J., Sikder J., Chakraborty S., Curcio S. and Drioli E. 2014 *Journal of Environmental Management* **147**. 55-72
[3] Sójka-Ledakowicz J., Koprowski T., Machnowski W. and Knudsen H.H. 1998 *Desalination* **119**. 1-9
[4] Chidambaram T., Oren Y. and Noel M. 2015 *Chemical Engineering Journal* **262**. 156-168
[5] Zahrim A.Y. and Hilal N. 2013 *Water Resour. Ind.* **3**. 23-34
[6] Hao O.J., Kim H. and Chiang P.C. 2000 *Crit. Rev. Env. Sci.Technol.* **30**. 449-505
[7] Akbari A., Remigy J.C. and Apte P. 2002 *Chem. Eng. Process.* **41**. 601-609
[8] Forgacs E., Cserhati T. and Oros G. 2004 *Environ. Int.* **30**. 953-971
[9] Tünay O., Kabdasli I., Erememktar G. and Orhon D. 1996 *Water Sci. Technol.* **34**. 9-16
[10] Robinson T., McMullan G., Marchant R. and Nigam P. 2001 *Bioresour. Technol.* 247-255
[11] Verma A.K., Dash R.R. and Bhunia P. 2012 *J. Environ. Manage.* **93**. 154-168
[12] Riera-Torres M., Gutiérrez-Bouzán C. and Crespi M. 2010 *Desalination* **252**. 53-59
[13] Bes-Piá A., Iborra-Clar M.I., Iborra-Clar A., Mendoza-Roca J.A., Cuartas-Uribe B. and Alcaina-Miranda M.I. 2005 *Desalination* **178**. 343-349
[14] Chen T., Gao B. and Yue Q. 2010 *Colloid Surf. A-Physicochem. Eng. Asp.* **355**. 121-129
[15] Zahrim A.Y., Tizaoui C. and Hilal N. 2011 *Desalination*. 1-16
[16] Lee K.E., Morad N., Teng T.T. and Poh B.T. 2012 *Chem. Eng. J.* **203**. 370-386
[17] Majewska-Nowak K. 1989 *Desalination* **71**. 83-95
[18] Cheng S., Oatley D.L., Williams P.M. and Wright C.J. 2011 *Adv. Coll. Interf. Sci.* **164**. 12-20
[19] Karisma D., Febrianto G. and Mangindaan D. 2017 *IOP Conf. Ser.: Earth Environ. Sci.* **109**. 012012
[20] Khulbe K.C., Feng C. and Matsuura T. 2010 *J. Appl. Polym. Sci.* **115**. 855-895
[21] Sutedja A., Josephine C.A. and Mangindaan D. 2017 *IOP Conf. Ser.: Earth Environ. Sci.* **109**. 012042
[22] Gunawan F.M., Mangindaan D., Khoiruddin K. and Wenten I.G. 2019 *Polym. Adv. Technol.* **30**. 360-367
[23] Han G., Zhang S., Li X., Widjojo N. and Chung T.S. 2012 *Chem. Eng. Sci.* **80**. 219-231
[24] Mulder M. 1992 *Basic Principles of Membrane Technology* (Netherlands: Kluwer Academic Publishers)
[25] Mangindaan D.W., Woon N.M., Shi G.M. and Chung T.S. 2015 *Chem. Eng. Sci* **122**. 14-23
[26] Mangindaan D.W., Shi G.M. and Chung T.S. 2014 *J. Membr. Sci.* **458**. 76-85
[27] Zhao Y., Li N. and Xia S. 2016 *Compos. Sci. Technol.* **132**. 84-92
[28] Zhou Y., Yu S., Liu M. and Gao C. 2006 *J. Membr. Sci.* **270**. 162-168

The Effect of Manganese Palmitate as Pro-Oxidant Additive on Mechanical Properties of Polypropylene

Reza Jati Pamungkas[*], Rachmawati and I Made Arcana[*]

Inorganic and physical Chemistry Research Group, Faculty Mathematics and Natural Sciences, Institut Teknologi Bandung, Jl. Ganesha 10, Bandung 40132, Indonesia

[*]Corresponding author: rezajatip@gmail.com, arcana@chem.itb.ac.id

Abstract. Polypropylene (PP) is one of the most widely used polymer materials, such as in textile, automotive spare part, furniture, household appliances, and packaging. However, PP is very difficult to degrade naturally. When the product from PP is not reused and discharged into the environment, this plastic can cause problems to the environment. One of the solution to improve degradability of PP is by adding pro-oxidant additives. The addition of pro-oxidant additives can make polypropylene to easily oxidize and degrade into shorter chains by producing carbonyl groups. The pro-oxidant additive used in this research is manganese palmitate. The additive was synthesized by reacting palmitic acid with sodium hydroxide and followed by reaction with manganese chloride tetrahydrate. The purpose of this research was to study the effect of manganese palmitate as a pro-oxidant additive to the structure and mechanical properties of PP before and after thermal treatment. The thermal treatment was carried out by providing heating in PP at various temperatures (25, 60, and 90°C) for 2, 4, 6, 8, and 10 days. The structure and mechanical properties of PP was characterized by functional group analysis (FTIR) and mechanical properties (Tensile tester). FTIR spectra of PP film after thermal treatment at temperature 60 and 90oC showed the presence of carbonyl group in 1700-1720 cm^{-1} region. After thermal treatment at 90°C, PP film could not be analyzed by mechanical properties because it was damaged and destroyed. The peak intensity of the carbonyl group increased with the increasing temperature of heating. PP films with the addition of manganese palmitate both before and after thermal treatment have a lower tensile strength and elongation compared to the pure PP, but their modulus young are higher, so that the films of PP with addition of pro-oxidant became more fragile and rigid. In addition, the PP film with heating treatment at 90°C cannot be analysed due to it was damaged and destroyed.

Keywords: manganese palmitate, polymer oxo-biodegradable, polypropylene, thermal degradation

1. Introduction

Polymer materials have been widely used in various fields such as packaging for food and electronics. Plastic is one of the most commonly used packaging materials in industry, from home industry to large industry. Plastic is a material that has properties such as flexibility, toughness, and another properties that support the product, and often used as packaging materials. The process of making plastic is easy and low cost than the other materials. Plastic is preferred as a packaging material[1]. Polyproylene (PP) is

Content from this work may be used under the terms of the Creative Commons Attribution 3.0 licence. Any further distribution of this work must maintain attribution to the author(s) and the title of the work, journal citation and DOI.
Published under licence by IOP Publishing Ltd

one of polymers the most widely used as packaging material. However, PP is difficult to decomposed in nature[2].

The use of many plastics can cause plastic waste to pollute the environment. Approximately 100 million tons of synthetic plastics are produced worldwide each year, and due to plastic can not be biodegraded by microorganisms, so it will continue to exist until hundreds of years[3].

One of the solutions to solve environmental pollution caused by plastic buildup is by making polymers that has oxo-degradation properties. The oxo-degradation polymer is formed by adding a pro-oxidant additive to the polymer[4]. The addition of pro-oxidant additives is a promising solution to solve the environmental pollution problems caused by plastic waste[5-6]. The process of oxo-biodegradation occurs in two stages. The first stage is to break the main chain of polymers into smaller fragments through the process of oxidation-reduction and the second stage is the fragment was attacked by microorganisms in environment[7]. These additives are transition metal ion complexes of carboxylic acids. The transition metal salt which have been widely used are stearate salts of iron, cobalt and manganese[8]. Previous study done by Maryudi, that using manganese carboxilic acid could increasing the degradation process of High Density Polyethylene (HDPE) films[9]. The effect of using Manganese palmitate (MnP) additive on the degradation of Polypropylene (PP) has not been investigated. In this work has been made to evaluate the effect of the addition of MnP on physical and mechanical properties of PP before and after thermal treatment.

2. Experimental

2.1. Materials
Manganese chloride tetrahydrate ($MnCl_2.4H_2O$), palmitic acid ($C_{15}H_{31}COOH$), sodium hydroxide (NaOH), aqua dm, ethanol (C_2H_5OH), xylene, were obtained commercially from Aldrich Chemical Co., and polypropylene was obtained from Polymer Industry (PT. Tripolyta).

2.2. Methods

2.2.1. Synthesis of manganese palmitate. Manganese palmitate was synthesized by reacting sodium hydroxide and palmitic acid with ethanol as a solvent to produce sodium palmitic. Manganese chloride tetrahydrate was heated at a temperature of 90 °C under nitrogen atmosphere and sodium palmitic was added into a solution of manganese chloride, then refluxed for 1,5 hours. After that the solution was left for 24 hours at room temperature until it occurs to precipitate. Manganese palmitic was filtered and washed with hot water, and ethanol, and then dried in a vacuum oven at 40 °C for 4 hours[7,10,11].

2.2.2. Preparation of polypropylene films. PP films were prepared by addition pro-oxidant additive with concentration from 0,2%, 0,6% and 1,0% (w/w) of the total mass, then it was dissolved in xylene. The mixture was placed in petri dish, its solvent was evaporated, and then pressed by using hot press at 150 °C for 5 minutes to form films.

2.2.3. Thermal degradation. PP films were heated for various days from 0, 2, 4, 6, 8 and 10 days at 60 °C and 90 °C. PP films before and after thermal treatment were characterization by using FTIR Bruker Alpha and Tensil tester faviraph Textechno41066.

3. Results and discussion
The most used parameter to evaluate the extent of thermal degradation of PP film is carbonyl index[1]. This was due the presence of autooxidation reaction mechanism of the transtition metal to catalysis the degradation of the PP films[12]. The presence of heat and oxygen, hydroperoxides (ROOH) as an impurity in polymer chains can be formed and further this hydroperoxides were decomposed by reduction-oxidation reaction to produce radicals by the presence of pro-oxidant additive in polymers. These

radicals then attack to other polymer chains to produce carbonyl group in polymer chains[3]. Figure 1 show the carbonyl group that was formed after thermal treatment at 60 and 90 °C for 10 days.

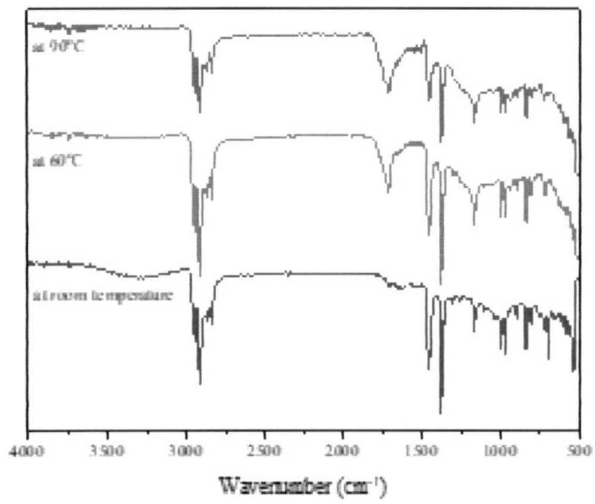

Figure 1. ATR spectrum of oxo-biodegradable PP films contains 1% MnP with various temperature for 10 days.

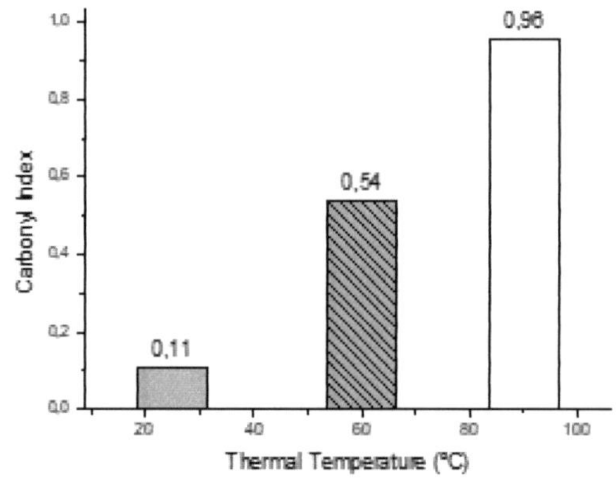

Figure 2. Carbonyl index of oxo-biodegradable PP films contains 1% of MnP with various temperature for 10 days.

Carbonyl index was calculated by comparing the absorbance of carbonyl group to absorbance of reference group[13]. The increasing of carbonyl index at higher temperature treatment indicates that the degradation of PP film was higher occured at higher temperature. The carbonyl index was smallest observed at room temperature, so this degradation of the PP films is slower than that of at higher temperature. This clearly indicates that the contribution of manganese palmitate (MnP) to accelerate the thermal degradation of the PP films at higher temperature. The increasing of temperature would effect the mobility of free radical to react with oxygen and polymer chain to produce new hydroperoxides[14]. The physical change of the PP films show that heating treatment at 60 °C and 90 °C, the PP films become more brittle than heating treatment at room temperature. In addition, the PP films with heating treatment at 90 °C was very brittle and can not be analyzed further. The previous work indicated that pro-oxidant additive will increase degradation process with using thermal treatment[15].

Figure 3 shows that with the various duration of thermal treatment could increase the degradation process showed with the increasing of carbonyl index value. This indicates that the presence of MnP could make the PP easier to degradate than pure PP[16].PP Oxo-biodegradation films was tested using tensile tester before and after thermal treatment to know the mechanical properties of the PP films.

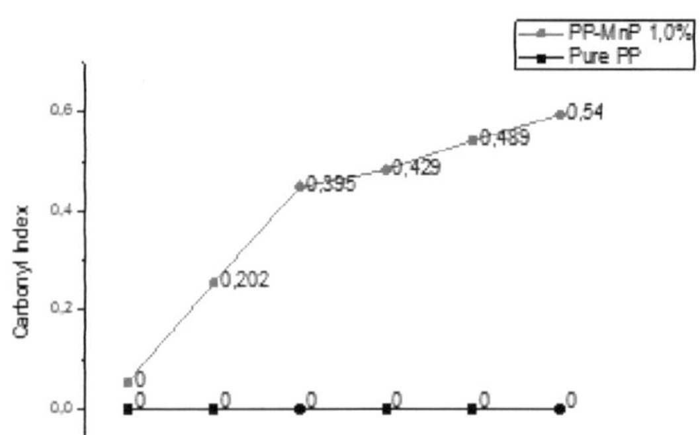

Figure 3. Carbonyl index of oxo-biodegradable PP films contains 1% of MnP with thermal treatment at 60 °C for various days.

The sample used for mechanical test were the pure PP films and the PP films with 1% MnP additive before and after thermal treatment at 60 ºC for 10 days. The PP film with 1% MnP additive after thermal treatment at 90 ºC for 10 days could not be analyzed the mechanical properties because the physically of film has been broken. This indicates that the PP film with 1% of MnP after thermal treatment at 90 ºC for 10 days had been thermally degraded. The results of tensile strength and elongation can be seen in Figure 4 and Figure 5.

The tensile strength and the elongation of the PP films containing 1% of MnP decrease after thermal treatment (Figure 4 and 5). This behaviour is explained due to phenomenon of the polymer chain reorganization with a lower degree of orientation. This is attributed by the presence of MnP. After thermal treatment, the MnP could accelarate to produce of free radical to produce macroradical and further react with oxigen and other PP chain to become smaller chain fragments and this is related with the increasing carbonyl index results. The increase in Modulus young indicates the brittleness of film PP increases (Figure 6), and so physically the PP films after thermal treatment is more brittle than pure PP.

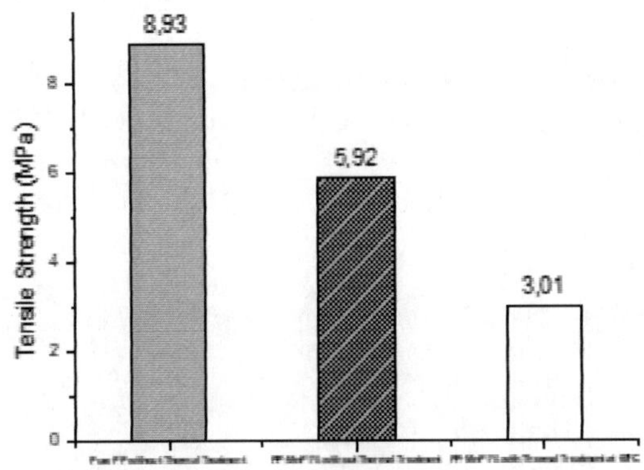

Figure 4. Tensile strength of PP films.

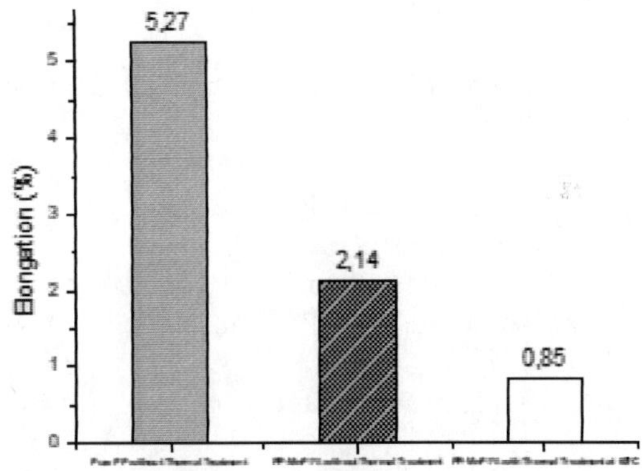

Figure 5. Elongation of PP films.

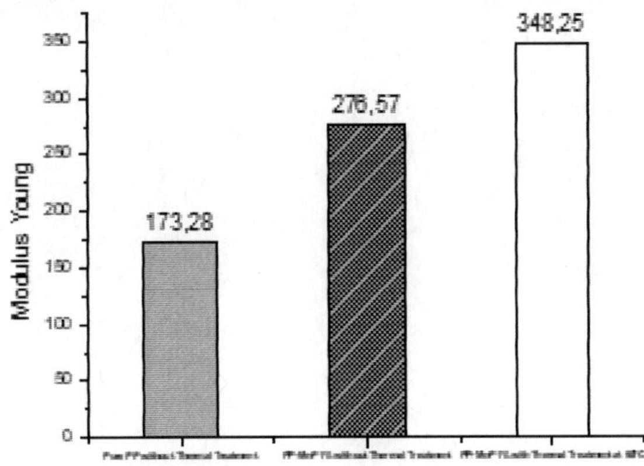

Figure 6. Modulus young of PP films.

4. Conclusion

The intensity of the carbonyl group peak increased with the increasing temperature of thermal treatment. The PP films with addition of manganese palmitate (MnP) both before and after thermal treatment have a lower tensile strength and elongation compared to the pure PP, but their modulus young are higher, so that the PP films with addition of pro-oxidant became more fragile and rigid. This result indicate that the presence of MnP and thermal treatment could make the PP easier to degradate than pure PP.

References

[1] Singh, B., and Sharma, N. 2007, *Polymer Degradation and Stability*, **92**, 876–885.
[2] Vijayakumar, C.T., Chitra, R., Surender, R., Pitchaimari, G., and Rajakumar, K. 2013 *Plastic and polymer technology* **2** 22-37.
[3] Adnan, Shah, J., and Jan, M.R. 2014, *Journal of Analytical and Applied Pyrolysis*, **109**, 196–204.
[4] Arutchelvi, J., Sudhakar, and Ambika, A. 2007 *Indian Journal of Biotechnology*. **7**, (1), 9–22.
[5] Koutny, M., Lemaire, J., and Delort, A.M. 2006, *Chemosphere*, **64**, 1243–1252.
[6] Gomes, L.B., Klein, J. P., Brandalise, R.N., Zani, M., Zoppas, B.C., and Grisa, A.M.C. 2014, *Material Research*, **17**, 1–6.
[7] Asriza, R. O. and Arcana, I M. 2015 *American Institute of Physics* 070007 1-4
[8] Chelliah, A., Subramaniam, M., Gupta, R., and Gupta, A. 2017, *Indian Journal of Science and Technology,* **10(6)**, 1-5.
[9] Maryudi, Hisyam, A., Yunus, R.M., dan Hossen Bag, M.D. 2013, *International Journal of Engineering Research and Applications*, **3**, 1156–1165.
[10] Maryudi, Yunus, R.M., Nour, A.H. and Abidin, M.H. 2009 *Journal of Applied Sciences* **17** 3156–3160.
[11] Aras, N. R. M. and Arcana, I M. 2015 *American Institute of Physics* 070024 1-5
[12] Rajakumar, K., Sarasvathy, V., Thamarai Chelvan, A., Chitra, R. and Vijayakumar, C.T. 2009 *Journal of Polymers and the Environment* **17** 191–202.
[13] Longo, C., Savaris, M., Zeni, M., Brandalise, R.N. and Coulon Grisa, A. M. 2011 *Materials Research*, **14(4)**, 442–448.
[14] Maalihan, R. D. 2016 *Material Science Forum*, **890**, 82–85.
[15] Roy, P.K., Surekha, P., Raman, R., and Rajagopal, C. 2009 *Polymer Degradation and Stability*, **94**, 1033–1039.
[16] Moo-Tun, N.M., Valadez-Gonzalez, A., and Uribe-Calderon, J.A. 2017 *Polymer Bulletin*, **75**, 1–21.

Comparison of tensile strength of continues-core method and 4 strand modifies method of after healing of repaired Kessler rabbit's Achilles tendon injury

P S Caksana[1], IKS Kawiyana[2], I W S Dusak[2], T G T Nindhia[3*]

[1] Resident of Studies Program of Orthopaedic Surgery and Traumatology, Medical Faculty of Udayana University, Denpasar, Indonesia
[2] Study Program of Orthopaedic Surgery and Traumatology, Medical Faculty of Udayana University, Denpasar, Indonesia
[3] Departmen of Mechanical Engineering, Engineering faculty, Udayana University, Jimbaran, Bali, Indonesia

[*]Corresponding author: tirta.nindhia@me.unud.ac.id

Abstract. Tendon injury causes significant morbidity during the productive age, and the number is increase in recent decades. Tensile strength is the most important factor in tendon function of style, and the results of repair should be able to withstand large current style of early mobilization. This research was an experimental research design with the post-test only control group design with the subject of the rabbit. A total of 18 research subjects who meet the requirements of the study inclusion were randomly divided into treatment and control groups. Control group was a group of rabbits that were repaired with 4 strand modified Kessler technique after Achilles tendon cut sharply. While the treatment group were a group of rabbits that were repaired with continuous-cores technique. The ratio of collagen I /III in rabbit Achilles tendon repair with the technique of continuous-cores are smaller compared to the control group with value of $P < 0.05$. Also Showed statistical analysis tensile strength of rabbit Achilles tendon repair with the technique of continuous-core is greater than the control group at $P < 0.05$.

Keywords: Achilles, tendon, injury, rabbits, tensile strength, continuous-cores method, 4 strand modifies method

1. Introduction

Injuries to the tendons become a significant cause of morbidity in the population of productive age. These injuries often experienced in sports injuries or traffic accidents, and addressed by a wide range of specialist and non-specialist personnel. The number is increasing in recent decades. The process of handling the tendon injury is still a challenge for orthopedics specialist doctors for healing tendon cause a variety of problems in terms of both anatomy and function [1].

Tendons are the connective tissue that lies between the bone and muscles that serves to transmit the force produced by skeletal muscles by maintaining posture and joint movement. Most injuries to the tendons occur in areas close to the joints, such as shoulders, elbows, knees, and ankles. Injury to the

Content from this work may be used under the terms of the Creative Commons Attribution 3.0 licence. Any further distribution of this work must maintain attribution to the author(s) and the title of the work, journal citation and DOI.
Published under licence by IOP Publishing Ltd

joints accompanied by partial or complete tear of a tendon occurs as much as 45% of all musculoskeletal injuries each year. Injuries to the tendon is quite often the case with the percentage of 30% - 50% of all injuries. These injuries mainly occur due to traffic accidents or sport injuries [1]

Injuries to the tendons vary from minor injuries that have relatively minimal risk to the daily activities, up to a tear in the tendon that requires surgery. Complications are common due to tendon injuries are disturbances spasticity, rigidity and reduced the power of the tendon itself. At the end of tendon function as a buffer as well as the nature of viscoelastic to reduce muscle damage will be decreased so that the movement and function of protection is limited [1-3].

The main purpose of tendon healing is to restore mechanical function where the tendons have a poor spontaneous regenerative capacity after injury, making it difficult to regain the biological and biomechanical function as before the injury. This is due to the formation of adhesions and scar tissue which inhibits the regenerative process of the tendon [3].

At the tendon healing process, there is a change in the type and distribution of collagen tissue that is formed. There is a decrease in the ratio of collagen type I / type III where an increase in type III collagen synthesis compared to Type I. Collagen type III formed more and much thinner than type I as well as the cross-linking of collagen fibres that form only 45% and most immature. The synthesis of type I collagen in tendon injury grade III will decline and be replaced by increased synthesis of collagen type III along with the expansion of granulation tissue and remodelling process so that the ratio of collagen type I / III decreased and the substitution of collagen towards the formation of scar tissue and fibrosis. Scarring of the tendon can cause a decrease in adhesion and tensile strength of up to 20% [4-7].

Tensile strength is the most important factor in the primary function of the tendon as a conductor of force, and therefore the results of repair should be able to withstand great force created when the initial mobilization. This leads to a lot of different repair techniques with a focus on increasing the tensile strength [8].

2. Experimentals

The experiment was conducted with the aim to prove different in tendon reconstruction technique that will reduce the collagen type I / type III ratio and increase of the tensile strength of the tendon after healing. This study is an experimental research design using post-test only control group design with the subject rabbits from eligible subject population with randomized sampling.

Thirty-six adult male New-Zealand white rabbits of the same age (12 weeks) weighing 2 - 3 kg were used in this experiment. All the animal was given adequate food and water in the laboratory. These rabbits were randomly divided into control and treatment groups. The control group was repaired with continuous-cores (Fig. 1), while the treatment group was compared with core repaired with 4 strand modified Kessler technique (Fig. 2) after Achilles tendon (Fig. 3) had been cut sharply and sewn with the non-absorbable nylon thread (4-0 monofilament), followed with immobilization of the affected limb with a cast. This technique uses six cores that divide the load at six anchor and multi strand locations. The addition of the number of core and node points can significantly increase the strength and resistance to the existence of gaps.

After that, the rabbits were sacrificed and the Achilles tendon was isolated for further examination. The rabbit tendon tissues were examined for tensile strength in the vertical direction in which the tendon is stretched until failure load with the universal testing machine. The immunohistochemistry examination is then performed to know the ratio of collagen type I/ III. This procedure is carried out with rabbit tendon tissue fixation using 10% neutral buffered formalin and processed for routine histological preparations with antimouse collagen I/III antibody and antimouse IgG secondary antibody / biotin which has been conjugated with horseradish peroxidase. The calculation of the amount of collagen, tendon using the Image J Network processing.

The obtained data were presented as mean SD of number of rabbits (n) that were used in the experiment. The mean difference was analyzed with independent t-test for statistical significant of Examined variables i.e. the tensile strength and the ratio of collagen I/ III. The difference was assumed to be significant at $P < 0.05$.

IOP Conf. Series: Materials Science and Engineering **622** (2019) 012021 doi:10.1088/1757-899X/622/1/012021

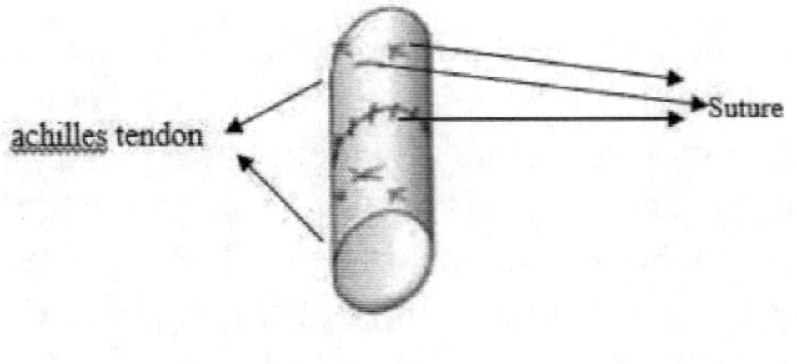

Figure 1. Continuous-cores technique

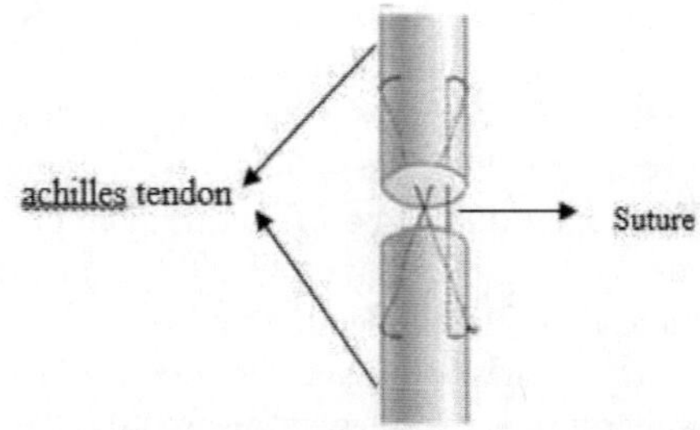

Figure 2. Four-strand modified Kessler technique

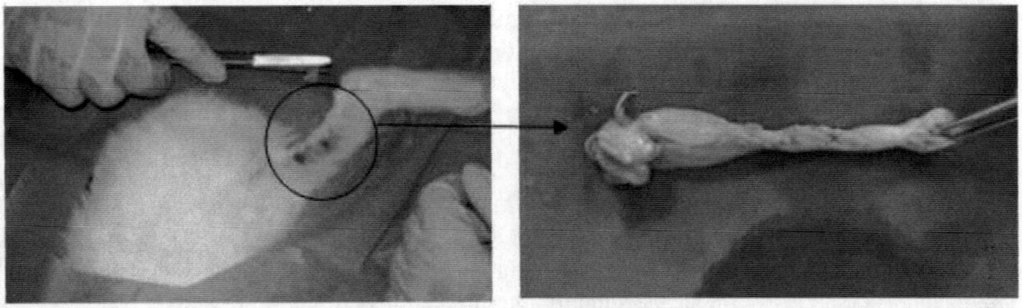

Figure 3. Rabbit Achilles tendon

3. Results and discussion

The distribution of the research subjects are as many as 32 subjects. The treatment group was compared with core repaired with 4 strand modified Kessler technique as much as 16 or 50.00% of the total subjects and the control group was repaired with continuous and cores as many as 16 or 50.00%.

The mean of the tensile strength in the treatment group was 62.34 ± 7.7173 N, while the average tensile strength in the control group was50.37 ± 3.8275 N. The highest tensile strength was 62.34 ± 7.7173 N in the treatment group and the minimum tensile strength was 50.37 ± 3,8275found in the control group. The mean percentage of type III collagen in the treatment group was 16.56 ± 1.9676 N, while the control group was 26.74 ± 2.5134 N.

The mean of collagen type I in the treatment group was 87.33 ± 1.1937, while in the control group was 73.42 ± 3.5091. The ratio of collagen I/ III in the treatment group was 5.3463 ± 0. 6934 compared with the control group 2.7668 ± 0.2795.

Fig. 3 Showed that the tensile strength is higher in the treatment group compared with the control group, and the mean difference between treatment and control group was statistically significant with p = 0,051 (P <0.05).While the ratio of collagen type I / III in the treatment group is smaller than the control group, and the mean difference between treatment and control group was statistically significant with p = 0,056 (p <0.05).

Based on the results of parametric analysis by using t-independent obtained the difference statistically significant between treatment groups with the control group, where the ratio of collagen I / III in the group treated with the technique of continuous-core is smaller compared to the control group by using 4 strand modified Kessler.

Based on the results of the parametric analysis with independent t-test found any statistically significant difference between the treatment group and control group, in which the tensile strength of the tendon in the group treated with the technique of Continuous and Core is greater than the control group at 4 Strand Modified Kessler technique.

The mean tensile strength of the Achilles tendon after treatment is still in the normal range of tensile strength in rabbits, but the strength after maximum loading decreased compared with normal tendon.

This is consistent with the study by Woo et. Al. [4] stating that the biomechanical component remains inferior tendon after injury grade III.

At the tendon healing process, there is a change in the type and distribution of collagen tissue that is formed. There is a decrease in the ratio of collagen type I / type III where an increase in type III collagen synthesis compared to Type I [7]. Collagen type III formed more and much thinner than type I as well as the cross-linking of collagen fibers and most immature [5]. The synthesis of type I collagen in tendon injury grade III will decline and be replaced by increased synthesis of collagen type III along with the expansion of granulation tissue and remodeling process so that the ratio of collagen type I / III decreased and the substitution of collagen towards the formation of scar tissue and fibrosis [4,6].

4. Conclusion

The ratio of type I/III I collagen in healing of achillestendon injuries in rabbit that are repaired by Continous-core techniques are lowered than the tendon repaired by 4 Strand Modified Kessler technique. Tensile strength that we measure in rabbit achilles tendon healing after injuriethat repair using Continous-core techniques is higher than achiles tendon that is repaired by 4 strand modified Kessler technique.

Acknowledgement

The authors wish to thank the Orthopaedic and Traumatology Department, Udayana University, Industry Manufacture and Machinery Research Group Udayana University, The Faculty of Veterinary Medicine Udayana University for all the support and facilitations during the research.

References

[1] Järvinen TAH, Kannus P, Maffulli N, Khan KM, 200, Achilles tendon disorders: Etiology and epidemiology, Foot Ankle Clin.,10(2), pp. 255-266.
[2] Hildebrand KA, Gallant-Behm CL, Kydd AS, Hart DA, 2005, The basics of soft tissue healing and general factors that influence such healing, Sports Med Arthrosc,13(3), pp.136-144.
[3] Giannotti S, Dell'Osso G, Bottai V, 2015, Treatment of Tendon Injuries of the Lower Limb with Growth Factors Associated with Autologous Fibrin Scaffold or Collagenous Scaffold, Surg Technol Int.,26, pp.324-328.
[4] Woo SL, Vogrin TM, Abramowitch SD, 2000, Healing and repair of ligament injuries in the knee, J Am Acad Orthop Surg., 8(6), pp.364-372.
[5] Young JS, Maffulli N, 2007, Etiology and Epidemiology of Achilles Tendon Problems, The Achilles Tendon, pp.39-49.
[6] Romani WA, Langenberg P, Belkoff SM, 2010, Sex, collagen expression, and anterior cruciate ligament strength in rats, J Athl Train., 45(1), pp. 22-28.
[7] Baugé C, Leclercq S, Conrozier T, Boumediene K,2015, TOL19-001 reduces inflammation and MMP expression in monolayer cultures of tendon cells, BMC Complement Altern Med., 15(1), pp. 1-10.
[8] Rawson S, Cartmell S, Wong J, 2013, Suture techniques for tendon repair; a comparative review, Muscles Ligaments Tendons J., 3(3), pp. 220-228.

On The Application of Diffusion Magnetic Resonance Imaging for Study of Agarose

Yanurita Dwihapsari*, **Salma Niatu Zakiah and Darminto**

Department of Physics, Institut Teknologi Sepuluh Nopember Surabaya, Indonesia

*Corresponding author: yanuritadh@physics.its.ac.id

Abstract. The precise and accurate measurements of material properties are needed in biomaterial and tissue engineering study to better describe the properties of material and ensure the product quality Diffusion Magnetic Resonance Imaging (MRI) is used in biomaterial study as a biomarker for composition and structure of biomaterial and a tool to observe mechanical properties of materials. The previous study predicted the mechanical properties of polyvinyl alcohol (PVA)-based material for biomedical phantom using diffusion MRI. The measurement was performed based on water diffusion inside gel and the method only worked for a certain concentration of PVA. In this study, Diffusion MRI was used to observe water diffusion of agarose as a potential hydrogel material for phantom development and tissue-mimicking materials.

Keywords: agarose, hydrogel, gel, implant, diffusion, magnetic resonance imaging, diffusion weighted imaging, diffusion coefficient

1. Introduction

The precise and accurate measurements of material properties are needed in biomaterial and tissue engineering study to better describe the properties of material and ensure the product quality. Diffusion Magnetic Resonance Imaging (MRI) is used in biomaterial study as a marker for composition and structure of material, such as membranes, permeability of membranes, packing density and compartment sizes. Diffusion MRI also allows differentiation between intracellular and extracellular parts of tissue. It provides information on water diffusion patterns in a material and its microsructure by sentisizing MRI signal to water diffusion inside the material.

Diffusion can be defined as random motion of particles due to thermal energy in uniform solution, known as self-diffusion. In other term, diffusion is also defined as the displacement of molecules due to concentration inhomogeneity between two non-equilibrium states, known as mutual diffusion [1]. Diffusion as a transport mechanism in hydrogel is described according to Fick's second law of diffusion, where C is concentration of hydrogel and D is diffusion coefficient of hydrogel.

$$\frac{\partial C}{\partial t} = D\nabla^2 C \tag{1}$$

Content from this work may be used under the terms of the Creative Commons Attribution 3.0 licence. Any further distribution of this work must maintain attribution to the author(s) and the title of the work, journal citation and DOI.
Published under licence by IOP Publishing Ltd

Diffusion coefficients are acquired using several techniques in MRI, such as relaxometry [2] and field gradient diffusometry [1] based on spin diffusion measurement by Stejskal-Tanner [3]. The diffusion coefficients acquired using field gradient diffusometry were discussed in this paper.

Diffusion MRI measurement utilized magnetic field which is applied for short period of time which causes phase coherence to the magnetic moment of each nucleus as a function of its position within the sample. Another magnetic field gradient with the same magnitude is applied after a certain period to let nuclei diffuse into the sample. The second gradient transmits a phase as a function of position, opposite that of the first gradient. The displacement rate of nuclei transport inside solution determines the net phase incoherence among magnetic moments within the sample. The net magnetization of magnetic moments will be lower due to the gradient and causes decreasing signal intensity.

The reduction of signal intensity due to higher diffusion-sensitizing magnetic field gradient compared to initial signal intensity without magnetic field gradient can be used to calculate diffusion coefficient according to [3].

$$S(t) = S_0 \exp - \left[\gamma^2 G^2 \delta^2 \left(\Delta - \frac{\delta}{3}\right)\right] D \qquad (2)$$

Where S(t) is signal intensity as a result of diffusion-sensitizing gradient, S_0 is signal intensity without diffusion-sensitizing gradient, □ is gyromagnetic ratio, G is diffusion-sensitizing gradient, □ is duration of diffusion gradients, □ is time between two diffusion-sensitizing gradient and D is diffusion coefficient. The factors of gyromagnetic ratio, diffusion-sensitizing gradient and the time between two diffusion gradient are defined as b-value.

$$b = \gamma^2 G^2 \delta^2 \left(\Delta - \frac{\delta}{3}\right) \qquad (3)$$

The b-value is one of the most important parameter in the acquisition of diffusion MRI. The diffusion coefficient, D can be obtained by fitting the normalized signal intensity $S(t)/S_0$ versus b-value in Eq. 2 according to

$$\frac{S(t)}{S_0} = exp - b\,D. \qquad (4)$$

The diffusion images acquired in MRI are usually in 3D as functions of position variable which encode the molecular displacement distribution in a particular voxel. Therefore the structural boundaries are important and must be considered in the calculation of diffusion coefficient. To simplify the effect of boundaries, apparent diffusion coefficient (ADC) term was introduced to describe the mean free diffusion coefficients in a certain voxel.

The previous study predicted the mechanical properties of polyvinyl alcohol (PVA)-based material for biomedical phantom using diffusion MRI [4]. The measurement was performed based on water diffusion inside gel and the method only worked for a certain concentration of PVA. Diffusion MRI was also assessed in agarose hydrogel phantom [5] and drug delivery study [6]. The previous study used linear and quadratic fitting between signal intensity and b-value to obtain diffusion coefficient of agarose phantoms [5].

In this study, diffusion MRI measurement was performed in agarose as a potential hydrogel material for phantom development and tissue implants. The diffusion coefficient was obtained and analyzed to describe water diffusion inside agarose gel with various concentrations. The result would be used to observe the potential of agarose as a reference material for the measurement of diffusion MRI and could be potentially explored for the development of phantom and tissue-mimicking materials.

2. Experimental

2.1. Sample Preparation
The samples were prepared as solutions consisting of agarose powder (Agarose, Vivantis, US) mixed with 250 ml of distilled water and formed the solution with various concentrations. The solution was heated to form gel before being poured into cylindrical tube of 5.7 cm diameter and 7.8 cm height. After reaching temperature of approximately 50°C and cloudy solution appeared, the solution was cooled to room temperature until gelation occurred. In this experiment, 5 variations of agarose gel concentration (in w/v%) were used: 0.5%, 1%, 1.5% and 2% respectively. In addition, a tube of pure water was used for reference.

2.2. Diffusion Measurement
Diffusion weighting images were obtained using method described in [4]. The measurement was performed using MRI 1.5 Tesla (General Electric, US) with parameters: TR 8000 ms, TE 88 ms, matrix size 256 × 256, and slice thickness 5 mm. The increasing diffusion sensitizing gradient, G increased b-value and caused signal intensity reduction (see Eq. 2 and Eq. 3).

Diffusion coefficient, D was obtained by fitting normalized signal intensity to b-value according to Eq. 4. The normalized signal intensity was obtained by comparing reduced signal intensity S(t) with initial signal intensity without diffusion-sentisizing gradient S_0. In this experiment, normalized signal intensity was displayed in normal logarithmic scale.

3. Results and discussion
The normal logarithmic plot of normalized signal intensity vs b-value in water as a reference showed linear relation and suggested mono exponential fitting. The observed diffusion coefficient of water as reference in this experiment was 2.3×10^{-9} m^2/s, consistent with result from other studies [5, 7].

The normalized signal intensity vs b-value in agarose was plotted in Fig. 1. The plot showed non-linear relation and suggested bilinear fitting instead of linear fitting. The non-linear relation was also found in the previous study, however the plot was fitted to linear and quadratic model [5].

The diffusion coefficient of agarose was obtained using bilinear curve fitting and yielded two diffusion coefficients which showed slow and fast components inside gel (Fig. 1). The slow and fast components can be interpreted either as intra- and extracellular part [8] or water and lipid proton inside gel [9]. The diffusion coefficient was expected to be stable for slow and fast component in any agarose concentrations using b-value of 1000 s/mm^2 and TR of 8000 ms in 1.5 Tesla MRI system [10].

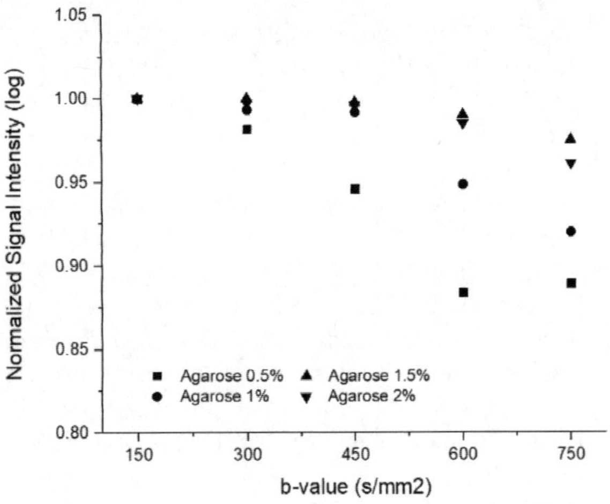

Figure 1. Normalized signal intensity vs b-value to calculate diffusion coefficient

The fast diffusion coefficient was found in the order of water diffusion coefficient (10^{-9} m^2/s), while the slow diffusion coefficient depended on agarose concentration. For slow components, the higher agarose concentration resulted in lower diffusion coefficient compared to pure water. The same result of lower diffusion coefficient was also found in the study of gelatin and starch system [11, 12]. It can be explained that more interaction between water molecules and hydroxyl group occurred in higher agarose concentration through hydrogen bonding and chemical exchange of protons. The higher interaction of water molecules with hydroxyl group in agarose obstructs translational motion of water and reduces diffusion coefficient of water inside agarose.

The increase of diffusion coefficient in higher agarose concentration are relatively linear compared to the result of PVA hydrogel in the previous study [4]. The result is in a good agreement with the lower diffusion coefficient for higher agarose concentration at TR = 8000 ms [9]. The exponentially lower diffusion coefficient in higher concentration of PVA hydrogel might be due to crosslinking effect in PVA. The linear relation between agarose concentration and diffusion coefficient could be used to predict and model water diffusion of agarose with various concentrations.

The previous studies also reported linear changes of MRI parameters, such as spin-lattice relaxation time (T1) and spin-spin relaxation time (T2) with increasing agar concentration using MRI relaxometry [13, 14]. For measurement of molecular structure of hydrogel and observation of its properties, agarose hydrogel is suitable for our purpose and can be used as the reference material for diffusion MRI measurement. Agarose can also be developed for phantom especially in MRI application and tissue-mimicking materials.

4. Conclusion

MRI diffusometry was used to describe water diffusion inside agarose hydrogel as a potential material for phantom development and tissue implants. The bi exponential diffusion coefficient was obtained and explained slow and fast components of diffusion coefficient inside gel. Agarose hydrogel is suitable for the reference material for diffusion MRI measurement. The other MRI methods can be performed for the next experiment to validate the result of this experiment in describing water movement inside agarose hydrogel.

Acknowledgement

Institut Teknologi Sepuluh Nopember Surabaya (ITS) is highly aknowledged for Research Grant Penelitian Lokal ITS.

References

[1] Price WS 2009 *NMR Studies of Translational Motion* (Cambridge: Cambridge University Press).
[2] Kruk D, Meier R and Rossler EA 2012 *Physical Review E* **85** 020201.
[3] Stejskal EO and Tanner JE 1965 *The Journal of Chemical Physics* **42** 288.
[4] Dwihapsari Y, Sari DP and Darminto 2010 *AIP Conference Proceeding* **1454(1)** 53.
[5] Portakal ZG, Shermer S, Jenkins C et al 2018 *Medical Physics* **45(6)** 2476.
[6] Al Faraj A, Shaik AP and Shaik AS 2015 *International Journal of Nanomedicine* **10** 157.
[7] Holz M, Heil SR and Sacco A 2000 *Physical Chemistry Chemical Physics* **2** 4740.
[8] Schwarcz A, Bogner P, Meric P, et al 2004 *Magnetic Resonance in Medicine* **51** 278.
[9] Ababneh Z, Haque M, Maier SE, et al 2004 *MAGMA* **17** 95.
[10] Ogura A, Hayakawa K, Miyati T and Maeda F 2011 *European Journal of Radiology* **77** 185.
[11] Farhat IA, Loisel E, Saez P et al 1997 International Journal of Food Science & Technology **32** 377.
[12] Kimmich R, Klammler F, Skirda VD et al 1993 *Applied Magnetic Resonance* **4** 425
[13] Dwihapsari Y, Maulidiyah N and Darminto 2018 *IOP Conference Series: Materials Science and Engineering* **395** 012025.
[14] Dwihapsari Y, Afriana RD and Darminto 2018 IOP Conference Series: Materials Science and Engineering **395** 012022.

Recent Development of Lactic Acid Production using Membrane Bioreactors

L. Aliwarga[1], A.K. Wardani[1], P.T.P. Aryanti[2], and I G. Wenten[1,3*]

[1] Chemical Engineering Department, Institut Teknologi Bandung, Jl. Ganesha 10, Bandung, 40132, Indonesia
[2] Chemical Engineering Department, Universitas Jenderal Achmad Yani, Jl. Terusan Jendral Sudirman, Cimahi, 40285, Indonesia
[3] Research Center for Nanosciences and Nanotechnology, Institut Teknologi Bandung, Jl. Ganesha 10, Bandung, 40132, Indonesia

*Corresponding author: igw@che.itb.ac.id

Abstract. Lactic acid has been widely used as flavour and preservative in the food, pharmaceutical, leather and textile industries. It can be produced by fermentation process of the substrates with high lactose content, such as cheese whey, soybean milk, corn, and potatoes. Among various existing technologies, membrane bioreactor is one of the promising methods to achieve high productivity of lactic acid. In addition, membrane bioreactor allows integration of fermentation and separation steps, thus it able to simultaneously maintain high cell density, recycle the cells for further use, and continuously remove lactic acid from the fermenter.

Keywords: lactic acid, membrane bioreactor, production yield

1. Introduction

Lactic acid (2-hydroxypropionic acid), CH_3-CHOHCOOH, is a simple organic compound containing both the hydroxyl and carboxylic acid groups. It was first isolated from sour milk by CW Scheele in 1780 and commercially produced by CE Avery in Littleton, MA, USA in 1881 [1-3]. The production demand of lactic acid has been increased over years due to due to its high potential of application in a wide range of fields [4-6]. In 2012, the production of lactic acid was around 259,000 metric tons with the worldwide growth 12–15% per year [7, 8]. In recent years, lactic acid manufacture is mostly based on carbohydrate fermentation. The major manufacturers are Archer Daniels Midland Company (USA), NatureWorks LLC (USA), Purac (The Netherlands), and Galactic S.A. (Belgium).

Lactic acid has been mainly used for food and food-related applications. It is due to the mild acidic taste of lactic acid. In addition, lactic acid is non-volatile, odourless, and classified as GRAS (generally recognized as safe) for use as a general purpose food additive [2]. Therefore, many industries choose lactic acid as a safety flavour and preservative in the food. Lactic acid also has been utilized in the cosmetic industry such as in the manufacture of hygiene and aesthetic products due to its moisturizing, antimicrobial and rejuvenating effects on the skin, as well as of oral hygiene products [8]. The other promising application of lactic acid lies on its polymer, the poly-lactic acid (PLA). It offers tremendous

Content from this work may be used under the terms of the Creative Commons Attribution 3.0 licence. Any further distribution of this work must maintain attribution to the author(s) and the title of the work, journal citation and DOI.
Published under licence by IOP Publishing Ltd

advantages like biodegradability, thermos-plasticity, high strength etc. [9]. PLA is considered as an environment-friendly alternative to substitute plastics derived from petrochemicals [10]. PLA can be applied in medical applications for filling the gaps in bones, producing sutures (stitching material), and joining membranes or thin skins in humans [11].

Another potential growth area for lactic acid derivatives is environmentally friendly solvents, particularly lactate esters of low molecular weight alcohols such as ethyl, propyl and butyl lactate. Oxygenated chemicals such as propylene glycol, propylene oxide, acrylic acid, acrylate esters, and other chemical intermediates such as lactate ester plasticizers also can be made from lactic acid [2, 3]. The schematic diagram of potential products from lactic acid derivatives can be seen in Figure 1.

The production of lactic acid has been dominated by carbohydrates fermentation processes. Various studies aimed to improve lactic acid productivity using membrane bioreactors. In membrane bioreactor, membrane module is integrated with conventional fermenters, thus permit simultaneous production and purification of lactic acid in the same unit. Membrane bioreactor can be operated in various modes, such as batch, semi-continuous, continuous, immobilized, and membrane recycle.

Membrane bioreactor offers advantages of great flexibility in scale of production depending on market demand as well as high levels of separation and purification. In addition, membrane bioreactor eliminates the requirement of separate purification units and results in compact design with reduced capital investment [12]. For better understanding, this paper then aims to give a brief review of recent development in lactic acid production using membrane bioreactors. The configuration and performance of membrane bioreactors are discussed.

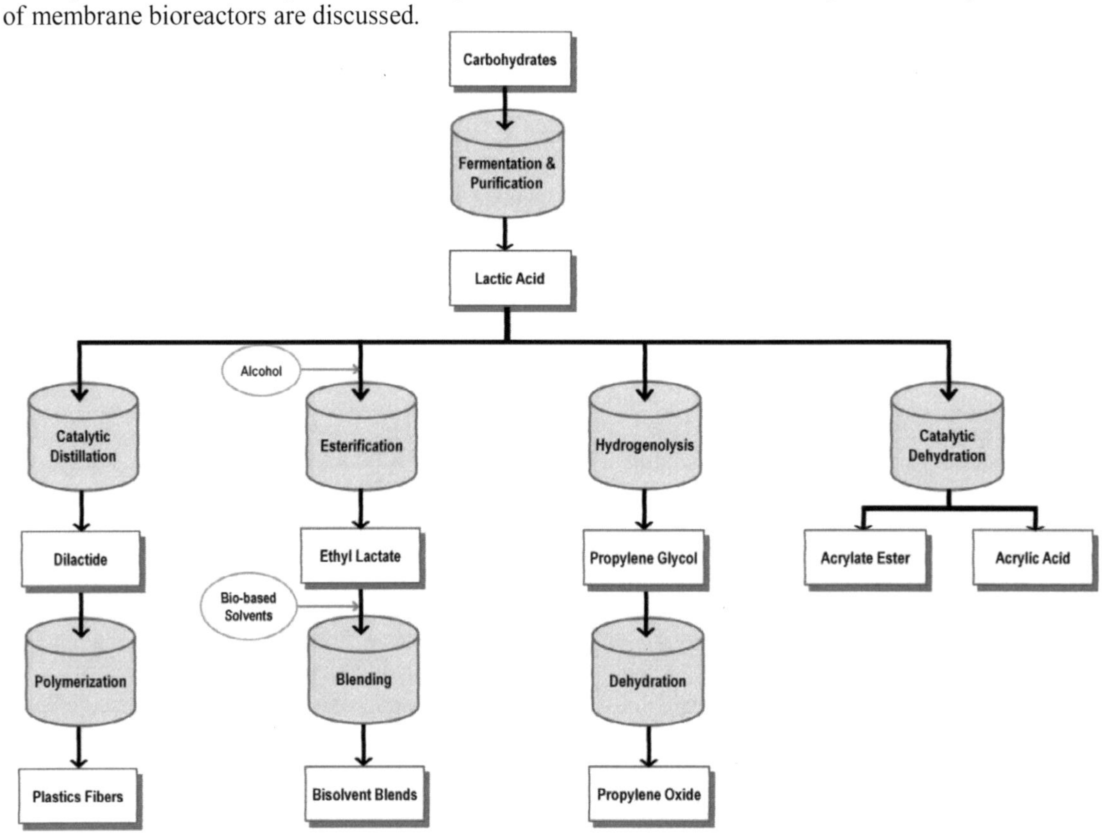

Figure 1. schematic diagram of potential products from lactic acid derivatives

IOP Conf. Series: Materials Science and Engineering **622** (2019) 012023 doi:10.1088/1757-899X/622/1/012023

2. Experimental

In general, lactic acid can be manufactured by either chemical synthesis or carbohydrates fermentation. The chemical synthesis is mainly based on the hydrolysis of lacto nitrile by a strong acid and produces only a racemic mixture of lactic acid [13]. Meanwhile, a fermentation process is able to produce a stereoisomer of lactic acid [7]. Fermentation process is more attractive in terms of its environmental impact, low production cost, decreased fossil-based feedstock dependency, reduction of CO_2 emission, and high product specificity [14]. Therefore, fermentation process is preferred for the production of lactic acid, approximately 90% from the total lactic acid production worldwide [13, 15].

Table 1. Comparison of different strains and substrates for lactic acid production

Substrate	Organism	Yield (g/g)	Ref.
Glucose	*Lactobacillus casei*	0.91	[16]
	Lactobacillus rhamnosus	0.74-0.93	[16-18]
	Lactobacillus delbrueckii	0.57-0.83	[19]
	Lactobacillus salivarius	0.92	[18]
	Lactobacillus zeae	0.71-0.98	[17, 20]
	Lactobacillus coryniformis	0.98	[20]
Lactose	*Lactobacillus rhamnosus*	0.38	[21]
	Lactococcus Lactis	1.50	[21]
Molasses	*Lactobacillus delbrueckii*	0.45-0.58	[22]
	Lactobacillus rhamnosus	0.40	[22]
Hydrolized barley flour	*Lactobacillus amylovorus*	0.52	[23]
	Lactobacillus casei	0.67	[23]
	Lactobacillus delbrueckii	0.85	[24]
	Lactobacillus rhamnosus	0.74	[24]
Hydrolized wheat flour	*Lactococcus Lactis*	0.76	[25]
	Lactobacillus delbrueckii	0.11-0.82	[25]
Paneer whey	*Lactobacillus kefir*	0.20	[26]
	Lactobacillus acidophilus	0.17	[26]
Whey	*Lactobacillus casei*	0.32	[27]
	Leuconostoc Lactis	0.22	[27]
Whey permeate	*Streptococcus Thermophilus*	0.35-0.50	[28-30]
	Lactobacillus delbrueckii	0.18-0.41	[28-30]
	Lactococcus Lactis	0.20-0.88	[28, 29, 31]
	Lactobacillus rhamnosus	0.71	[31]
Solid waste	*Lactobacillus plantarum*	0.42-0.46	[32]
	Lactobacillus pentosus	0.43-0.51	[32]
	Lactococcus Lactis	0.16	[32]

2.1. Carbohydrate Resources

Lactic acid can be produced from various substrates, either sugar in pure form such as glucose, sucrose, lactose etc. or sugar-containing materials such as whole-wheat powder [33], starch [13, 34], cucumber juice [35], cheese whey [36, 37], molasses [38], and sugarcane juice [39]. The selection of carbohydrate feedstock depends on the price, availability and its purity. The purified sugars are mostly expensive as the feedstock for lactic acid production. Therefore, many researchers utilized agricultural by-products such as corn starch, cassava starch, cottonseed hulls, wheat bran, sugarcane press mud, barley starch, carrot processing waste, corn fiber hydrolyzates, and potato starch as potential substrates for lactic acid production [9]. Meanwhile, proteinaceous and other complex nutrients that required by the organisms can be provided from corn steep liquor, yeast extract, soy hydrolysate, etc. [3, 39, 40].

2.2. Microorganisms

The characteristic of the produced lactic acid depends much on the type of organism. In general, microorganisms for lactic acid production can be classified into homofermentative strain and heterofermentative strain. A homofermentative strain produces a single product, lactic acid only, while heterofermentative strain produces other products such as ethanol, diacetyl, formate, acetoin or acetic acid, and carbon dioxide along with lactic acid [9]. The comparisons of homofermentative strain and heterofermentative strain organisms for lactic acid production are shown in Table 2. The existing commercial lactic acid production processes mainly use homofermentative organisms, especially from the genera *Lactobacillus, Streptococcus, Enterococcus, and Pediococcus*. These organisms exhibit maximum productivity only within a very narrow pH range, preferably in the range of 5.5 to 6.5 [41].

The choice of microorganisms is mainly determined by the resource of carbohydrate. For substrate from glucose, any homofermentative strains of the genus *Lactobacillus* can be used, however the preferred organism is *Lactobacillus delbrueckii* [42]. For lignocellulosic substrates, *Lactobacillus pentosus* is required to maximize the yield [32]. Meanwhile, starch is able to be both hydrolyzed and fermented by certain amylase-producing *Lactobacillus* strains, such as *Lactobacillus amylovorus* [43]. *Lactobacillus bulgaricus* is the most used organism for fermentation of whey. This organism is able to produce lactic acid with yields in the range of 80-90% [44, 45]. Besides, fungal strains such as *Mucor, Monilia*, and *Rhizopus* can also be utilized to produce lactic acid. The best-known fungal source of lactic acid is *Rhizopus oryzae* that able to obtain 63–69% yields of lactic acid from chemically defined media containing 15% glucose [46, 47].

2.3. Methods

Lactic acid is a metabolic product of simple carbohydrates produced by many species of organisms mainly through the fermentative metabolic pathway. After supplementation of nutrients, sugar contained substrates are inoculated with the selected microorganism, and the fermentation takes place [8]. The stoichiometry for homofermentative fermentation of lactic acid from hexose and pentose can be expressed as equation (1) and (2), respectively [48].

$$C_6H_{12}O_6 \rightarrow 2\ C_3H_6O_3 \quad (1)$$

$$C_5H_{10}O_5 \rightarrow C_3H_6O_3 + C_2H_4O_2 \quad (2)$$

The carbohydrate fermentation operates most efficiently and effectively at near neutral pH, which requires neutralization and produces the salt of the acid instead of the acid itself. Meanwhile, pH of the fermentation broth goes on lowering as lactic acid is formed and accumulated. Therefore, pH of the fermentation must be maintained, usually by addition of lime, thus lactic acid (partly) is converted to be calcium lactate. The precipitation and acidification of calcium lactate then generates huge quantity of calcium sulphate (gypsum). Approximately one ton of gypsum by-product is produced for every ton of lactic acid produced by the conventional fermentation, which poses an environmental problem [2]. In addition, the fermentation broths of lactic acid also contain impurities such as a residual sugars, nutrients, and microorganisms, and other organic acids [7, 41]. Therefore, the complex separation steps to recover and purify the lactic acid from the fermentation broths become one of the major economic hurdle and process cost of conventional carbohydrate fermentation process.

Table 2. Comparison of different strains and substrates for lactic acid production

Characteristic	Homofermentative strain	Heterofermentative strain
Products	• Lactic acid	• Lactic acid • Ethanol • Diacetyl • Formate • Acetic acid • Carbon dioxide
Genera	• *Lactococcus* • *Streptococcus* • *Pediococcus* • *Enterococcus* • *Lactobacillus* • *Sporolactobacillus*	• *Leuconostoc* • *Oemococcus* • *Bifidobacterium* • *Lactobacillus*
Selectivity of lactic acid	High selectivity	Low selectivity (high by-product formation)
Availability for commercial production	Available	Not available

Various technologies have been introduced for lactic acids recovery from the fermentation broth, such as precipitation, extraction, adsorption, ion-exchange system, membrane, etc. [51-59]. However, those existing processes still have many obstacles that need to be addressed. Precipitation and liquid-liquid extraction need a large number of chemical agents and energy. They produce a large amount of water effluent as well as solid residue and involve phase changes that lead to quality degradation of citric acid [60-62]. Meanwhile, adsorption has a short lifetime of adsorbents and low capacity [54, 60]. In addition, fouling still becomes the most significant hurdle in membrane processes [63, 64].

3. Results and discussion

The increasing demand of lactic acid and increasing concern over environmental impact of gypsum accumulation as a by-product leads to development of alternative technologies for lactic acid production. Batch, fed-batch, repeated batch, and continuous fermentations have been developed to increase production efficiency of lactic acid. Various studies showed that higher lactic acid concentration could be obtained in batch and fed-batch fermentations, however higher productivity was achieved by the use of continuous fermentations [15]. It is due to the low concentration of microbes in the batch system.

To solve the problem of production efficiency as well as gypsum formation, membrane bioreactor can be a promising alternative. Membrane bioreactor integrates the fermentation and separation step using membrane such as microfiltration, ultrafiltration, Nano filtration, and electro dialysis, thus simultaneously maintaining high cell density, recycling the cells for further use, and continuously removing lactic acid from the fermenter [65-68]. In continuous fermentation processes using membrane bioreactor, components like microbial cells, proteins, nutrients (yeast extract, salts of ammonium, potassium, phosphorus, etc.), unconverted carbon sources, and water are separated by a filtration unit and returned to the fermenter , while the lactic acid is concentrated in the permeate [4, 12, 69]. Therefore, the use of membrane bioreactor in lactic acid production does not produce any harmful bio-product. The detail comparison of membrane bioreactor and conventional technologies for lactic acid production are shown in Table 3.

In general, both polymer (i.e. polysulfone [65] and polyacrylonitrile [70]) and ceramic (alumina [71]) can be used as membrane bioreactor materials to produce lactic acid. Ceramic membrane tends to have better mechanical and chemical resistance than polymer membrane. Meanwhile, polymer membrane offers advantages of flexible configuration and low production cost [72, 73]. There are only a few researchers who mention the material of membrane bioreactor for lactic acid production. They mostly focused on the operation mode and product efficiency.

Table 3. Comparison of membrane bioreactor and conventional technologies for lactic acid production [12, 50, 74]

Characteristic	Membrane bioreactor	Conventional technology
Unit operations	• Fermentation tank • Membrane module	• Fermentation tank • pH adjustment through hydrated lime • Rotary drum filtration • Carbon bleaching • Plate and frame filtration • Calcium lactate evaporation • Crystallization • Extraction • Distillation unit • Dilution
Flexibility parameters • Modules • Production capacity • Steps of operations	 Flexible in size and numbers Flexible Few steps	 Fixed configuration Fixed Large number of steps
Effect to the environment and economic aspects	• Does not produce any harmful bio-product • Free from pre- and post-treatment • Low energy consumption • Use no harsh chemicals	• Generates million tons of gypsum/year • Heat generation due to exothermic reaction • Highly energy intensive • Use harsh chemicals, especially acids and alkalis

In lactic acid production, the operation mode of membrane bioreactor can be divided into batch, semi-continuous, continuous, immobilized, and membrane recycle. Among those modes, membrane recycle bioreactor has the highest efficiency in production of lactic acid, as shown in Table 4. However, the high productivity is not the only requirement for the economic feasibility of the process. Timmer and Kromkamp [44] investigated that the process might be primarily influenced by production capacity and product concentration and to a lesser extent by the volumetric productivity when annual lactic acid production capacity rose to as high as 4540 metric tons. When lactic acid concentration is significantly low, the energy cost for water removal in the downstream process offsets the benefits of the increased productivity.

In 1986, Mehaia and Cheryan [37] produced lactic acid from acid whey permeate in a membrane recycle bioreactor. Whey permeate was obtained by ultrafiltration of cottage cheese whey and supplemented with yeast extract. The lactose in the permeate was converted into lactic acid by Lactobacillus bulgaricus in a high-performance membrane bioreactor configured in the cell recycle mode. At a cell concentration of 10 g/L, optimum productivity of lactic acid was only 35 g/L.h. Meanwhile, lactic acid productivity of 80 g/L.h only could be obtained at cell concentration of 60 g/L. Other studies [89-93] also showed that lactic acid concentration produced by membrane bioreactor was still below 60 g/L.

To enhance the economic advantage of the membrane bioreactor process, methods that increase the lactic acid concentration along with the high-cell density are required. Ohashi et al. [84] studied continuous production of lactic acid by retaining cells at a high density of Lactococcus lactis in a stirred ceramic membrane reactor (SCMR). The mass concentration and productivity of lactic acid reached 40 g/L and 10.6 g/L.h, respectively. In 1998, Kamoshita et al. [94] developed the improved SCMR system. Using the improved SCMR system, a cell mass concentration of 178 g/L and viability of 98 % were obtained after 198 h of fermentation. They demonstrated that the improved permeability of the SCMR with the use of a membrane cleaning system influenced a rapid increase in the concentration and viability of cells, and accordingly, the increased production rate of lactic acid in proportion to the concentration of viable cells.

Table 4. Comparison of bioreactors for lactic acid production

Bioreactor	**Substrate**	**Organism**	**Lactic acid (g/L)**	**Productivity (g/L.h)**	**Ref.**
Batch	Banana wastes	*Lactobacillus casei*	-	0.13	[75]
	Cassava bagasse	*Lactobacillus delbrueckii*	81.9	1.36	[76]
	Sugar cane baggage	*Lactococcus lactis*	10.9	0.17	[77]
	Cheese whey	*Lactobacillus bulgaricus*	9.6	4.8	[78]
Semi-continuous	Whey permeate	*Lactobacillus helveticus*	55	3.54	[69]
Continuous	Whey permeate	*Lactobacillus helveticus*	64	22	[79]
	Glucose	*Lactobacillus lactis*	210	2.2	[80]
	Cheese whey	*Lactobacillus helveticus*	38	19-22	[81]
Immobilized	Glucose	*Lactobacillus lactis*	115	2.25	[82]
	Liquid distillery stillage	*Lactococcus rhamnosus*	42.2	1.22	[83]
Membrane recycle	Whey permeate	*Lactobacillus bulgaricus*	43	85	[37]
	Molasses	*Lactococcus lactis*	40	10.6	[84]
	Glucose	*Lactococcus rhamnosus*	88	35.2	[85]
	Glucose	*Lactococcus paracasei*	91	31.5	[86]
	Glucose	*Lactococcus paracasei*	120	150	[87]
	Glucose	*Lactobacillus delbrueckii*	35	76	[88]

Another effort to increase lactic acid productivity was developed by combining the advantage of both membrane bioreactor and multi-staged bioreactor. Kulozik et al. [95] investigated the performance of a seven-staged cascade reactor with cell recycle. In comparison with a single-stage membrane bioreactor, the cascade reactor resulted 4 times higher productivity, 28 g/L.h, in which the cell concentrations were maintained at 20 g/L and the lactic acid concentrations were around 72 g/L. In 2001, Kwon et al. [96] produced lactic acid by a two-stage cell-recycle culture of Lactococcus rhamnosus. The membrane cell-recycle bioreactors were arranged in a series and successfully obtained 92 g/L of lactic acid with a productivity of 57 g/L.h. Meanwhile, Xu et al. [86] developed a membrane cell-recycle bioreactor that equipped with a diaphragm pump and tangential flow-rate controller to produce lactic acid by Lactococcus paracasei. The maximum productivity of this system was 31.5 g/L.h.

Later, Danner et al. [97] developed UF membrane bioreactor coupled with on-line monopolar electro dialysis to recover, pre-purify, and concentrate lactic acid. The results showed that the volumetric productivity was low (1.38 g/L h). The lactic acid concentration was 35 g/L with lactic acid yield on consumed glucose appeared stable at around 80%.

4. Conclusion

Fermentative production of lactic acid has roused interest among researchers in the recent years due to its high potential of application in a wide range of fields. High purity lactic acid could be produced and separated continuously in a fully membrane-integrated fermentation process using a cheap and renewable carbon source. Various types of substrate such as whole-wheat powder, starch, cheese whey, molasses, and sugarcane juice have been investigated to produce lactic acid with the help of organisms, especially from the genera *Lactobacillus*.

In general, the efficiency of lactic acid production by membrane bioreactor is higher than conventional technologies. However, several studies showed that membrane bioreactor was only able to produce lactic acid with concentration below 60 g/L. Therefore, further development of methods that increase the lactic acid concentration was required. The improvement in continuous fermentation could be done by increasing the cell density in the bioreactor, thus enhances the substrate-to-product conversion rate and resulting in higher productivity. The increase of cell densities in the fermentation process can be conducted either by the use of cell immobilization or by cell-recycle using membrane

References

[1] Padmini E and Miranda L R 2013 Chem. Eng. J 232 249-258
[2] Vick-Roy T 1985 *Dic Pergamon, Toronto*
[3] Datta R, Tsai S-P, Bonsignore P, Moon S-H and Frank J R 1995 *FEMS Microbiol. Rev.* **16** 221
[4] Rathin D and Michael H 2006 *J. Chem. Technol. Biotechnol.* **81** 1119
[5] Li Y, Shahbazi A, Williams K and Wan C 2008 (Totowa, NJ: Humana Press) 369
[6] González M I, Álvarez S, Riera F and Álvarez R 2007 *J. Food Eng.* **80** 553
[7] Olmos-Dichara A, Ampe F, Uribelarrea J L, Pareilleux A and Goma G 1997 *Biotechnol. Lett.* **19** 709
[8] Joglekar H G, Rahman I, Babu S, Kulkarni B D and Joshi A 2006 *Sep. Purif. Technol.* **52** 1
[9] Castillo Martinez F A, Balciunas E M, Salgado J M, Domínguez González J M, Converti A and Oliveira R P d S 2013 *Trends Food Sci. Technol.* **30** 70
[10] John R P, Nampoothiri K M and Pandey A 2007 *Appl. Microbiol. Biotechnol.* **74** 524
[11] Senthuran A, Senthuran V, Mattiasson B and Kaul R 1997 *Biotechnol. Bioeng.* **55** 841
[12] Shikinami Y, Kawarada H and Nishi C 2002 Google Patents)
[13] Pal P, Sikder J, Roy S and Giorno L 2009 *Chem. Eng. Process., Process Intensification* **48** 1549
[14] Lasprilla A J R, Martinez G A R, Lunelli B H, Jardini A L and Filho R M 2012 *Biotechnol. Adv.* **30** 321
[15] Lunelli B H, Andrade R R, Atala D I P, Wolf Maciel M R, Maugeri Filho F and Maciel Filho R 2010 *Appl. Biochem. Biotechnol.* **161** 227
[16] Hofvendahl K and Hahn–Hägerdal B 2000 *Enzyme Microb. Technol.* **26** 87
[17] Hujanen M and Linko Y-Y 1996 *Appl. Microbiol. Biotechnol.* **45** 307
[18] Vaccari G, y González-Vara A R, Campi A L, Dosi E, Brigidi P and Matteuzzi D 1993 *Appl. Microbiol. Biotechnol.* **40** 23
[19] Siebold M, P.v F, R J, D R, K S and H R 1995 *Process Biochem.* **30** 81
[20] Demirci A and Pometto A L 1992 *J. Ind. Microbiol.* **11** 23
[21] Antonio G-V R, Pinelli D, Rossi M, Fajner D, Magelli F and Matteuzzi D 1996 *J. Ferment. Bioeng.* **81** 548
[22] Taniguchi M, Kotani N and Kobayashi T 1987 *J. Ferment. Technol.* **65** 179
[23] Tiwari K P, Pandey A and Mishra N 1979 *Zentralblatt für Bakteriologie,* **134** 544
[24] Giraud E, Champailler A and Raimbault M 1994 *Appl. Environ. Microbiol.* **60** 4319
[25] Melzoch K, Votruba J, Hábová V and Rychtera M r 1997 *Journal of Biotechnology* **56** 25
[26] Hofvendahl K, van Niel E W J and Hahn-Hägerdal B 1999 *Appl. Microbiol. Biotechnol.* **51** 669
[27] Gupta R and Gandhi D 1995 *Indian J. Dairy Sci.* **48** 636
[28] Roukas T and Kotzekidou P 1998 *Enzyme Microb. Technol.* **22** 199
[29] Audet P, Paquin C and Lacroix C 1988 *Appl. Microbiol. Biotechnol.* **29** 11
[30] Audet P, Paquin C and Lacroix C 1989 *Appl. Environ. Microbiol.* **55** 185
[31] Audet P, Paquin C and Lacroix C 1990 *Appl. Microbiol. Biotechnol.* **32** 662
[32] Mulligan C and Gibbs B 1991 *Biotechnol. Appl. Biochem.*
[33] McCaskey T A, Zhou S D, Britt S N and Strickland R 1994 *Appl. Biochem. Biotechnol.* **45** 555
[34] Akerberg C, Hofvendahl K, Zacchi G and Hahn-Hägerdal B 1998 *Appl. Microbiol. Biotechnol.* **49** 682
[35] Yin P, Nishina N, Kosakai Y, Yahiro K, Pakr Y and Okabe M 1997 *J. Ferment. Bioeng.* **84** 249

[36] Passos F V, Fleming H P, Ollis D F, Felder R M and McFeeters R 1994 *Appl. Environ. Microbiol.* **60** 2627
[37] Ghaly A, Tango M A and Adams M 2003 *Agric. Eng. Int.: CIGR J.*
[38] Mehaia M A and Cheryan M 1986 *Enzyme Microb. Technol.* **8** 289
[39] Dumbrepatil A, Adsul M, Chaudhari S, Khire J and Gokhale D 2008 *Appl. Environ. Microbiol.* **74** 333
[40] Timbuntam W, Sriroth K and Tokiwa Y 2006 *Biotechnol. Lett.* **28** 811
[41] Hofvendahl K and Hahn-Hägerdal B 1997 *Enzyme Microb. Technol.* **20** 301
[42] Vaidya A N, Pandey R A, Mudliar S, Kumar M S, Chakrabarti T and Devotta S 2005 *Crit. Rev. Env. Sci. Technol.* **35** 429
[43] Hammes W P and Hertel C 2006 *The Prokaryotes: Volume 4: Bacteria: Firmicutes, Cyanobacteria,* ed M Dworkin, *et al.* (Springer US) 320
[44] Zhang D X and Cheryan M 1991 *Biotechnol. Lett.* **13** 733
[45] Timmer J and Kromkamp J 1994 *FEMS Microbiol. Rev.* **14** 29
[46] Giorno L, Chojnacka K, Donato L and Drioli E 2002 *Ind. Eng. Chem. Res.* **41** 433
[47] Yu R-c and Hang Y D 1989 *Biotechnol. Lett.* **11** 597
[48] Ward G E, Lockwood L B and May O E 1938 Google Patents)
[49] Tsao G T, Cao N J, Du J and Gong C S 1999 *Recent Progress in Bioconversion of Lignocellulosics,* ed G T Tsao, *et al.* (Springer Berlin Heidelberg) 243
[50] Vijayakumar J, Aravindan R and Viruthagiri T 2008 *Chem. Biochem. Eng. Q.* **22** 245
[51] Abdel-Rahman M A, Tashiro Y and Sonomoto K 2013 *Biotechnol. Adv.* **31** 877
[52] Aljundi I H, Belovich J M and Talu O 2005 *Chemical Engineering Science* **60** 5004
[53] Boonkong W, Sangvanich P, Petsom A and Thongchul N 2009 *Chem. Eng. Technol.* **32** 1542
[54] Cao X, Yun H S and Koo Y-M 2002 *Biochem. Eng. J.* **11** 189
[55] da Silva A H and Miranda E A 2013 *J. Chem. Eng. Data* **58** 1454
[56] Zhang K, Zhang L and Yang S-T 2014 *Industrial & Engineering Chemistry Research* **53** 12802
[57] Gao M-T, Shimamura T, Ishida N, Nagamori E, Takahashi H, Umemoto S, Omasa T and Ohtake H 2009 *Enzyme and Microbial Technology* **44** 350
[58] Keshav A, Wasewar K L and Chand S 2009 *Chemical Engineering Communications* **197** 606
[59] Li S J, Chen H L, Xu J Y and Zhang L 2007 *Sep. Sci. Technol.* **42** 2347
[60] Marinova M, Kyuchoukov G, Albet J, Molinier J and Malmary G 2004 *Sep. Purif. Technol.* **37** 199
[61] Huang C, Xu T, Zhang Y, Xue Y and Chen G 2007 *J. Membr. Sci.* **288** 1
[62] Cheng K-K, Zhao X-B, Zeng J, Wu R-C, Xu Y-Z, Liu D-H and Zhang J-A 2012 *Appl. Microbiol. Biotechnol.* **95** 841
[63] Wang Q, Cheng G, Sun X and Jin B 2006 *Process Biochem.* **41** 152
[64] Cho Y H, Lee H D and Park H B 2012 *Industrial & Engineering Chemistry Research* **51** 10207
[65] Wenten I G, Khoiruddin K, Aryanti Putu T P, Victoria Agnes V and Tanukusuma G 2018 *Rev. Chem. Eng.* https://doi.org/10.1515/revce-2017-0117
[66] Tejayadi S and Cheryan M 1995 *Appl. Microbiol. Biotechnol.* **43** 242
[67] Handojo L, Wardani A, Regina D, Bella C, Kresnowati M and Wenten I 2019 *RSC Adv.* **9** 7854
[68] Kresnowati M T A P, Regina D, Bella C, Wardani A K and Wenten I G 2019 *Food Bioprod. Process.* **114** 245
[69] Mangindaan D, Khoiruddin K and Wenten I G 2018 *Trends Food Sci. Technol.* **71** 36
[70] Jeantet R, Maubois J L and Boyaval P 1996 *Enzyme Microb. Technol.* **19** 614
[71] Zhang Y, Chen X, Qi B, Luo J, Shen F, Su Y, Khan R and Wan Y 2014 *Bioresour. Technol.* **163** 160
[72] Moueddeb H, Sanchez J, Bardot C and Fick M 1996 *J. Membr. Sci.* **114** 59
[73] Himma Nurul F, Prasetya N, Anisah S and Wenten I G 2019 *Rev. Chem. Eng.* **35** 211
[74] Himma N F, Wardani A K, Prasetya N, Aryanti P T P and Wenten I G 2018 *Rev. Chem. Eng.* https://doi.org/10.1515/revce-2017-0094

[75] Pal P and Dey P 2013 *Chem. Eng. Process., Process Intensification* **64** 1
[76] Chan-Blanco Y, Bonilla-Leiva A R and Velázquez A C 2003 *Appl. Microbiol. Biotechnol.* **63** 147
[77] John R P, Madhavan Nampoothiri K and Pandey A 2006 *Appl. Biochem. Biotechnol.* **134** 263
[78] Laopaiboon P, Thani A, Leelavatcharamas V and Laopaiboon L 2010 *Bioresour. Technol.* **101** 1036
[79] Plessas S, Bosnea L, Psarianos C, Koutinas A A, Marchant R and Banat I M 2008 *Bioresour. Technol.* **99** 5951
[80] Boyaval P, Corre C and Terre S 1987 *Biotechnol. Lett.* **9** 207
[81] Bai D-M, Wei Q, Yan Z-H, Zhao X-M, Li X-G and Xu S-M 2003 *Biotechnol. Lett.* **25** 1833
[82] Schepers A W, Thibault J and Lacroix C 2006 *Enzyme Microb. Technol.* **38** 324
[83] Zhang Y, Cong W and Shi S Y 2011 *Bioprocess. Biosyst. Eng.* **34** 67
[84] Djukić-Vuković A P, Mojović L V, Jokić B M, Nikolić S B and Pejin J D 2013 *Bioresour. Technol.* **135** 454
[85] Ohashi R, Yamamoto T and Suzuki T 1999 *J. Biosci. Bioeng.* **87** 647
[86] Xavier A M R B, Gonçalves L M D, Moreira J L and Carrondo M J T 1995 *Biotechnol. Bioeng.* **45** 320
[87] Xu G q, Chu J, Wang Y h, Zhuang Y p, Zhang S l and Peng H q 2006 *Process Biochem.* **41** 2458
[88] Kuznetsov A, Beloded A, Derunets A, Grosheva V, Vakar L, Kozlovskiy R and Shvets V 2017 *Clean Technol. Environ. Policy* **19** 869
[89] Vick Roy T B, Mandel D K, Dea D K, Blanch H W and Wilke C R 1983 *Biotechnol. Lett.* **5** 665
[90] Cheryan M 1998 *Ultrafiltration and microfiltration handbook* (CRC press)
[91] Litchfield J H 1996 *Advances in Applied Microbiology,* ed S L Neidleman and A I Laskin (Academic Press) 45
[92] Ohleyer E, Blanch H W and Wilke C R 1985 *Appl. Biochem. Biotechnol.* **11** 317
[93] Zhang D X and Cheryan M 1994 *Process Biochemistry* **29** 145
[94] Choudhury B and Swaminathan T 2006 *Appl. Biochem. Biotechnol.* **128** 171
[95] Kamoshita Y, Ohashi R and Suzuki T 1998 *J. Ferment. Bioeng.* **85** 422
[96] Kulozik U, Hammelehle B, Pfeifer J and Kessler H G 1992 *J. Biotechnol.* **22** 107
[97] Kwon S, Yoo I K, Lee W G, Chang H N and Chang Y K 2001 *Biotechnol. Bioeng.* **73** 25
[98] Danner H, Madzingaidzo L, Thomasser C, Neureiter M and Braun R 2002 *Appl. Microbiol. Biotechnol.* **59** 160

Simulation of Feed Modulation on Dynamic Fixed-Bed Reactor for CO Methanation over Ni-based Catalyst

Utari Nuravifah, Sarah Eka Putri, Yogi Wibisono* Budhi and Jenny Rizkiana

Depatment of Chemical Enggineering, Faculty of Industrial Enggineering, Institut Teknologi Bandung, Jl. Ganesha 10, Bandung 40132, Indonesia

*Corresponding author: Y.Wibisono@che.itb.ac.id

Abstract. The hydrogenation of CO to produce synthetic natural gas (SNG) is highly exothermic and usually catalyzed by nickel as an active site. These reactions are typically conducted under elevated pressures and low temperatures to shift the reversible reactions to the products. However, conducting reaction under such low temperature is kinetically limited. An alternative method that can be applied to ameliorate this limitation is by conducting a dynamic operation. This study focused on model development and reactor approach for dynamic fixed-bed operation intended for CO methanation. One dimensional pseudo-homogeneous reactor model was developed for a typical laboratory scale by neglecting internal and external diffusion based on Weisz-Prater, Anderson, and Mears criteria. The gas phase model was governed for compounds in the bulk phase. The model consisted of the dynamic term, convective term, diffusive term, and source term. The design criteria involving pressure drop, ratio of the height of catalyst bed to particle diameter (L_B/d_p), ratio of reactor diameter to particle diameter (d_r/d_p), ratio of bed length to reactor diameter (L_B/d_r) and axial dispersion were taken into consideration. A kinetic model to complement the simulation was taken from literature. The reactor model was simulated for steady-state and unsteady-state operation with optimum feed composition. The result of steady-state model simulation was considered as a base case and comparison to judge the reactor performance under unsteady-state operation. Modulating the value of the inlet CO fraction in step function was introduced to the unsteady-state model in order to enhance methane production. The simulation results showed that the highest methane production could be achieved by modulating CO inlet fraction between 0.45 and 0.4 with the overall switching time of 25 s.

Keywords: energy storage, CO methanation, Ni-based catalyst, feed modulation, reactor design

1. Introduction

The current effort to surmount the lack of alternative clean energy source is to generate synthetic natural gas (SNG) having characteristic closes to natural gas [1]. This can be achieved by reacting carbon monoxide (CO) with hydrogen (H_2), or so-called CO methanation reaction, in the presence of a heterogeneous catalyst. This process has been evaluated as a means of storing non-continuous electrical power from wind and solar devices, in which the energy is produced at locations where it is not directly consumed [2,3]. The accumulated electrical power is first converted to hydrogen (H_2) through electrolysis process. Then CO gas coming from coal or biomass gasification is injected into the

Content from this work may be used under the terms of the Creative Commons Attribution 3.0 licence. Any further distribution of this work must maintain attribution to the author(s) and the title of the work, journal citation and DOI.
Published under licence by IOP Publishing Ltd

methanator unit together with hydrogen input [4,5]. Carbon monoxide methanation is typically comprised of a series of parallel reactions [6]. There are two reactions involved as follows [5]:

$$CO + 3\,H_2 \rightleftharpoons CH_4 + H_2O \qquad \Delta H_R^{298K} = -206 \text{ kJ mol}^{-1} \qquad (1)$$

$$CO + H_2O \rightleftharpoons CO_2 + H_2 \qquad \Delta H_R^{298K} = -41 \text{ kJ mol}^{-1} \qquad (2)$$

Methanation of CO as in equation (1) is the key reaction and water-gas shift (WGS) reaction as in equation (2) is one of many side reactions that may occur during methanation. In industrial application, the WGS reaction is usually carried out before methanation to raise the H_2/CO ratio for high yield of methane (CH_4) [7]. According to thermodynamic analysis, methanation strongly depends on the chemical equilibrium. To obtain high yield and selectivity of methane (CH_4), low temperatures and high pressures operating conditions are recommended. However, it would be kinetically limited in regard to the deployed catalyst [6,8]. At temperatures higher than 350°C, the reverse reaction of CO methanation, steam reforming of methane, becomes thermodynamically favorable, thus resulting in limited amounts of methane and an increase in CO and CO_2 [2]. These conditions must be avoided as much as possible thermodynamically and kinetically.

A catalyst with high activity below 350°C is required to maximize the yield of methane [2]. Transition metal such as nickel has been widely examined as an active catalyst for CO methanation [9] despite being easily deactivated in low-temperature operation [10,11]. The deactivation is resulted from the interaction of the metal particles with CO and formation of mobile nickel carbonyl species [11]. In the catalysis, catalytic performance is influenced by the nature and properties of the applied support. A support acts as a stabilizer for the active phase as well as an enhancer for the adsorption of key reactants [9,12-14]. For the promotion of CO methanation, metal oxide supports are widely applied (Al_2O_3, SiO_2, zeolites, TiO_2, CeO_2, MgO, etc) and alumina supported catalysts are the commercial ones [9,14-16].

Beside catalyst support, high amount of nickel content is also giving positive effect on CO conversion and CH_4 selectivity [17]. Catalyst with a higher Ni-content provides a higher activity at low-temperature [12,16,18]. Increasing reaction temperature from 200 to 350°C [14,18] as well as conducting dynamic operation could improve the catalytic performance significantly [19]. Substituting one or more operating variables such as feed composition periodically could increase conversion and selectivity to the desired product [20-23]. In this study, we aimed to investigate a way to enhance the yield of methane in synthetic natural gas production by modulating one of the feed components in a step function form. The simulation results were then compared to the value obtained in steady-state operation.

2. Methods

2.1. Mathematic modeling of fixed-bed reactor

To simulate the dynamic behavior of the methane synthesis a reactor description is needed. An unsteady-state adiabatic fixed-bed reactor (FBR) was then constructed. A one-dimensional pseudo-homogeneous FBR with axial mixing was applied to model the reactor [24]. The effect of catalyst bed inside a reactor results in axial mixing. For design purposes, a Bodenstein number, Bo, of 150 was applied to calculate the value of effective diffusivity of mixture, D_{ax}, as used in mass balance equation. The chosen Bo number still ensures plug flow behavior [25]. The effective conductivity of catalyst bed in axial direction used in energy balance equation was adapted from literature [26]. The unsteady-state mass and energy balance equations as well as the steady-state pressure drop correlation are presented in table 1.

The effective conductivity of catalyst bed in axial direction used in energy balance equation was adapted from literature [26]. The unsteady-state mass and energy balance equations along with pressure drop criteria [27] are presented in the table 1. Ergun's equation [24] was applied to calculate the pressure drop along the catalyst bed. In this present work, the obtained pressure drop calculation was in agreement with the specified criteria. Thus it has been assumed to be negligible. The physical properties of gases were extracted from literature [28].

IOP Conf. Series: Materials Science and Engineering **622** (2019) 012024 doi:10.1088/1757-899X/622/1/012024

2.2. Design criteria

In designing the pseudo-homogeneous PFR model, one must ascertain that mass and heat transports do not limit the overall reaction rate. This is crucial because transfer phenomena between two phases in heterogeneous creates a film layer resistance. Consequently, causing concentration and heat gradients between the gas phase and the film covering the catalyst surface [29]. To investigate any significant temperature and concentration gradients either inside the catalyst bed or within the catalyst particles, criteria [30] listed in table 2 are employed. The external mass and heat transfer coefficients needed in table 2 are presented in table 3. The gas phase properties and transfer parameters required for all criteria were taken from literature. The following assumptions are considered during modeling the FBR:

- Temperature gradients both in axial and radial directions within the reactor are neglected according to criterion listed in table 4.
- Reactor is operated adiabatically, hence heat loss is neglected.
- Plug flow pattern inside FBR is also considered according to criteria listed in table 4.
- Pressure drop were calculated based on Ergun's equation for fixed-bed reactor.

Table 1. Mass and energy balances.

Definition	Expression	Equation
Mass balance	$\frac{\partial P_i}{\partial t} + u_z \frac{\partial P_i}{\partial z} = D_{ax} \frac{\partial^2 P_i}{\partial z^2} \pm \rho_B RT \sum v_{ij} r_i$ i = gas component; j = reaction number.	(3)
Energy balance	$(1-\varepsilon_B) C_{p,s}\, \rho_b \frac{\partial T_G}{\partial t} + \varepsilon_B\, C_{p,g}\, \rho_g u_z \frac{\partial T_G}{\partial z} =$ $\rho_B \sum(-\Delta H_{R,i} r_i) + \lambda_{ea} \frac{\partial^2 T_G}{\partial z^2}$	(4)
Initial conditions	at t = 0 and $0 \leq z \leq L_B$ $P_i = y_i{*}P_{in}$ $T_G = T_{G,in}$	(5)
Boundary conditions	at z = 0 $P_i = y_i{*}P_{in}$ $T_G = T_{G,in}$ at z = L_B $\frac{dP_i}{dz} = 0$ $\frac{dT_G}{dz} = 0$	
Pressure drop criterion	$\frac{150(1-\varepsilon_B)^2 \mu u_s L_b}{p_0 \varepsilon_B^3 d_p^2} + \frac{1.75(1-\varepsilon_B)\rho_g u_s^2 L_b}{p_0 \varepsilon_B^3 d_p} \leq 0.03$	(6)

Table 2. Heat and mass transfer limitation criteria.

Criteria	Expression	Transport process
Weisz-Prater	$\frac{r^{eff}\rho_b d_p^2}{4C_{CO}D_{CO}^{eff}} < 1$	Intraparticle mass transport.
Anderson	$\frac{r^{eff}\rho_b d_p^2 \lvert \Delta Hr \rvert}{4\lambda^{eff}T} < \frac{0.75RT}{E_A}$	Intraparticle heat transport.
Mears	$\frac{r^{eff}\rho_b d_p}{k_C C_{CO}} < 0.3$	External mass transport between bulk phase and catalyst surface.
Mears	$\frac{r^{eff}\rho_b d_p \lvert \Delta Hr \rvert}{hT} < \frac{0.3RT}{E_A}$	External heat transport between bulk phase and catalyst surface.

Table 3. Correlations used for estimating heat and mass transfer coefficients used in table 2.

Parameter	Correlation
External mass transfer coefficient	$Sh = 2 + 0.6\, Re^{1/2}\, Sc^{1/3}$ $Sc = u_z/D_{CO}^{eff}$ $Re = \rho_g u_z d_p/\mu_g$ $k_c = S_h D_{CO}^{eff}/d_p$
External heat transfer coefficient	h = 500 W/m^2.K [26]

Table 4. Aspect ratio criteria to maintain plug flow pattern in FBR.

Criteria	Limitation	Reference
$L_B/d_p \geq 50$	Radial mixing can be neglected against fluid velocity	[24, 27]
$d_r/d_p > 10$	Gas-solid wall effects can be neglected	[24]
$L_B/d_r > 0.04\ (u_z\, d_r/\varepsilon_B D_{ar})$	Neglected intrareactor mass gradient in radial direction	[31]

2.3. Chemical reaction model

A set of detailed kinetic equations and constants is compulsory for complementing the simulation process. Since catalysts are reaction-specific, they own unique kinetic mechanism and lead to certain reaction rate equations. Choosing an applicable reaction rate equation for a wide range of operating conditions is of fundamental importance to model, simulate and optimize a conventional or novel reactor concept. The scarcity of published kinetic data for commercial nickel catalysts and the manifold differences of the applied catalysts are the other obstacles in finding the right equations. A guide to choose an appropriate kinetic approach especially for methanation might be based on catalyst's nickel content as reported from literature [5].

Rönsch *et al.* [5] had modified two methanation rate equations derived by Klose and Baerns [32] and Zhang *et al.* [33]. Both authors used different catalyst's nickel content, 18% wt and 50% wt, respectively. However, only Zhang *et al.* [33] who had considered involving water-gas shift reaction in their CO methanation study.The modified reaction rate equations contain term of the reverse reactions, methane reforming and reverse water-gas shift. In this paper, the modified reaction rate equation which rooted from Zhang *et al.* [33] is adopted. The rate equations were valid for operating conditions with temperature window ranging from 250 to 360°C, while pressure between 1 and 5 bar. Moreover, the catalytic reaction kinetic was then incorporated into the reactor model developed earlier.

$$r_1 = \frac{k_1 K_C K_H^2 P_{CO}^{0.5} P_{H_2}}{\left(1 + K_C P_{CO}^{0.5} + K_H P_{H_2}^{0.5}\right)^3} - \frac{k_1 K_C K_H^2 P_{CH_4} P_{H_2O} P_{CO}^{-0.5} P_{H_2}^{-2} \left(\frac{1}{K_M}\right)}{\left(1 + K_C P_{CO}^{0.5} + K_H P_{H_2}^{0.5}\right)^3} \quad (7)$$

$$r_2 = \frac{\frac{k_2}{P_{H_2}}\left(P_{CO} P_{H_2O} - \frac{P_{H_2} P_{CO_2}}{K_{WGS}}\right)}{\left(1 + K_{CO} P_{CO} + K_{H_2} P_{H_2} + K_{CH_4} P_{CH_4} + \frac{K_{H_2O} P_{H_2O}}{P_{H_2}}\right)^2} \quad (8)$$

The rate coefficients k_1 and k_2 as in equation (7) and (8) denote the rate coefficient of CO methanation and water-gas shift reaction, respectively. The kinetic parameters of both reactions are showed in table 5. Meanwhile, the equilibrium constants for CO methanation (K_M) and water-gas shift reaction (K_{WGS}) as a function of temperature are adapted from different literature [8]. Properties of methanation reactor catalyst are listed in table 6.

Table 5. Kinetic parameters of equation (7) and (8) [5].

Symbol	Value	Unit
E_1	103,000	J/mol
E_2	62,000	J/mol
ΔH_C	-42,000	J/mol
ΔH_H	-16,000	J/mol
ΔH_{CO}	-70,650	J/mol
ΔH_{H2}	-82,900	J/mol
ΔH_{CH4}	-38,280	J/mol
ΔH_{H2O}	88,680	J/mol
k_1	$(7.0/3.6)\text{x}10^{10}$ exp (E_1/RT_G)	mol/kg cat.s
k_2	$(7.83/3.6)\text{x}10^{6}$ exp $(-E_2/RT_G)$	mol/kg cat.s.bar
K_C	$5.8\text{x}10^{-4}$ exp $(-\Delta H_C/RT_G)$	$\text{bar}^{-0.5}$
K_H	$1.6\text{x}10^{-2}$ exp $(-\Delta H_H/RT_G)$	$\text{bar}^{-0.5}$
K_{CO}	$8.23\text{x}10^{-5}$ exp $(\Delta H_{CO}/RT_G)$	bar^{-1}
K_{H2}	$6.12\text{x}10^{-9}$ exp $(\Delta H_{H2}/RT_G)$	bar^{-1}
K_{CH4}	$6.65\text{x}10^{-4}$ exp $(\Delta H_{CH4}/RT_G)$	bar^{-1}
K_{H2O}	$1.77\text{x}10^{3}$ exp $(\Delta H_{H2O}/RT_G)$	-
K_M	exp $(25.8+24883/T_G)$	bar^2
K_{WGS}	exp $(4.35+4593/T_G)$	-

Table 6. Properties of methanation reactor catalyst.

Property	Value	Reference
Catalyst material	Ni/Al_2O_3 (50 wt% Ni)	[33]
Pellet shape	Sphere	Assumed
Void fraction, ε_B	0.4	[34]
CO effective diffusivity, D_{ax}	1.36 x 10^{-6} m^2/s	[26]
Effective thermal conductivity, λ^{eff}	2.90 x 10^{-4} kW/m.K	[26]

3. Results and discussion

3.1. *Simulation frame conditions*

The simulation was performed on a fixed-bed reactor designed earlier. The reactor physical dimensions and the appropriate catalyst loading for the typical laboratory scale were shown in table 7. The unsteady-state heat and material balances for five gas components were solved using a software package FlexPDE version 6.5.

Table 7. Reactor physical dimensions and catalyst loading used in the simulation model.

Parameter	Value	Unit
Catalyst bed height, L_B	0.07	m
Reactor diameter, d_R	0.008	m
Catalyst particle diameter, d_p	0.6	mm
Catalyst load, m_{cat}	0.2	mg
Bulk catalyst density, ρ_{cat}	56.87	kg/m^3

For the simulation of CO methanation process, the feed composition ratio of H_2/CO was set to be stoichiometric with refering to equation (1). The gas feed volume flowrate was fixed and derived from design criteria and assessment of transport limitations. Turning to temperature and pressure operating conditions, both were chosen based on the conditions used by [33] when investigating the kinetics of CO hydrogenation. Table 8 gives an overview of those operating variables.

Table 8. Operating conditions of CO methanation used in the simulation model.

Parameter	Value	Unit
H_2/CO	3	-
Volume flowrate	0.02	m^3/h
Temperature	300	°C
Pressure	2	bar

3.2. *Steady-state operation*

Since the nature of methane production from syngas is highly exothermic, the purpose of this part of simulation was to investigate the appropriate feed gas composition which gave lower adiabatic temperature rise along the reactor length. A steady-state simulation with various feed gas composition was conducted and the results were presented in figure 1.

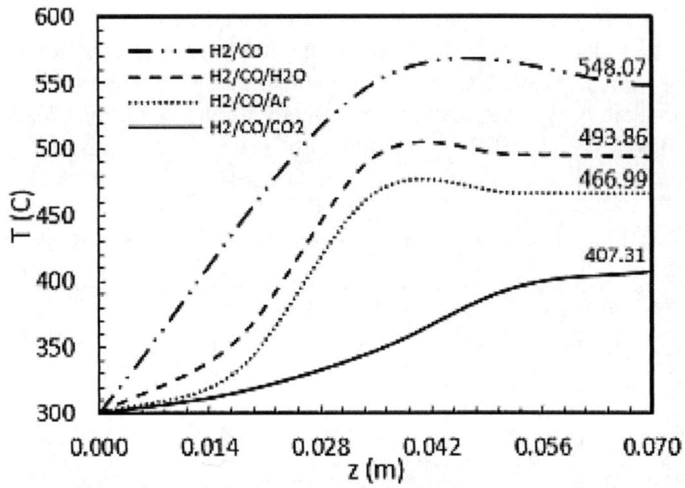

Figure 1. Adiabatic temperature rise distribution along axial direction of reactor length under steady-state simulation with T_{in} = 300°C and P_{in} = 2 bar. The ratio of H_2/CO for each variation was retained equal to 3 (67.5% H_2, 22.5% CO, 10% Ar or CO_2 or H_2O).

As can be seen from figure 1, the presence of CO_2 in the feed gas results in lower adiabatic temperature rise. The CO_2 gas might act as a thermal flywheel. At temperature above 350°C the water-gas shift reaction occured, hence limiting the methanation rate [15]. It is also worthy of notice that adding inert gas to the reactor benefited the exothermic reactions as the inert gas dilutes the reactant and absorbs the heat released during reaction [36]. Adding steam to the reactant gas mixture did not do much with lowering adiabatic temperatur rise. Even higher adiabatic temperature rise might indicate higher methanation activity, the latter variation was not favorable. The product composition at reactor outlet was dominated by H_2 and H_2O (see figure 2). According to the results above, the option of adding CO_2 to the inlet feed gas mixture was chosen for the next simulations.

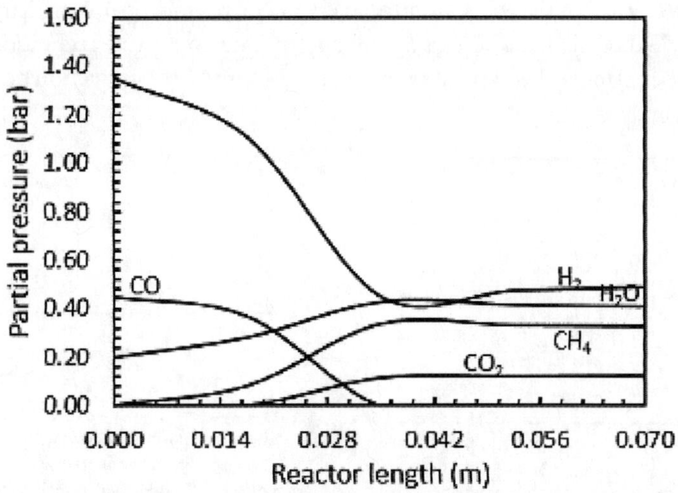

Figure 2. Gas partial pressure distribution along the reactor length under steady-state operation. Feed gas composition: 67.5% H_2, 22.5% CO, 10% H_2O. Inlet temperature and total pressure: 300°C and 2 bar, respectively.

Under steady-state operation within this simulation frame, the CO gas was almost completely consumed (98.14%) along the reactor length and the obtained CH_4 was 85.78% (see figure 3). The results are in agreement with Lazdans et al. [3] who reported that highest CO conversion rates can be achieved at temperature ranging from 300 to 400°C. Referring to the previous temperature profiles in figure 1, the obtained adiabatic temperature rise was also in agreement with the work done by Su et al. [37] that the rise in temperature did not exceed 550°C for reactant with H_2/CO ratio equals to 3. Hence, it was safe to say that our simulation was reliable enough to describe the methanation process.

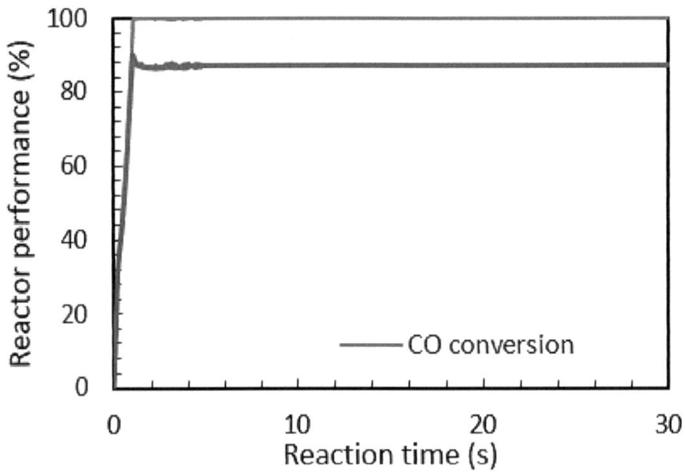

Figure 3. Ideal CO conversion (red line) and CH_4 yield (blue line) at reactor outlet within simulation time frame during start-up and steady-state operation. Feed gas composition: 67.5% H_2, 22.5% CO and 10% CO_2, inlet temperature and total pressure: 300°C and 2 bar, repectively.

3.3. Dynamic operations

From section 3.2, the optimum composition for methanation process was 67.5% H_2, 22.5% CO and 10% CO_2. This composition was then used as a starter to conduct dynamic operation. The aim of this section was to increase the yield of methane by examining appropriate contacting technique that will influence the reactor performance. Under this type of operation, the quantity of production becomes time-average quantity.

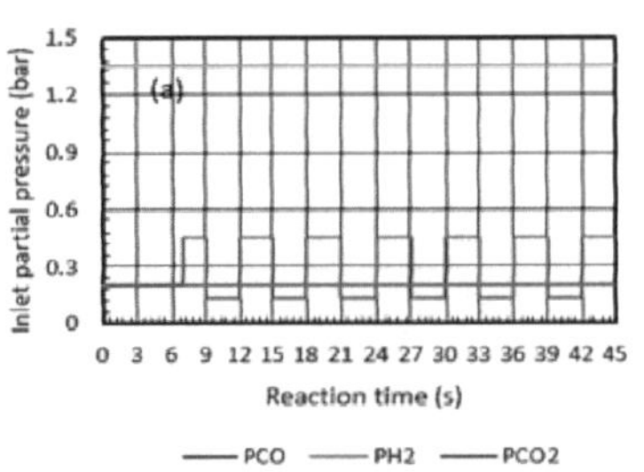

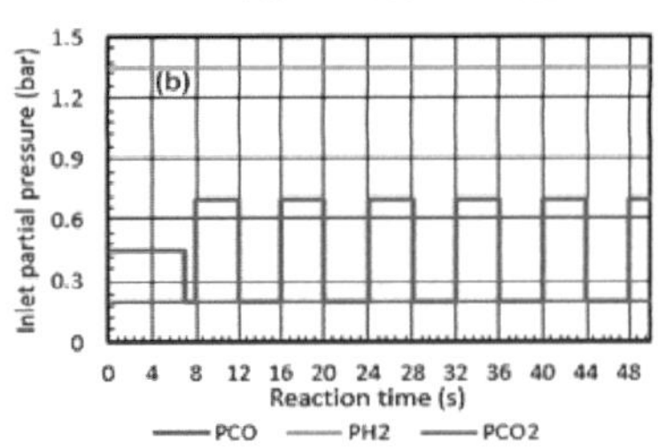

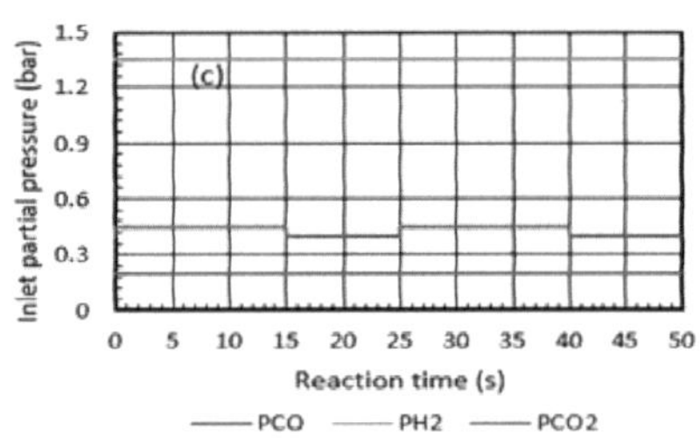

Figure 4. Single step change of CO feed ratio for various composition and switching time at inlet reactor, inlet temperature and total pressure= 300°C and 2 bar, respectively. Inlet CO pressure in bar (a) minimum= 0.135, maximum= 0.45 and switching time= 3 s, (b) minimum= 0.2, maximum= 0.7 and switching time= 4 s and (c) minimum= 0.4, maximum= 0.45 and switching time= 25 s.

IOP Conf. Series: Materials Science and Engineering **622** (2019) 012024 doi:10.1088/1757-899X/622/1/012024

The choices of manipulation for dyamic operation are so wide. In this simulation, the amplitude of CO composition in the feed gas as the input variable has been modulated periodically around its steady-state values at different switching time. The modulated variable is varied in the form of a square wave function. While the CO ratio in the feed was performed to be varied, the H_2 and CO_2 ratios were hold to be constant. The total inlet pressure in the reactor was kept at 2 bar. However, modeling a modulation with constant pressure in the reactor has a consequence that the mass flow rates into the reactor will vary during the cycle period as the changing composition causes changing consumption rates [21]. The visualization of the overall various processes were shown in figure 4.

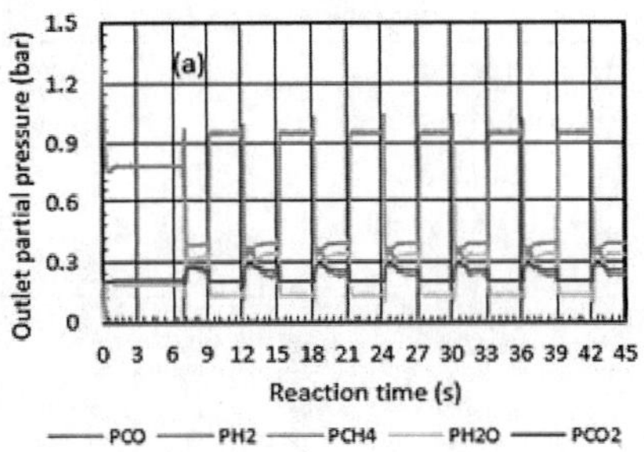

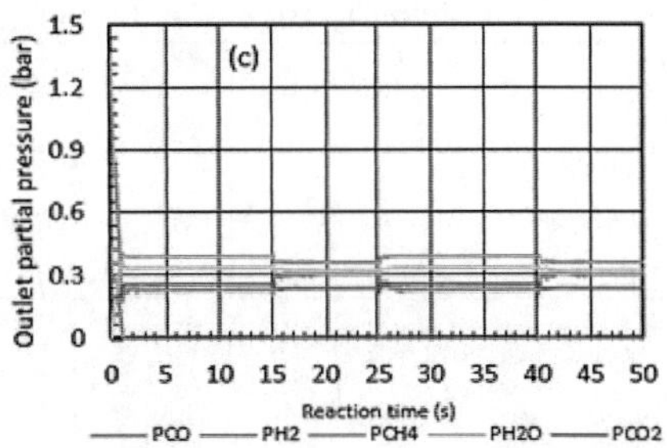

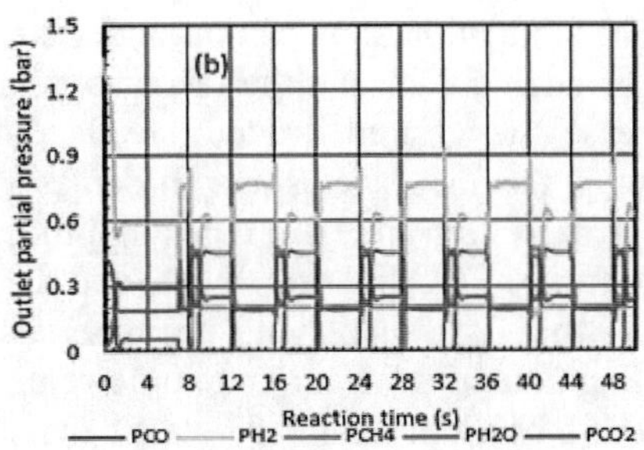

Figure 5. Transient system response of CO feed ratio for various composition and switching time at outlet reactor. Inlet CO pressure in bar (a) minimum= 0.135, maximum= 0.45 and switching time= 3 s, (b) minimum= 0.2, maximum= 0.7 and switching time= 4 s and (c) minimum= 0.4, maximum= 0.45 and switching time= 25 s.

The effect of switching time variation on reactor performance was also studied. The switching time is an importance parameter in periodic operation, which strongly influences the system dynamics. Short switching time results in the system having difficulty adjusting to the sudden change. As can be seen from figure 5, the reactor performance was constant for the following cycles, indicated by almost identical area formed during the cycle period. As for scenario (a), the CO conversion and CH_4 yield were indeed higher than the steady-state value. However, the obtained CH_4 product at the end of reactor was less than the other scenarios. This was due to the lower time-average of CO inlet value. Hence, this scenario was not suitable. The lower performance of scenario (b) compared to the steady-state value was caused by short switching time. Because of the short switching time, the process might not adjust to the sudden change. Within this ration and cycle of time, it cannot reach steady condition.

Modulating CO composition with scenario (c) as seen both in figure 4 and 5 did give slight improvement on reactor performance. Scenario (c) was performed with total switching time of 25 s. The selected CO maximum value was introduced to the reactor for 15 s and followed by adding the minimum value for the next 10 s. This was to ensure that every step change for each cycle approached steady-state condition. For the lower time-average CO pressure, the CH_4 product closed to the value of steady-state. The given CO conversion and methane yield was also satisfied. Our work results are also in agreement with [38] that modulating CO composition through step change function can give significant impact to the methanation performance. Table 9 summarized the resulted reactor performance for all variation.

Table 9. Summary of time-average reactor performance for all variation.

Steady-state		Scenario (a)	
P_{CO} in (bar)	4.5 x 10^{-1}	P_{CO} in (bar)	2.9 x 10^{-1}
P_{CO} out (bar)	8.4 x 10^{-3}	P_{CO} out (bar)	1.9 x 10^{-4}
P_{CH4} out (bar)	3.8 x 10^{-1}	P_{CH4} out (bar)	2.6 x 10^{-1}
CO conversion (%)	98.14	CO conversion (%)	99.93
CH_4 yield (%)	86.13	CH_4 yield (%)	88.73
Scenario (b)		**Scenario (c)**	
P_{CO} in (bar)	4.5 x 10^{-1}	P_{CO} in (bar)	4.3 x 10^{-1}
P_{CO} out (bar)	1.8 x 10^{-1}	P_{CO} out (bar)	4.7 x 10^{-3}
P_{CH4} out (bar)	2.5 x 10^{-1}	P_{CH4} out (bar)	3.8 x 10^{-1}
CO conversion (%)	59.69	CO conversion (%)	98.90
CH_4 yield (%)	54.61	CH_4 yield (%)	87.51

4. Conclusion

The methanation of CO has been studied at temperatures ranging from 300°C to 400°C since it is limited by methane steam reforming at elevated temperature. The simulation showed that under steady-state process, methanation reaction can be enhanced slightly by adding CO_2 into the feed gas to retain low adiabatic temperature rise along the reactor. To increase the desired product, feed composition modulation was introduced. In this work, CO composition in the inlet reactor was chosen as a variable to be modulated. Modulating CO inlet fraction between the maximum and minimum amplitude value of 0.45 and 0.4, respectively, gives a minimum time-average value of CO fraction as 0.43 in order to reach the desired reaction performance. The overall switching time is also playing an important role in determining the dynamic operation regimes. The contacting time of different amplitude value within a total time of 25 s of each value benefited to keep the process being in the steady-state behavior whilst the system adjusts to the periodic change. By performing feed modulation with different contacting time, the obtained minimum time-average value of CO inlet fraction was able to produce CH_4 with amount closed to the steady-state value. Thus, this strategy is quietly applicable to CO methanation.

References

[1] Spazzafumo G 2016 *Int J Hydrogen Energy* **41** 19492-19498
[2] Janke C, Duyar M S, Hoskins M and Farrauto R 2014 *Appl Catal B: Environ* **152-153** 184-191
[3] Lazdans A, Dace E and Gusca J 2016 *Energy Procedia* **95** 540-545
[4] Götz M, Lefebvre J, Mörs F, Koch A M, Graf F, Bajohr S, Reimert R and Kolb T 2016 *Renew Energy* **85** 1371-1390
[5] Rönsch S, Köchermann J, Schneider J and Matthischke S 2016 *Chem Eng Technol* **39 (2)** 208-218
[6] Ghaib K, Nitz K and Ben-Fares F 2016 *ChemBioEng Rev* **3 (6)** 1-11
[7] Chein R Y, Chen W Y, and Yu C T 2016 *J Nat Gas Sci Eng* **29** 243-251
[8] Molino A and Braccio G 2015 *Fuel* **139** 425-429
[9] Kuznecova I and Gusca J 2017 *Energy Procedia* **128** 255-260
[10] Chen D, Lødeng R, Anundskås A, Olsvik O and Holmen A 2001 *Chem Eng Sci* **56** 1371-1379
[11] Agnelli M, Kolb M and Mirodatos C 1994 *J Catal* **148** 9-21
[12] Le T A, Kim M S, Lee S H, Kim T W and Park E D 2017 *Catal Today* **293-294** 89-96
[13] Frontera P, Macario A, Ferraro M and Antonucci P 2017 *Catalysts* **7 (12)** 59
[14] Muroyama H, Tsuda Y, Asakoshi T, Masitah H, Okanishi T, Matsui T and Koichi E 2016 *J. Catal.* **343** 178-184

[15] Zhang R, Xia G, Li M, Wu Y, Nie H and Li D 2015 *Journal of Fuel Chemistry and Technology* **43 (11)** 1359-1365
[16] Garbarino G, Belloti D, Riani P, Magistri L and Busca G 2015 *Int J Hydrogen Energy* **40** 9171 9182
[17] Hwang S, Lee J, Hong U. G., Seo J. G., Jung J. C., Koh D. J., Lim H., Byun C and Song I. K 2011 J. Ind. Eng. Chem. **17** 154-157
[18] Rahmani S, Rezaei M and Meshkani F 2014 *J. Ind. Eng. Chem.* **20** 1346-1352
[19] Silveston P, Hudgins R. R and Renken A 1995 *Catal. Today* **25** 91-112
[20] Klusácek K and Stuchlý V 1995 *Catal. Today* **25** 169-174
[21] De Deugd R. M., Ypma S. M., Kapteijn F, Meeuse F. M., Moulijn J. A and Verheijen P. J. T 2001 Stud. Surf. Sci. Catal. **133** 255-262
[22] Chibane L, Djellouli B and Benguerba Y 2011 *Chem Eng J* **178** 398-406
[23] Kreitz B, Wehinger G D, Turek T 2019 *Chemical Engineering Science* **195** 541-552
[24] Froment G F and Bischoff K B 1990: Chemical Reactor Analysis and Design, 2nd ed., John Wiley, New York
[25] Güttel R 2013 *Chem Eng Technol* **36 (10)** 1675-1682
[26] Hwang S and Smith R 2009 *Chem. Eng. Comm.* **196** 616-642
[27] Campman M A J. 1996: Kinetics of carbon monoxide oxidation over supported platinum catalysts: The role of steam in presence of ceria, Technische Universiteit Eindhoven, Eindhoven
[28] Kind M and Martin H 2010: VDI Heat Atlas 2nd edition, Springer, Berlin
[29] Iranshahi D, Pourazadi E, Paymooni K, Rahimpour M R 2011 *Int J Hydrogen Energy* **36** 2076 2085
[30] Schlereth D 2015: Kinetic and reactor modelling for the modelling of carbon dioxide, Technische Universitat Munchen, Munchen, Germany
[31] Bischoff K B and Levenspiel O 1962 *Chem. Eng. Sci.* **17 (4)** 245-255
[32] Klose J and Baerns M 1984 *J. Catal.* **85** 105-116
[33] Zhang J, Fatah N, Capela S, Kara Y, Guerrini O and Khodakov A. Y 2013 *Fuel* **111** 845-854
[34] Zhang W, Thompson K. E, Reed A. H and Beenken L 2006 *Chem. Eng. Sci.* **61** 8060-8074
[35] Li X, Li J, Yang B and Zhang Y 2015 *Chin. J. Chem. Eng.* **23** 389-397
[36] Riaz A, Farsi A, Zahedi G and Manan Z A 2011 *World Acad Sci Eng Technol* **80** 53-56
[37] Su X, Xu J, Liang B, Duan H, Hou B and Huang Y 2016 *J Energy Chem* **25** 553-565
[38] Darmawan I R, Ivan, Sutjiono H 2017 *Journal of Applied and Physical Sciences* **3(3)** 81-88

Ionic Liquids (ILs)-based Electrolytes System for Lithium Ion Batteries

Astrid .O._Nandika, Achmad Rochliadi, Aep Patah*

Inorganic and physical Chemistry Research Group, Faculty Mathematics and Natural Sciences, Institut Teknologi Bandung, Jl. Ganesha 10, Bandung 40132, Indonesia

*Corresponding author: aep@chem.itb.ac.id

Abstract. Electrolytes system plays an important part in Lithium Ion Batteries (LIBs) due to its role in lithium ion (Li^+) transport between anode and cathode. Commercial LIBs use organic carbonates-based electrolytes system, but these electrolytes are flammable and volatile. For these reasons, replacement with non-flammable, non-volatile and high conductive compounds become recent research focus. Owing to the excellent properties, ILs are expected to cover the limitation of the organic-based electrolytes system in LIBs. The conductivity of electrolytes system which consists of dimethyl carbonate (DMC)/diethyl carbonate (DEC) (1:1, v/v) as organic solvent (OS) and imidazolium-based ILs is measured at various temperatures. Furthermore, the thermal stability of the electrolytes and the redox properties of lithium ion in IL-based electrolytes system are also investigated. [bmim][BF_4]-based electrolyte yields a conductivity of ~13 mS cm^{-1} at 20 °C which is 10,000 times higher than DMC/DEC based only. Thermal analysis shows that the IL-based electrolytes decompose at 360 °C, much higher than the organic-based only (~110 °C). Cyclic voltammetry measurement of the ILs-based electrolytes with [bmim][BF_4]: DEC: DMC (3:1:1) + $LiBF_4$ 0.2 M compositions displays reversible redox reaction. The presence of [bmim][BF_4] affects the redox reaction of Li^+ ion, in both current and potential. Based on these results, a mixture of ILs and organic solvent has a potential as a new electrolytes system in the next LIBs.

Keywords: Conductivity, Electrolytes system, Imidazolium, Ionic Liquids, Lithium ion

1. Introduction

Nowadays, lithium ion batteries (LIBs) are one of the most important rechargeable batteries. LIBs have been applied in various electronic devices due to their high energy density, long life cycle and less toxic compared to other commercial batteries.[1,2] In general, LIBs consist of electrodes (cathode and anode), separator membrane and electrolytes system. Electrolytes system plays an important role in lithium ion transport. Commercial LIBs usually use organic carbonates-based electrolytes system such as ethylene carbonate (EC), dimethyl carbonate (DMC), diethyl carbonate (DEC) or a mixture of EC/DMC. However, their volatile and flammable properties have been a safety issue of LIBs which can initiate into an explosion.[3]

Some researchers proposed to replace the commonly used organic carbonates compound with polymer or ionic liquids (ILs) as the electrolytes system. Wang et al. reported a Li/$LiFePO_4$ cell with a

Content from this work may be used under the terms of the Creative Commons Attribution 3.0 licence. Any further distribution of this work must maintain attribution to the author(s) and the title of the work, journal citation and DOI.
Published under licence by IOP Publishing Ltd

mixture of [bmim][BF_4]/γ-butyrolactone and vinylene carbonate resulted a high discharge capacity of 123 mAh g^{-1} and stable up to 20 cyles.[4] Moreover, ILs can be an alternative electrolyte due to their excellent properties such as non-flammable, high thermal stability and conductivity.[5,6] Electrolytes system with a high conductivity, low viscosity and high thermal stability is needed to develop LIBs. To the best of our knowledge, the study of imidazolium-based ILs, DEC/DMC and $LiBF_4$ mixture in Graphite/$LiFePO_4$ batteries has not been reported. Here, we report the investigation of physical, electrical, thermal stability and electro-chemical properties of ILs-based electrolytes system.

2. Experimental Method

2.1. Materials
All chemicals were commercially purchased. 1-butyl-3-methylimidazolium tetrafluoroborate [bmim][BF_4] 99.5% (Io-li-tec), 1-butyl-3-methylimidazolium bis(trifluoromethylsulfonyl)imide [bmim][NTf_2] 99.0% (Io-li-tec), 1-butyl-3-methylimidazolium hexafluorophosphate [bmim][PF_6] 99.5% (Io-li-tec), lithium tetrafluoroborate ($LiBF_4$) 98.0% (Sigma-Aldrich), DMC 99.0% (Merck) and DEC 99.0% (Merck).

2.2. Electrolytes System
Electrolytes system consisted of imidazolium-based ILs: DEC: DMC in various ratios of volume/volume. The different ratio between organic solvent and ILs was employed to investigate the optimal condition for the electrolytes system. The ratio between imidazolium-based ILs: DEC: DMC was (4:3:3), (2:1:1), (3:1:1) and (14:3:3) v/v. Lithium salt $LiBF_4$ 0.2 M was used in this electrolytes system.

2.3. Density and viscosity of ILs
The density of electrolytes system was measured using pycnometer, meanwhile the viscosity using Ostwald viscometer 200 at various temperatures of 20, 30, 40, 50, and 60 °C.

2.4. Conductivity of ILs
The conductivity of electrolytes system was determined using Hanna instrument HI 9033. Prior to use, the conductometer was calibrated with KCl solution 0.01 M, 0.1 M and 1 M at room temperature. The conductivity of electrolytes system was measured three times at various temperatures of 20, 30, 40, 50, and 60 °C.

2.5. Conductivity of ILs
Thermal stability of electrolytes system was obtained using Thermal Analyser NETZSCH STA 449F1. Sample (± 3 mg) was heated at temperature range of 20−500 °C under a N_2 atmosphere with a flow rate of 20 mL min^{-1} at a heating rate of 10 °C min^{-1}.

2.6. Cyclic Voltammetry
Redox reaction of electrolytes system under voltammetric cyclic method was investigated using Gamry reference 3000 potentiostat/Galvanostat/ZRA. Glassy Carbon (GC), Platina (Pt) and Tungsten (W/WO_3) were used as working electrode (WE), counter electrode (CE) and reference electrode (RE), respectively. Range of voltage from -3.5 until 3 V, scan rate of 100 mV s^{-1} and equilibrium time of 600 s were employed in this measurement.

3. Results and discussion
An ideal electrolytes system should have low viscosity, high ionic conductivity, high thermal stability and reversible redox reaction. Those properties are important for LIBs electrolytes system.

3.1. Density and viscosity of ILs

The viscosity of liquid affects the ion mobility between anode and cathode, but on the other hand it depends on the density. Therefore, we measure the density of three types of ILs such as [bmim][BF_4], [bmim][NTf_2] and [bmim][PF_6]. Figure 1 shows the density and viscosity of ILs, [bmim][BF_4] yields the lowest density affected by its molecular mass. Molecular mass of [bmim][BF_4], [bmim][NTf_2] and [bmim][PF_6] is 226.03 g mol^{-1}, 284.18 g mol^{-1} and 419.37 g mol^{-1}, respectively.

These results show good linear trends as reported by L. E. Barrosse-Antle et al.[7] Furthermore, the increase of temperature measurement leads to the decrease of ILs density. Molecular mobility increases at a higher temperature and enlarges the volume which is inversely proportional to the density. Viscosity of ILs affects the lithium ion (Li^+) transport between anode and cathode. ILs with a lower viscosity enhance the Li^+ ion transport to electrode, thus facilitating the redox reaction process on the surface of electrode. Figure 3.1 also shows that the viscosity increases with an increase in temperature. These phenomena are caused by the thermal energy from environment that induce the molecules to move faster than at a low temperature.[8,9] As a result, ion requires a relatively short time to move and decreases the viscosity of ILs. Moreover, the bulkiness of anion also affects the viscosity of ILs, a bulky anion requires a long time to move and reach the low part of viscometer. [bmim][NTf_2] yields the lowest viscosity compared with [bmim][BF_4] and [bmim][PF_6] because [NTf_2]$^-$ anion has a less bulky structure among other ILs.

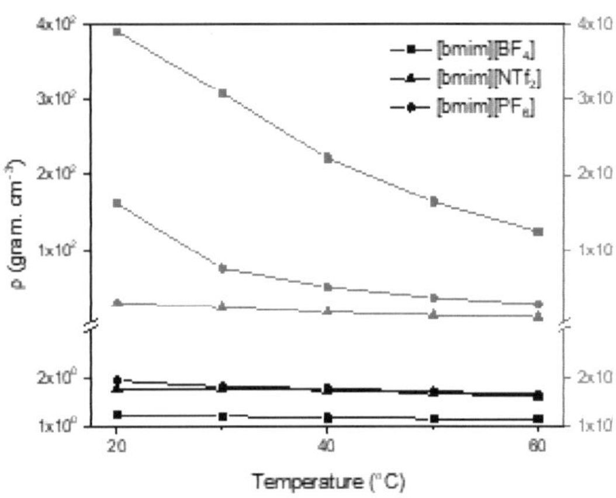

Figure 1. Density (black line) and viscosity (red line) of ILs at various temperatures.

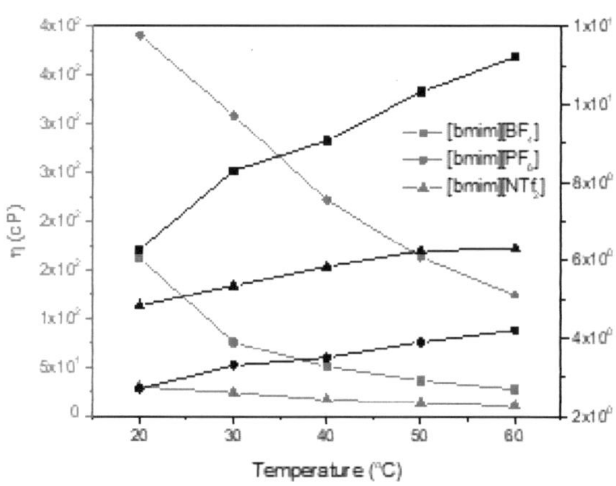

Figure 2. Viscosity (red line) and conductivity (black line) of ILs at various temperatures.

3.2. Conductivity of ILs

The conductivity of ILs at various temperatures shows that [bmim][BF_4] has the highest conductivity compared with [bmim][NTf_2] and [bmim][PF_6]. Besides the viscosity, other factors such as ion pairing (H-bonding) between cation-anion and the environment condition (water or Cl^- ion impurities content) also influence the conductivity. Generally, ILs with low viscosity exhibit a higher conductivity due to the faster ion mobility. However, in this measurement [bmim][BF_4] yields the highest conductivity among other ILs because of the ability of ion pairing between [bmim]$^+$ and [BF_4]$^-$. Kundong et al. reported the order of ion pairing stability (H-bonding in imidazolium ring and anion) as follows: [BF_4]$^-$ > [NTf_2]$^-$ > [PF_6]$^-$ using DFT calculation and molecular dynamic simulation. [10] The more stable H-bonding with anion, the more difficult to dissociate. The conductivity of ILs depends on the cation, anion and also the impurities such as Cl^- ion and water content.[10,11] It is already known that [bmim][BF_4] was a hydrophilic ILs,[12,13] thus makes a further distance between anion and cation because of the water content in ILs. The presence of low amount of water in [bmim][BF_4] starts to weaken or break down the hydrogen bonding between the proton in imidazolium ring and anion.[14] Therefore, this property enables [bmim][BF_4] to have a higher conductivity than [bmim][NTf_2] which has the lowest viscosity. The result of conductivity measurement is shown in Figure 2.

Electrolytes system contains two crucial things, there are lithium salt and electrolytes. ILs have been investigated to replace the organic carbonates-based electrolytes system, but on the other hand ILs have a higher viscosity compared with common organic solvent. To overcome the low solubility of lithium salt in ILs, combination of organic solvent and ILs is used. Organic carbonates-based solvent is used to dissolve lithium salt and decrease the viscosity of ILs.

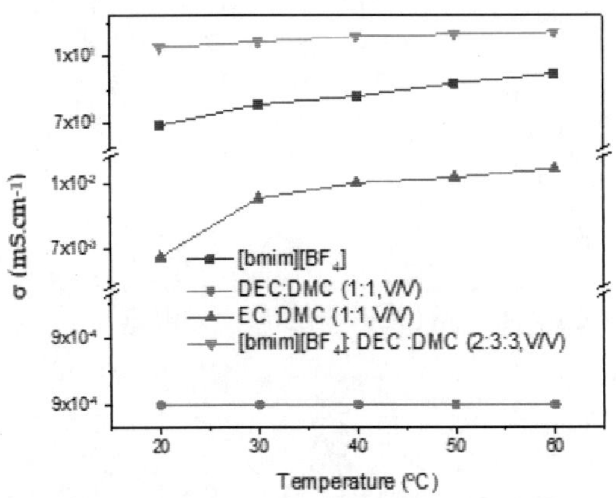

Figure 3. Conductivity of organic solvent at various temperatures..

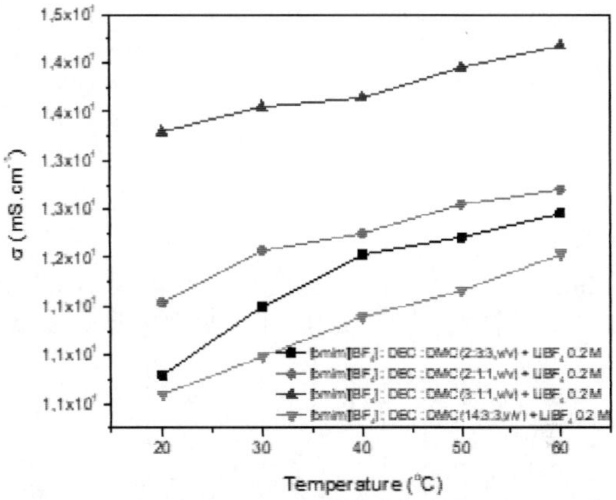

Figure 4. Conductivity of electrolytes system.

A mixture between [bmim][BF_4] and organic solvents DEC: DMC (1:1, v/v) has the highest conductivity among others, as shown in Figure 3. The conductivity increases proportionally with the number of ions involved in a mixture system. This system yields the highest conductivity due to the effect of organic solvents addition such as DEC and DMC that decreases the viscosity of [bmim][BF_4], therefore the ion mobility and conductivity increase.

Furthermore, in this study we use the ratio of organic solvents DEC: DMC (1:1, v/v) to optimize the electrolytes system. The optimization of electrolytes was conducted at 5 different ratios between ILs and organic solvent. The optimal condition of electrolytes system is [bmim][BF_4]: DEC: DMC (3:1:1, v/v), as shown in Figure 4. When the ILs ratio is higher than the organic solvents, it slows down the ion mobility due to the viscosity of ILs. Therefore, the conductivity decreases proportionally with the less organic solvent ratio.

3.3. Cyclic Voltammetry of Electrolytes system

Cyclic voltammetry (CV) is an electrochemical technique generally used to investigate the reduction and oxidation processes of molecular species.[15] Voltammogram of electrolytes system that consists of [bmim][BF_4]: DEC: DMC (3:1:1, v/v) + $LiBF_4$ 0.2 M shows three reversible reduction and oxidation peaks. The first and second reduction peaks that reveal at -0.5 V and -1.5 V represent the reduction of [bmim]$^+$ cation. Kroon et al.[16] reported a formation of 1-butyl-3-methylimidazolium radical from [bmim]$^+$ cation. [bmim] radical undergoes several reactions such as radical-radical coupling (dimer) or disproportionation reaction. The most possible mechanism was proposed by Kroon et al.

IOP Conf. Series: Materials Science and Engineering **622** (2019) 012025 doi:10.1088/1757-899X/622/1/012025

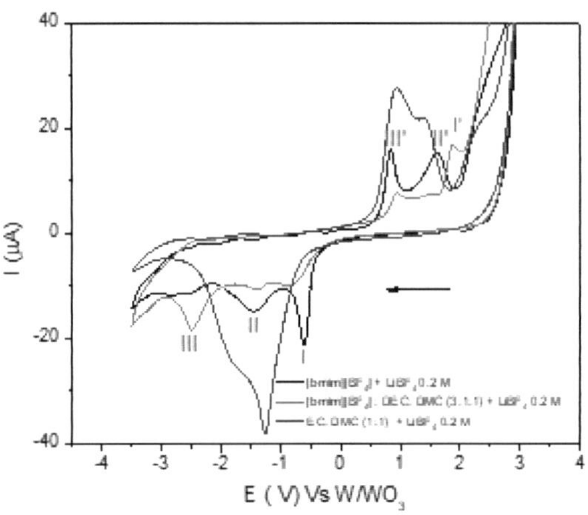

Figure 5. Voltammogram of [bmim][BF_4] + $LiBF_4$ 0.2 M (black line), [bmim][BF_4]: DEC: DMC (3:1:1, v/v) + $LiBF_4$ 0.2 M (red line) and EC: DMC (1:1) + $LiBF_4$ 0.2 M (blue line).

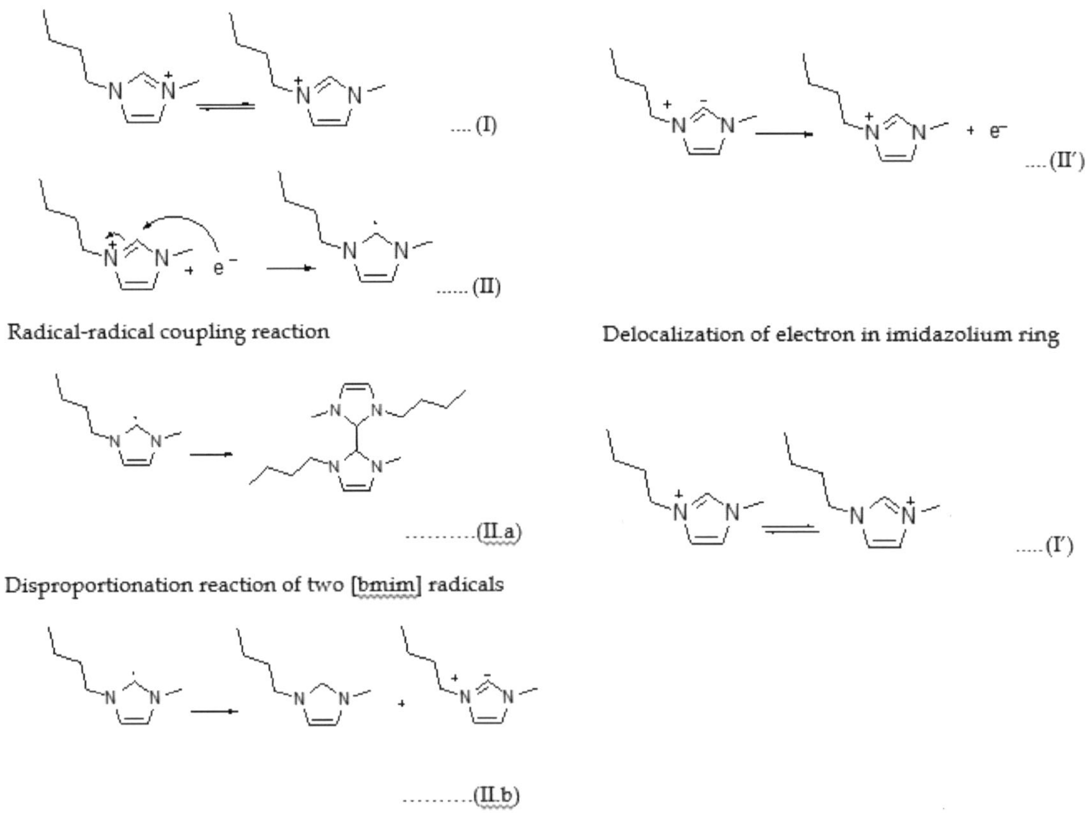

Figure 6. Radical mechanism proposed by Kroon et al.

Reduction peak at -2.5 V (Figure 6 (III)) indicates the intercalation of Li^+ ion, meanwhile oxidation peak at 1.1 V (Figure 6 (III')) belongs to the de-intercalation of Li^+ ion. Redox reaction of Li^+ ion is found shifted when dissolved in ILs. The main factor of this result is suggested by the different viscosity of the solvents. In a high viscous solvent like ILs, the diffusion coefficients of Li^+ are smaller than in organic solvent. Patah et al.[17] reported electro-active species like complex ion have various diffusion coefficients in different medium that depend on their viscosities. Moreover, the disproportionation reaction product (alkene) can be oxidized to form [bmim]$^+$, this oxidation occurs at +1.5 V (Figure 6 (II')).

3.4. Thermal stability of ILs

Thermal stability characterization was conducted to investigate the decomposition temperature of electrolytes system. [bmim][BF_4]: DEC: DMC (3:1:1, v/v) + $LiBF_4$ 0.2 M electrolytes system exhibits a higher thermal stability than DEC: DMC (1:1) + $LiBF_4$ 0.2 M. TGA profile of this electrolytes system shows two steps of decomposition. The decomposition at 110 °C and 360 °C corresponds to the organic solvent and [bmim][BF_4], respectively. This electrolytes system stability is indicated by there is no significant weight loss at temperature range of 100−300 °C. This result explains that [bmim][BF_4]: DEC: DMC (3:1:1, v/v) + $LiBF_4$ 0.2 M electrolytes system has a great thermal stability even after heating process above 500 °C, yields residue about 33.9%. Addition of ILs to electrolytes system increases the thermal stability and the similar result has been reported by Wang et al.[4]

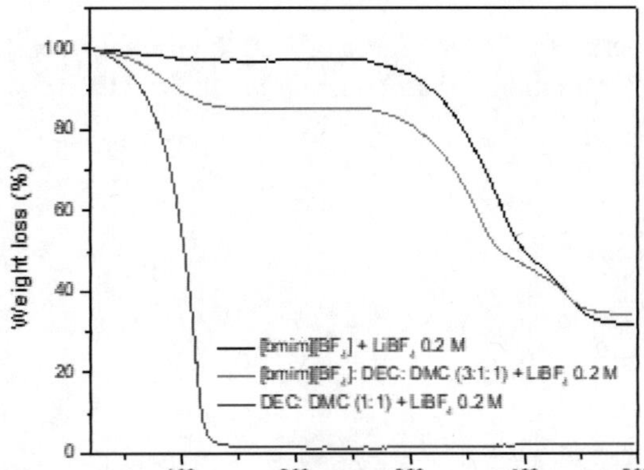

Figure 7. Thermal stability of electrolytes system

4. Conclusion

Electrolytes system of [bmim][BF_4]: DEC: DMC (3:1:1, v/v) has high conductivity, reversible lithium ion redox reaction and high thermal stability. This electrolytes system provides promise as an alternative to replace the organic carbonates-based in lithium ion batteries.

Acknowledgement

This work was financially supported by Direktorat Riset dan Pengabdian Masyarakat, Kemenristek Dikti under reseach program Peneltian Dasar Unggulan Perguruan Tinggi (PDUPT) 2018 and Program Penelitian, Pengabdian kepada Masyarakat dan Inovasi (P3MI) ITB 2018.

References

[1] Liu HW, Yang HM and Li JL 2010 *Electrochim. Acta.* **55** 1626
[2] Owen JR 1997 *Chem. Soc. Rev.* **26** 259
[3] Balducci A 2017 *Top Curr Chem (Z).* **375** 20
[4] Wang H, Liu S, Wang N and Liu Y 2012 *Int. J. Electrochem. Sci.* **7** 7579
[5] Armand M, Endres F, MacFarlane DR, Ohno H and Scrosati B 2009 *Nat. Mater.* **8** 621
[6] Diogo Cabral PCH, Pringle JM, Zhang X and MacFarlane DR 2015 *Electrochim. Acta.* **180** 419
[7] Barrosse-Antle LE, Bond AM, Compton RG, O'Mahony AM, Rogers EI and Silvester DS 2010 *Chem. Asian J.* **5** 202
[8] Guerfi A, Dontigny M, Charest P, Petitclerc M, Lagacé M, Vijh A and Zaghib K 2010 *J. Power Sources.* **195** 845
[9] Bonhôte P, Dias AP, Papageorgiu N, Kalyanasundaram K and Grätzel M 1996 *Inorg. Chem.* **35** 1168
[10] Dong K, Zhang S and Wang J 2016 *Chem. Commun.* **52** 6744
[11] Alfassi ZB, Huie RE, Milman BL and Neta P 2003 *Anal. Bioanal. Chem.* **377** 159
[12] Fadeev AG and Meagher MM 2001 *Chem. Commun.* **1** 295
[13] Seddon KR, Stark A and Torres M 2000 *Pure Appl. Chem.* **72** 2275
[14] Cha S, Ao M, Sung W, Moon B, Ahlström B, Johansson P, Ouchi Y and Kim D 2014 *Phys. Chem. Chem. Phys.* **16** 9591
[15] Elgrishi N, Rountree KJ, McCarthy BD, Rountree ES, Eisenhart TT and Dempsey JL 2018 *J. Chem. Educ.* **95** 197
[16] Kroon MC, Buijs W, Peters CJ and Witkamp GJ 2006 *Green Chem.* **8** 241
[17] Patah A, Bächle J and Grampp G 2016 *Electrochim. Acta.* **219** 305

Effect of Electroplating Current on Surface Morphology and Thickness of Plated-Through Hole (PTH) in Multilayer Printed Circuit Board (PCB)

Muhammad Arifin[1,2*], Kamarudin[1], Vivin Octowinandi[1,2], and Riki Ria[1,2]

[1] Department of Electrical Engineering, Politeknik Negeri Batam, 29461 Batam, Indonesia

[2] Teaching Factory Manufacturing of Electronics, Politeknik Negeri Batam, 29461 Batam, Indonesia

*Corresponding author: arifin@polibatam.ac.id

Abstract. Multilayer printed circuit board (PCB) consists of many layers in PCB. In order to connect one layer to other layers is needed plated-through-hole (PTH). PTH is commonly formed from the copper metal using some methods such as electroplating or electroless plating. Electroplating is method which there is adhering process the copper metal using the electrical current. This method is easy and cheap in PCB manufacture. One of the important part of electroplating result is surface morphology and thickness of PTH in multilayer PCB. The objective of this study is to im the electroplating current in electroplating process to learn more about the effect of electroplating current against the surface morphology and thickness of PTH.. The electroplating current applied in this experiment is at 5 A, 10 A, 15 A, and 20 A. In this research is also used Scanning Electron Microscopy (SEM) and Energy Dispersive X-Ray Spectroscopy (EDS) to characterize the formed PTH in multilayer PCB. The results shows that the PTH was successfully formed in through hole of multilayer PCB. The thickest PTH is about 37.8 μm for the applied current 20A and the smoothest morphology of PTH cross section is at the electroplating current 15 A.

Keywords: electroplating current, PTH thickness, multilayer PCB.

1. Introduction

One of the hardest surface adjustment technologies that it is applied for pre-treatment product is the electroplating process. The electroplating process generally utilizes the electochemical reaction principle [1], so the electric current also is applied in this process to make metal coating. Some metal materials which they are used in this electroplating process are copper, tin, silver, gold, chromium, nickel, and so on [2]. The electronic product coated by metal using electroplating process has many advantages such as lower resistivity, better anti-electromigration ability, higher thermal exchange rate [3], corrosion prevention, aesthetic finishes, and coating component [4]. Furthermore, surface smooth, copper grain, and resistivity of coated metal using the electroplating process are affected by many parameters such as the electric current density, the electric field distribution, the stirring style, and

Content from this work may be used under the terms of the Creative Commons Attribution 3.0 licence. Any further distribution of this work must maintain attribution to the author(s) and the title of the work, journal citation and DOI.
Published under licence by IOP Publishing Ltd

temperature [5,6]. This electroplating proses is widely used in aerospace, electronics, automotive, and so on [3]. One of important part in electronic products utilizing the electroplating process is printed circuit board (PCB). Copper is metal which it is commonly used as coating material in manufacturing printed circuit board (PCB) [7,8].

An important technology to manufacture multifunctional electronic products is the high-density interconnection (HDI) of printed circuit board (PCB), so making multi-layered PCB is substantial to be the high-density interconnection (HDI) [9]. One of parts on multilayer PCB is plated-through hole (PTH) [9,10]. PTH has the purpose to connect one layer to other layers in multilayer PCB [11-13]. Therefore, PTH is essential part in multilayer PCB. The objective of this study is to learn more about the effect of electroplating current against the surface morphology and thickness of PTH. The optimal electroplating current can increase the PTH performance in A4 size PCB. The high PTH performance can improve the connection between one layer and other layers in multilayer PCB as well as the adhesiveness between PTH and component.

2. Experimental Method

In order to form PTH in multilayer PCB is needed some steps such as designing, preparing material, pressing the FR-4 materials, drilling, brushing, cleaning, immersing in the conductive carbon solution (black hole solution), drying, electroplating process, cleaning, drying, testing and characterization [14]. Some steps was conducted sequentially from designing till characterization to obtain the good PTH in multilayer PCB.

The first steps is started by designing PTH in A4 size PCB using Altium Designer. The next step is preparing material. The materials include FR-4 materials (A4 size), conductive carbon solution (black hole solution), distilled water, ethanol 98%, and copper (II) sulphate pentahydrate ($CuSO_4.5H_2O$, meltex). Making the PCB stack from three FR-4 materials was conducted when preparing material was complete. The PCB stack was made by gluing the FR-4 materials using solid glue (A4 size) and pressing them to make the four layers of PCB. The pressing process was conducted with the constant pressure 12 bar and temperature 170 ^{o}C for 30 minutes. The next process is the drilling process. The drilling process is to make the holes in multilayer PCB with hole diameter 0.6 mm [13] and it works following the PCB design. Brushing and cleaning were conducted to clean the drilled PCB. The distilled water was used during brushing and cleaning. The next step is immersing the PCB surface in the conductive carbon solution for 5 minutes and drying at temperature 80 ^{o}C for 10 minutes. The next process is the electroplating. The electroplating process was conducted for four multilayer PCBs, but the electroplating current was different for each multilayer PCB in which the electroplating currents were 5 A, 10 A, 15 A, and 20 A. Each electroplating process needed about 45 minutes. Cleaning and drying of multilayer PCB were done after the electroplating process was finished. The formed PTH in multilayer PCB is characterized using Energy Dispersive Spectroscopy X-Ray (EDS) and Scanning Electron Microscopy (SEM) (Phenom ProX Desktop SEM).

3. Results and discussion

Plated-Through Hole (PTH) in multilayer PCB is generally formed using electroplating method [11]. In this electroplating method is needed some materials such as copper plate as anode, FR-4 material as cathode, copper (II) sulphate pentahydrate ($CuSO_4.5H_2O$) solution, and solution chamber. In this electroplating process is also applied the direct current (DC). The illustration of electroplating process for multilayer PCB is as shown in figure 1. There is an oxidation-reduction reaction or redox reaction when the DC current is applied in this electroplating process.

$$Cu^{2+} + 2e \longrightarrow Cu \quad \text{(reduction in cathode)}$$

$$Cu \longrightarrow Cu^{2+} + 2e \quad \text{(oxidation in anode)}$$

IOP Conf. Series: Materials Science and Engineering **622** (2019) 012026 doi:10.1088/1757-899X/622/1/012026

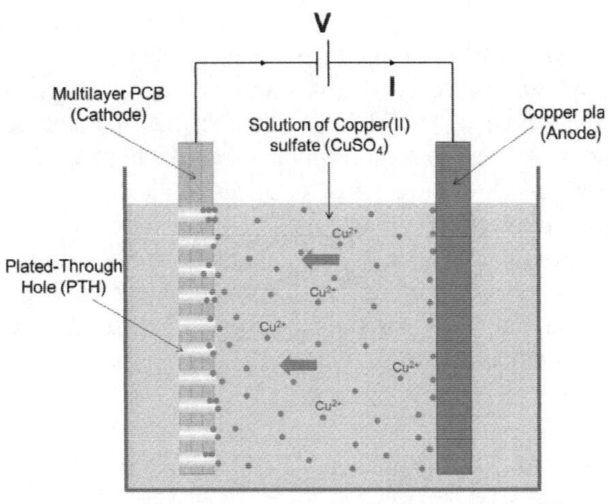

Figure 1. The illustration of electroplating process for multilayer PCB.

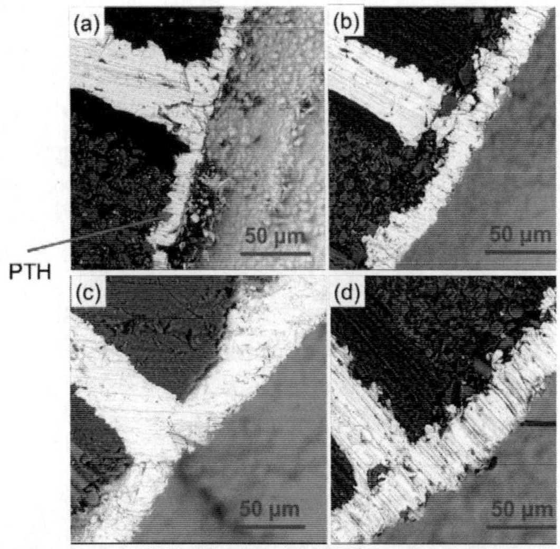

Figure 2. . SEM images of PTH cross section in multilater PCB for electroplating current at (a) 5 A, (b) 10 A, (c) 15 A, and (d) 20 A.

Plated-through holes (PTHs) were successfully formed using the electroplating method. The SEM results show that the copper metal adhered on the hole surface of multilayer PCB as shown in figure 2. The copper metal adhered on hole surface of multilayer PCB was formed by applying the electroplating current at 5 A, 10 A, 15 A, and 20 A for each multilayer PCB. Further, the SEM images exhibit that copper metal of PTH attached and bonded with FR-4 material. The copper metal of PTH also connect one conductive layer to other conductive layers of multilayer PCB. The SEM images also show the morphology of PTH cross section. The morphology of PTH cross section for electroplating current at 15 A is seen smoother than others.

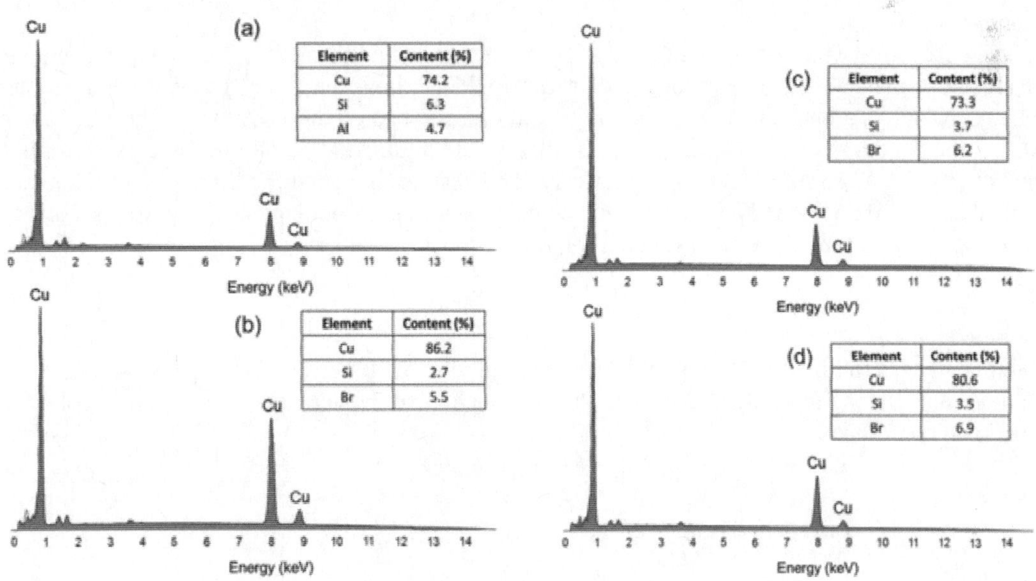

Figure 3. EDS results of PTH

IOP Conf. Series: Materials Science and Engineering **622** (2019) 012026 doi:10.1088/1757-899X/622/1/012026

The element contents of PTH are characterized by using Energy Dispersive X-Ray Spectroscopy (EDS). The EDS results show that the major content of PTH is copper as shown in figure 3. The copper content of PTH is more than 73 %. The copper contents of PTH for the electroplating current at 5 A, 10 A, 15 A, and 15 A are 74.2 %, 86.2 %, 73.3 %, and 80.6 %, respectively. These results show that the formed PTH in multilayer PCB has the copper content more than 73 %. It indicates that the electrical conductivity of formed PTH is good because the copper is one of metals which it own the high electrical conductivity. The high electrical conductivity of PTH is one of the criteria for the good PTH. The higher electrical conductivity of PTH, the better PTH in multilayer PCB.

The thickness of PTH is measured by using Phenom ProX Desktop SEM. The results show that the thickness of PTH is 12.15 μm, 21.20 μm, 30.90 μm, and 37.38 μm for the electroplating current at 5 A, 10 A, 15 A, and 20 A, respectively. The PTH thickness at 5 A is the thinnest about 12.15 μm whereas the PTH thickness at 20 A is the thickest about 37.38 μm as shown in figure 4.

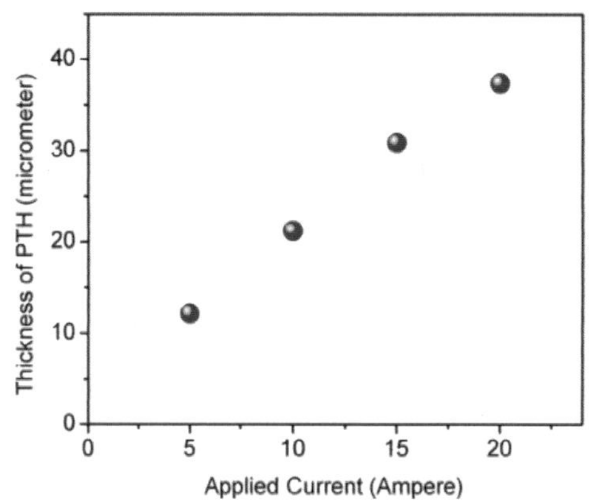

Figure 4. The effect of applied current on the PTH thickness.

4. Conclusion

PTH was successfully formed in multilayer PCB using electroplating method. The copper contents of PTH are more than 73 %. These copper contents of PTH show that PTH has the good electrical conductive which it allows the current flowing from one layer to other layers well. The SEM images show that the thickness of PTH is 12.15 μm, 21.20 μm, 30.90 μm, and 37.38 μm for the electroplating current at 5 A, 10 A, 15 A, and 20 A, respectively. This results show that the thickest PTH is obtained at applied current 20 A about 37.38 μm as well as the smoothest morphology of PTH cross section is at the electroplating current 15 A. So, the electroplating current affected the morphology and thickness of PTH.

Acknowledgement

The authors thanks to Program Penelitian Dosen Pemula research fundfrom DRPM Ristekdikti in 2018. The author also thanks to Politeknik Negeri Batam which gave the accomodation funds to follow the international conference.

References

[1] Arifin M, Rus Y B, Aimon A H, Iskandar F, Winata T, Abdullah M and Khairurrijal K 2017 *Mater Res Express* **4** 034005
[2] Selhi A and Nikhil S 2014 *Proc Int and All India Manufacturing Technology, Design and Research Conf* (Assam: Indian Institute of Technology Guwahti) p 439-1
[3] Chen Y, Li X, Chen S Q and Zhang H 2008 *ASME Proceedings* 619
[4] Masuku E S, Mileham A R, Hardisty H, Bramley A N, Johal C and Detassis P 2002 *CIRP Ann Manuf Technol* **51** 168
[5] Hu C-C and Wu C-M 2003 *Surf Coat Technol* **176** 75
[6] Seah C H, Mridha S and Chan L H 1999 *J Mater Process Technol* **89** 432
[7] Khandpur R S 2006 *Printed Circuit Board: Design,* Fabrication, and Assembly (New Delhi: McGraw-Hill) p 313
[8] Lin G-Y, Yan J-J, Yen M-Y, Dow W-P and Huang S-M 2013 *J Electrochem Soc* **160** D3028
[9] Wang C, Zhang J Q, Yang P X and An M Z 2012 *Int J Electrochem Sc* **7** 10644
[10] Chen C H, Lu C W, Huang S M and Dow W P 2011 *Electrochim Acta* **56** 5954
[11] Kim C, Espalin D, Liang M, Xin H, Cuaron A, Varela I, Macdonald E and Wicker R B 2017 *IEEE Access* **5** 25286
[12] Liu J, Chen J, Zhang Z, Yang J, He W and Chen S 2016 *Circuit World* **42** 141
[13] Cobley A J, Comeskey D J, Paniwnyk L and Mason T J 2010 *Circuit World* **36** 9
[14] Arifin M, Atmaja ABK and Octowinandi V 2018 *IOP Conf Ser : Mater Sci Eng* **395** 012016

IOP Conf. Series: Materials Science and Engineering **622** (2019) 012027 doi:10.1088/1757-899X/622/1/012027

Spin Current Generation in One – Dimension Spin – Chain Model of Insulating Magnetic Interface

Adam B. Cahaya[*] and M. A. Majidi

Department of Physics, Faculty of Mathematics and Natural Science, Universitas Indonesia, Depok 16424, Indonesia

[*]Corresponding author: adam@sci.ui.ac.id

Abstract. In a system of a thin ferromagnetic layer sandwiched by non-magnetic metals, the precession of magnetization of the ferromagnetic layer has been shown to generate spin current in the adjacent non-magnetic metals. The spin current arises from the spin-dependent scattering that is originated by the exchange interaction between the magnetic moment and the spin of adjacent normal metal's electron. While the theory was originally studied in magnetic multilayer that consists of a thin ferromagnet sandwiched by non-magnetic materials, to be able to describe a more realistic ferromagnetic layer, we need to study how magnetic moments far from the interface can influence the spin current generation. We use a simple model one dimension ferromagnetic spin chain to show the criteria in which the collective movements of the spin moments can enhance the spin current generation.

Keywords: spin pumping, spin current, spin chain

1. Introduction

While conventional electronics focuses on charges degree of freedom, field of study of spintronics also includes the spin degree of freedom. As the size of electronic devices reduces to nanosize, spin properties become important, since the electron scattering in this scale still conserve its spin angular momentum. Because of the conservation of the spin angular momentum, we need to consider spin current when studying the electronics transport of nanoscale materials [1,2,3].

Spin current can be generated by spin pumping effect. In a bilayer of ferromagnet and non-magnetic metal, movement of the magnetization of the ferromagnet pumps spin current into the adjacent non-magnetic metal. The pumped spin current is characterized by spin-mixing conductance parameter [4,5]. The spin-mixing conductance $g_{\uparrow\downarrow} = g_r + ig_i$is complex-valued, with $g_i \ll g_r$. Spin current is a tensor, which has current direction and polarization direction. In the case of spin pumping, the current direction is from ferromagnet to non-magnetic metal, while the polarization direction J_s is characterized by spin-mixing conductance parameter.

$$J_s = g_r m \times \dot{m} - g_i \dot{m} \quad (1)$$

Here $\boldsymbol{m}$ is the normalized magnetization vector and spin-mixing conductance $g_{\uparrow\downarrow} = g_r + ig_i$is complex-valued, with $g_i \ll g_r$.

Content from this work may be used under the terms of the Creative Commons Attribution 3.0 licence. Any further distribution of this work must maintain attribution to the author(s) and the title of the work, journal citation and DOI.
Published under licence by IOP Publishing Ltd

Although the theory was developed for ferromagnetic metal, spin pumping experiment showed that ferromagnetic insulators also generate spin current with similar magnitude [6,7]. The theory was originally studied in magnetic multilayer that consists of a thin ferromagnet sandwiched by non-magnetic materials. When the ferromagnet is thin enough, its magnetic moments can be considered as a uniform (coherently moving) magnetic moments [4,8]. We determine the criteria in which the coherent magnetic moments approximation held. The spin current transport in the bulk and the interface of one-dimension spin chain, the simplest case ferromagnet with one dimensional symmetry, sandwiched by non-magnetic metal, are discussed in section 2 and 3, respectively.

2. Spin current in insulating one dimensional spin chain

One dimensional ferromagnet can be modeled by an array of spins. The exchange interaction of the spin chain can be represented in the following Hamiltonian.

$$H = -J\sum S_n \cdot S_{n+1}. \quad (2)$$

where J is the exchange constant and S_n is the n-th spin, which interact only with the nearest spins $S_{n-1}$and S_{n+1}. The collective movement of this spin chain is characterized by the magnon dispersion relation

$$\hbar\omega = 2JS(1 - coska) \quad (3)$$

where a is the distance between each spin. $\hbar\omega$ and k are the energy and wave number that characterized the collective movement of the spin chain. The magnon dispersion relation is illustrated in Fig 1.

The collective moment of the spin magnetic moment, *i.e.* magnon, generates a magnonic (spin wave) spin current in the bulk of the spin chain. The polarization of the magnonic spin current that flow in x direction (see Fig 2) is

$$J_s = M_s J a m \times \frac{\partial m}{\partial x} \quad (4)$$

where M_s is the saturated magnetization [9,10]. In the next section we will determine the boundary condition that should be satisfied by this magnonic spin current.

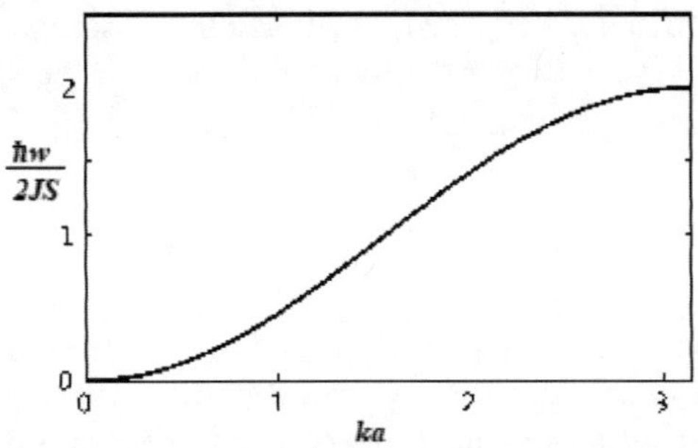

Figure 1. Energy and wave number that characterized the collective movement of the magnetic moments in the spin chain follow the magnon dispersion relation.

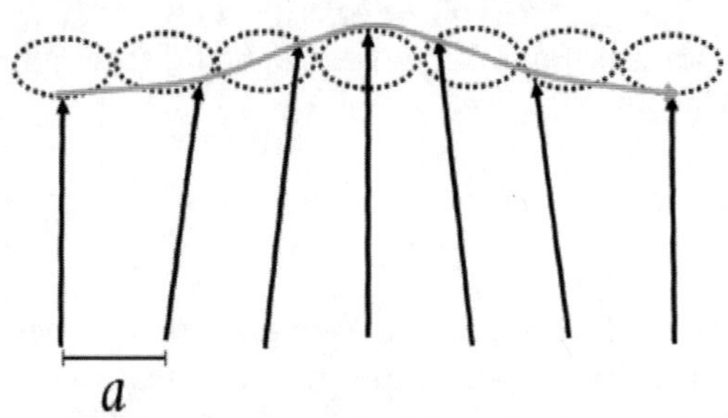

Figure 2. Collective movement of spin magnetic moments carry magnonic spin current in the bulk of the ferromagnetic insulator.

3. Boundary condition of the spin current at interface with non-magnetic metal

The spin current generated at the non-magnetic metal immediate to the interface with ferromagnetic insulator is characterized by the spin-mixing conductance. For $g_i \ll g_r$, the spin polarization is

$$J_s = g_r m \times \dot{m} \tag{5}$$

The origin of spin mixing conductance is the spin-dependent scattering due to the exchange interaction between spins of the conduction electron and the spin of the ferromagnetic insulator [9,10].

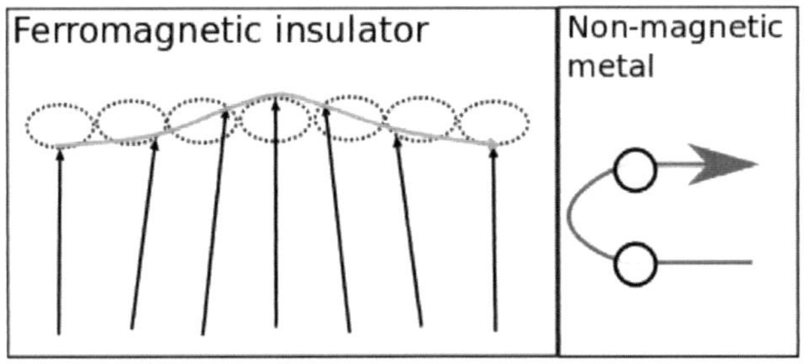

Figure 3. A bilayer of ferromagnetic insulator and non-magnetic metal. Spin current transport across the interface of ferromagnetic insulator and non-magnetic metal. When the scattering at the interface conserves spin, there should be spin current continuity across the interface.

Assuming the conservation of spin angular momentum in the interface scatterings, there should be a spin current continuity across the interface, which suggested the following boundary condition [10]

$$M_s J a \frac{\partial m}{\partial x} + g_r \dot{m} = 0 \tag{6}$$

$$\omega = \frac{M_s J a}{g_r} k$$

Eq (6) suggest the following criteria for combination of ω and k that satisfy the boundary condition

Drawing Eq. (6) into the magnon dispersion relation, we arrive at the minimum value of the spin mixing conductance

$$g_0 \simeq 0.68 \frac{\hbar M_s}{S} \tag{7}$$

in which the coherent-magnetic-moments model is a good approximation. It should be noted here that the frequency of this coherent movement in our case is zero because we neglected the influence of applied magnetic field and anisotropic fields. If these fields are included, the coherent frequency will correspond to the precession frequency due to these fields.

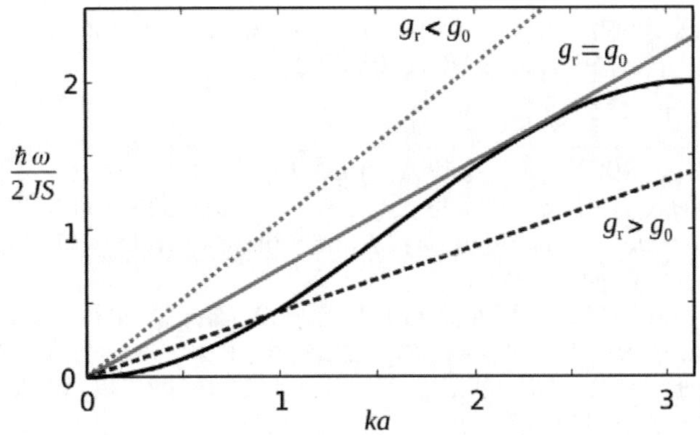

Figure 4. Red line represents the critical value of spin mixing conductance. A coherent magnetic moments approximation is a good approximation when the spin mixing conductance is small than this critical value (green line). However, other modes of collective movements, k and ω that satisfy Eqs. (3) and (6), should be taken into account when the spin mixing conductance is larger than the critical value (blue line).

4. Conclusion

We discussed the distribution of spin current in the bulk of the ferromagnetic insulator and its interface with non-magnetic metal. In the bulk, the spin current carried by the collective movement of the spin magnetic moment, *i.e.* magnon. On the other hand, in the metal side, spin current is carried by moving charges (see Fig 3). The spin pumping effect transfers the spin angular momentum across the interface and transfer the spin current from the ferromagnetic insulator to the non-magnetic metal.

We found the critical value of the spin mixing conductance, in which a uniform magnetic moment is a good approximation for treating the spin pumping effect from magnetic insulator to non-magnetic metal. It imply that when the spin mixing conductance is larger than the critical value, other modes, k and ω that satisfy Eqs. (3) and (6), contributes to the spin current transfer. While we discuss the simplest model of one-dimension ferromagnetic insulator, the result can be generalized to a more realistic model. Furthermore, by parameterizing external and anisotropic fields, the discussion can be extended to a more general case.

Acknowledgement
This work was supported by Indonesian Ministry of Research, Technology and Higher Education PDUPT Research Grant No. 374/UN2.R3.1/HKP05.00/2018

References
[1] Naef, F., & Zotos, X., 1998, *J. Phys. Cond. Mat.*, **10**, L183.
[2] Murakami, S., Nagaosa, N., & Zhang, S. C., 2003, *Science*, **301**, 1348-1351.
[3] Maekawa, S., Valenzuela, S. O., Kimura, T., and Saitoh, E. (Eds.). 2017. *Spin current* (Vol. 22). Oxford University Press.
[4] Tserkovnyak Y., Brataas A., and Bauer G.E.W., 2002, *Phys. Rev. Lett.* **88**, 117601.
[5] Weiler, M., Althammer, M., Schreier, M., Lotze, J., Pernpeintner, M., Meyer, S., Huebl, H., Gross, R., Kamra, A., Xiao, J., Chen, Y.T., Jiao, H.J., Bauer, G.E.W. & Goennenwein, S.T.B., 2013, *Phys. Rev. Lett.* **111**, 176601.
[6] Uchida, K., Xiao, J., Adachi, H., Ohe, J. I., Takahashi, S., Ieda, Ota, J.T., Kajiwara Y., Umezawa H., Kawai H., Bauer, G.E.W., Maekawa S., & Saitoh, E., 2010, Spin seebeck insulator. *Nature mater.*, **9**, 894.
[7] Uchida, K. I., Adachi, H., Ota, T., Nakayama, H., Maekawa, S., & Saitoh, E., 2010, *Appl. Phys. Lett.*, **97**, 172505.
[8] Šimánek, E., and Heinrich, B., 2003, *Phys. Rev. B*, **67** 144418.
[9] Alvarez, J. V., and Gros, C., 2002, *Phys. Rev. Lett.*, **88**, 077203.
[10] Hoffman, S., Sato, K., & Tserkovnyak, Y., 2013, *Phys. Rev. B*, ***88***, 064408.
[11] Cahaya, A. B., Leon, A. O., and Bauer, G. E. W., 2017, *Phys. Rev. B*, **96**, 144434.

The Porosity and Hardness of Fe-18Al Alloy Added ZrO_2 Nanoparticles

P. Pawawoi[13], D.H. Prjitno[2,3], R. Adhitya[3], A. Manaf[1*]

[1] Post Graduate Program Department of Physics, Faculty of Mathematics and Natural Sciences, Universitas Indonesia Depok, 16424, Indonesia
[2] PTNBR- BATAN, Jln. Tamansari 71 Bandung, 40317 Bandung, Indonesia
[3] Metallurgical Engineering, Faculty of Engineering, Universitas Jenderal Achmad Yani, Bandung 40285, Indonesia

[*]Corresponding author: azwar@ui.ac.id

Abstract. The additive effect of ZrO_2 nanoparticles in Fe-18Al based alloys through the mechanical alloying process to the porosity and hardness was investigated. Results of microstructure observations showed that the only intermetallic Fe_3Al present as the major phase in the samples along with other phases which are very minor together with the presence of porosity that indicated by objects of dark color. The porosity tends to diminish with the addition of ZrO_2 nanoparticles. However, the porosity is still visible on the Fe-Al alloy with the addition of 2% additive and the higher compaction load especially in the surface area. The results of phase study by XRD confirmed the presence of the intermetallic Fe_3Al. It is shown that ZrO_2 nanoparticles are dispersed in the metal matrix. The presence of ZrO_2 also confirmed by SEM-EDS. The fraction of pores that estimated based on the microstructure observation showed a decrease with an increase in volume fraction of ZrO_2 nanoparticles and the compaction load. The study concluded that the hardness of Fe-18Al alloys increased with a decrease in porosity which achieved in Fe-18Al alloy added with 2% ZrO_2 nanoparticles. The lowest porosity level is 0.17% and the highest hardness value is 352.59 HV

Keywords: Mechanical alloying, Intermetallic, Nanoparticle ZrO_2, Fe-Al alloys, Porosity

1. Introduction

High temperature resistant alloys which meet specific needs characterized by a capability of able to be operated continuously without protection at temperatures above 1200°F (650°C) [1]. Metallurgically, such high temperature resistant alloys are characterized by high creep and oxidation resistances property [2]. The need for high temperature resistant alloys is now increasing, while the type of material that potential to have such specific properties is very limited. The limitation is due to high temperature operating conditions that require specific property to overcome the oxidation. So far, the best material, the so called super alloys are reliable to fulfil such requirement [1, 3, 4, 5]. Broadly speaking, there are three groups of metal alloys that can be used in high temperature applications respectively super alloys, refractory alloys, and intermetallic alloys [1, 6, 7]. However, the latter alloys are being attracted great

Content from this work may be used under the terms of the Creative Commons Attribution 3.0 licence. Any further distribution of this work must maintain attribution to the author(s) and the title of the work, journal citation and DOI.
Published under licence by IOP Publishing Ltd

attention by many researchers because the alloys are technologically important to be produced at the reduced cost [8] [9].

One of intermetallic alloys which has good prospect for use at high temperature is iron-aluminium (Fe-Al) based alloys. Two types of intermetallic phase respectively Fe-Al (B_2) and Fe_3Al (DO_3) in Fe-Al based system has received widespread attentions which are expected that the intermetallic alloy can replace the stainless steels for high temperature applications. The alloy has a high temperature resistance through the formation of a thin layer on the surface when oxidized. The oxide thin layer protects the alloys which cause resistant to high temperature and good creep resistance coupled with reliable mechanical properties turning the alloy favourable for high temperature applications. It has been widely known that the Fe-Al alloys have a relatively low density, high melting point, possess a high strength property at high temperatures, good oxidation and crack resistances. Not surprisingly, the Fe-Al Alloy is one of the preferred materials applicable as the structural parts applied at high operating temperature with corrosive environment. [6, 10].

One of the fabrication process for the Fe-Al alloys are mechanical alloying in which powder of alloy components is milled, welded and fractured in a high energy ball milling apparatus and then subsequently annealed at sufficient temperatures, allowing the solid state reaction to take place for phase formation in the alloys. [11, 12, 13]. The alloy which produced from this process will have mechanical properties which still can be modified by the addition of tertiary metal oxides. Because of having a very high melting temperature and strength, the properties of alloys will improve. However, this process has drawbacks such as the porosity level that is high enough that will affect the alloy resistance to oxidation or the formation of an oxide layer and can reduce the mechanical properties, especially the strength of the material [14, 15, 16, 17, 18, 19, 20, 21].

In this paper, results of investigation on the hardness and porosity of Fe-Al alloys synthesized through mechanically alloying of the iron (Fe) and aluminium (Al) powders with the addition of 1 and 2 vol.% ZrO_2 nanoparticles are presented and discussed. To study the microstructure and mechanical properties of the alloys, Scanning Electron Microscope, Vickers hardness testing was used. An additional tool an image analyser is used to evaluate the porosity in the alloys.

2. Experimental methods

Alloy samples were prepared from an analytical grade of Fe and Al powders. ZrO_2 nanoparticles of 30-60 nm and 99.99% purity were used as during mechanical alloying the Fe and Al feedstock. The designated composition for the Fe-Al alloys are respectively $Fe_{82}Al_{18}$, $Fe_{81}Al_{18}$-ZrO_2, and $Fe_{80}Al_{18}$-$(ZrO_2)_2$ in weight percent (wt. %). Samples of such compositions are coded respectively F18, F18Z1 and F18Z2.

The mechanical alloying process was carried out in a planetary ball mill with a charge to ball mass ratio 5:1 in a stainless cup of 285 ml capacity. The ball was made of zirconium oxide with 10 mm in diameter. The planetary ball mill was operated under a jog speed maintained at a speed of 600 rpm for 60 and 90 minutes. The mechanically milled materials produced planetary ball milling was charged into a cylindrical die and pressed under a compressive strength of 90 and 100 kg/cm^2 using a hydraulic machine at room temperature. The green compacts were then sintered in a ceramic tube furnace with preheating temperature 200 °C for 30 minutes and then heated up to 1000 °C for 2 hours followed by a furnace cooling to room temperature.

Phase identification of all samples was characterized by XRD PANalytical Empyrean with an X'Pert MPD diffractometer (Cu Kα radiation, λ = 1.5406 Å, and generator settings 30 mA, 40 kV). The observation was carried out in the angle range 2θ = 20-80° and a step size of 0.017°. The microstructure of samples was observed under an optical microscope and SEM JEOL JSM-6360LA.

3. Results and discussion

Plot of XRD data for samples F18 and F18Z1 is compared in Figure 1. The identification study made to the strong diffraction peak of the two plots suggest that the peaks are well matched with that of intermetallic Fe_3Al phase referred to Jia Lia et al. [22]. Furthermore, when referring to the phase diagram of Fe-Al system [16], composition of the two samples coded F18 and F18Z1contained 18 wt.% Al must be a single phase alloy with Fe_3Al phase the only phase in the samples. No other minor phases may be identified confidently through their respective x-ray traces.

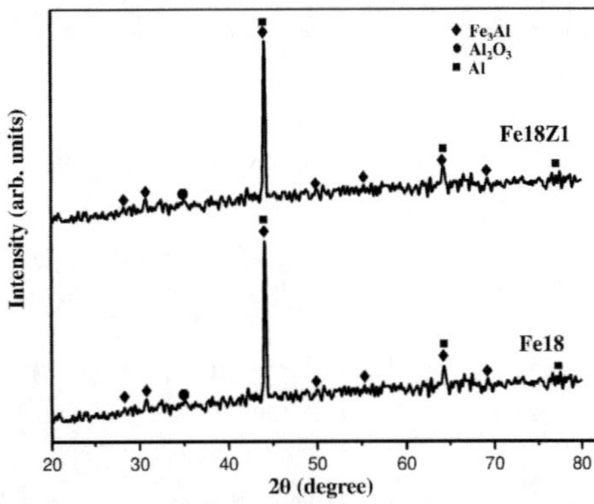

Figure 1. X-ray diffraction patterns of samples coded (a) F18 and (b) F18Z1 after sintered at 1000^0 C for 2 hours

Hence, if other phases exist in the samples like Al_2O_3 in sample coded F18 and an additional minor phase ZrO_2 in sample coded F18Z1, the fraction must be extremely small below the detection limit of the x-ray diffractometer. The presence of Al_2O_3 is most probable since Al tends to oxidize when standing alone and treated at a high temperature. While, the presence of ZrO_2 is also possible because the ZrO_2 component was added in the preparation of sample coded F18Z1. Fortunately, the additional of oxide phase, especially Al_2O_3 is needed by material for high temperature applications as protective layers. U. Prakash [9] reported that the sintering treatment of mechanically alloyed powders of Fe and Al changes the crystal structure of α-Fe (ferrite) to an intermetallic Fe_3Al phase at high fraction of Al.

A set of photo images resulted from microstructure observation under an optical microscope is presented in Figure 2. The images were taken in two different directions respectively surface section and cross section. Samples are made of mechanically alloyed powder of two different milling time and two different compaction loads. Basically, there is no difference in grain morphology of the two directions indicating that the grains in the sample are equiaxed grain and randomly orientated. No prefer orientation was observed in the samples.

The feature of the microstructure of the whole samples consisted of an intermetallic $FeAl_3$ as the main phase appears dominantly in the whole area of observation, another phase □-Al_2O_3 was also observed as the second phase. The microstructure of samples also shows small dark areas which identified as the porosity. The porosity was found less in the ZrO_2 added Fe-Al samples and almost porosity free as the fraction of additive increased to 2 % by weight. The compaction load seems gave no significant improvement to reduce the porosity.

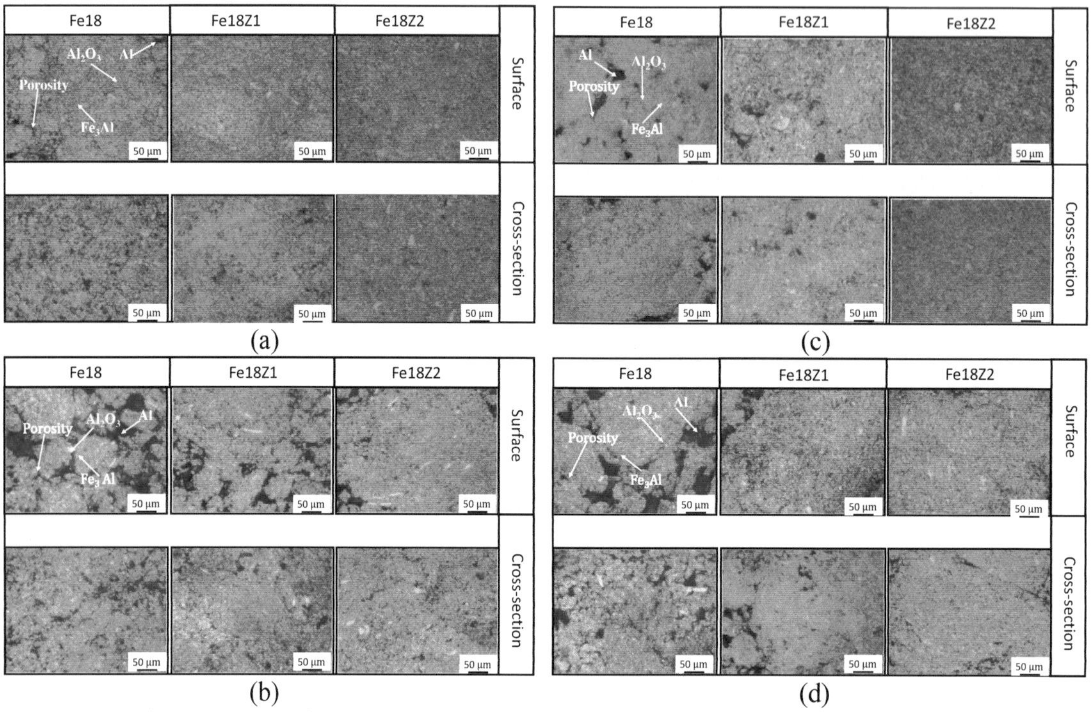

Figure 2. Microstructure of samples obtained after mechanically alloyed for: a) 60 minutes, b) 90 minutes both under compaction load 90 kg_f/cm^2, c) 60 minutes and d) 90 minutes under compaction load 100 kg_f/cm^2

While SEM image for sample F18 is compared with that of F18Z1 in Figure 3 along with EDS spectrums of microanalysis. Porosity is clearly seen in the two types of samples. The present of ZrO_2 in the samples is also shown in the micrograph appear as fine particles. The particle was treated as the object for microanalysis. Result of EDS analysis to the spot on the surface of sample F18 shows that only Fe and Al are present with the weight fraction, respectively 87.67% and 12.33 % which close to the designated composition. Almost the same result of EDS analysis of the surface of sample F18Z1 was obtained, but with an additional energy peak of Zr which is assumed present as an oxide form of ZrO_2. The mass fraction of ZrO_2 is only 0.79 % closer to the designated composition in which the ZrO_2 was added 1 wt. %.

The formation of intermetallic Fe_3Al phase together with densification toward pore free samples were achieved through sintering at a temperature 1000 ^{0}C. The addition of ZrO_2 nanoparticles in sample coded F18Z1 is expected to enhance the hardness. Figure 4 presents result of image analysis for porosity in the samples ZrO_2 added Fe-Al samples compared with that of ZrO_2 free Fe-Al sample.

(a)

Element	Weight (%)
Al K	12.33
Fe K	87.67
Total	100

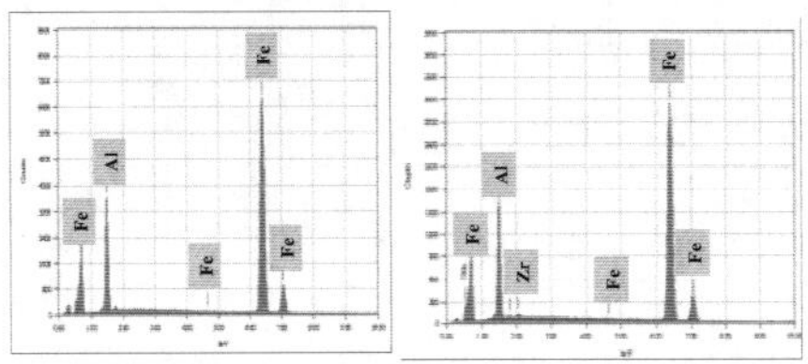

Element	Weight (%)
Al K	12.33
Fe K	87.67
ZrO_2K	0.79
Total	100

(b)

Figure 3. (a) SEM micrograph and (b) EDS for samples F18 and F18Z1 after sintering at 1000^{0}C for 2 hours

Images in Figure 4 confirm the suppression of pores in ZrO_2 added Fe-Al samples. The fraction of pores which originally 7.757 % in sample F18 was suppressed down to 2.641 % with the addition of 1 wt. % of ZrO_2 particles. Further suppression was achieved with the addition of 2 wt. % ZrO_2 particles. Results of porosity evaluation of the Fe-Al alloys as summarized in the Figure 4 confirmed that zirconia can minimize the porosity in the sample which prepared under consolidated the mechanically alloying powders. Hence, hardness of the sample can be expected is enhanced with the increase of zirconia content in the sample.

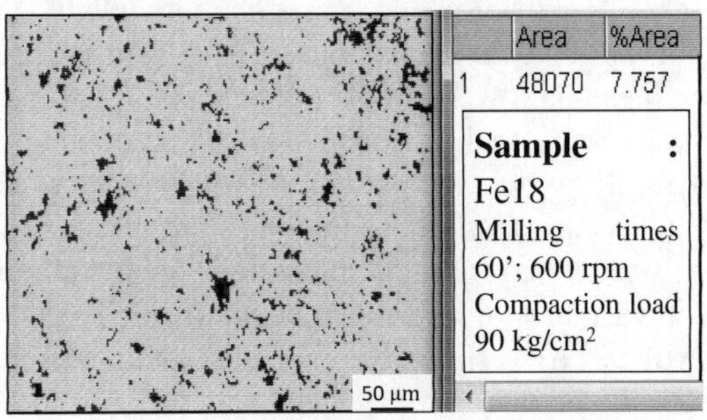

(a)

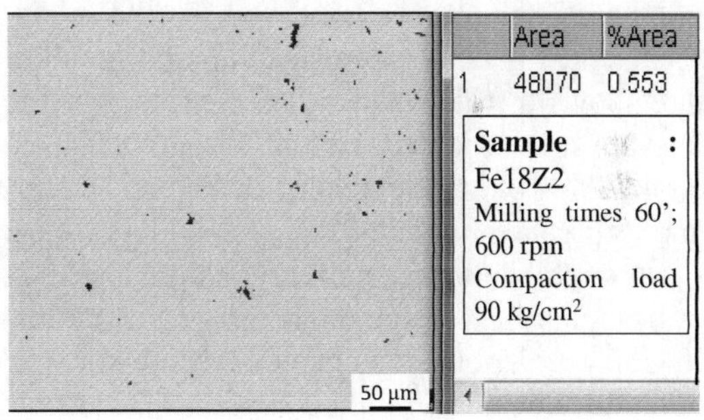

(c)

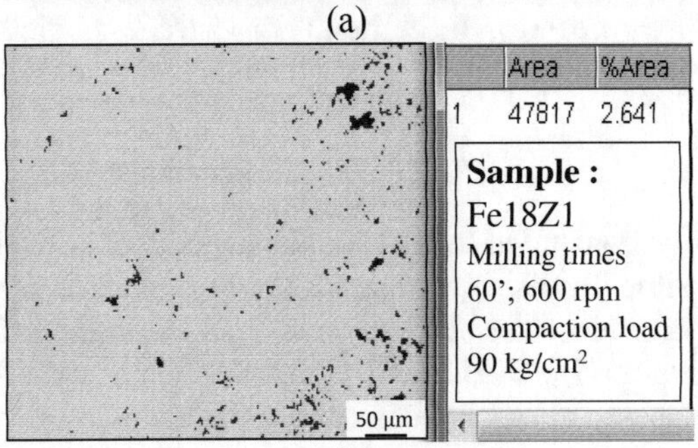

(b)

Figure 4. Photo images showing porosity in samples for: a) F18, b) F18Z1 and c) F18Z2 along with result of the calculation the fraction of porosity in the samples.

The ZrO_2 nanoparticles must be dispersed throughout grains of Fe_3Al in the sample to form a ceramic metal matrix composite. It is reasonable to expect the hardness of ZrO_2 added Fe-Al sample increases.

The higher of the ZrO_2 content in the sample, the dimensions and the weight changes is down. Based on hardness data, the hardness value increases along with the increase of zirconia content in the sample. Supported by microstructural, and porosity calculation data as seen in figure. 5.

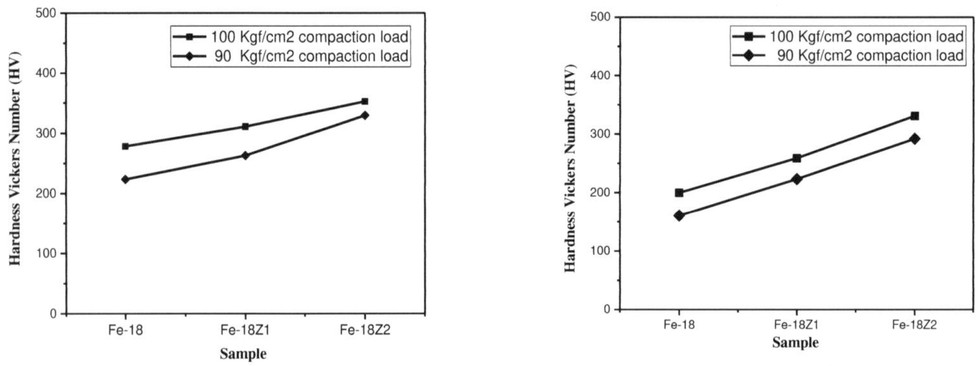

(a) (b)

Figure 5. Plot of hardness value of various samples prepared under a) 60 minutes and b) 90 minutes milling time. Compaction load of 90 kg_f/cm2 and 100 kg_f/cm2 were applied.

Hardness value of Fe-Al based samples consolidated by different compaction loads and the influence of milling time to the hardness value were summarized in figure 5. The hardness value of Fe-18Al prior to sintering coded F18BS is also inserted for comparison purpose. It is clear that the hardness of Fe-18Al based samples is improved by the addition of ZrO_2 particles and sintering. Obviously, the improvement is due to densification and the diminishing of pores in the samples. The hardness curve value shows that the F18BS has the lowest hardness value. The low hardness because grains of main phase are still un-joined. Hence, the sample has a relatively low density compared with that of sintered sample. Plots of hardness value in Figure 5 indicate that milling time also affect the hardness in which samples prepared under 60 minutes milling time have higher hardness values than those of 90 minutes.

Although, the longer milling time reduces the size of welded particles but in contrary upon sintering this brought effect to the reduction of the hardness. The variation of micro hardness from 37.5 HV to 220 HV was also observed in pure Al and Al-10, 20, 30 vol. % Fe3Al sintered composite system reported by D. Panda et.al [16]. Wento Han et.al [23] also reported the hardness of Fe9Al18Cr was 87 HV for an aging time 5000 hours at 4750C temperature. As seen in figure 5 (a), the highest micro hardness is 352.9 HV coded Fe18Z2 which mean that the addition of nanoparticles ZrO2 increased the hardness of Fe18 alloy compared with Fe_3Al intermetallic phase itself (338 HV).

4. Conclusion

In the light of the results and above discussion, it can be concluded that Fe_3Al phase is the only major phase of Fe-18Al alloys under studied. The other possible phases are suspected very minor present in the samples. The addition of ZrO_2 to the Fe-18Al alloys reduced the porosity from 7.757 % in the ZrO_2 free Fe-18Al sample to the lowest 0.53 % in the 2 wt.% ZrO_2 added Fe-18Al sample. Along with the diminishing of porosity, the hardness value of Fe-18Al based alloys increased from the lowest value of 160.38 HV to the high hardness value of 352.59 HV.

Acknowledgement
The authors gratefully acknowledge the support of Postgraduate Program of Materials Science, Universitas Indonesia (UI) for the research facilities. We are also thankful for the financial support provided by the Directorate of Research and Community Service - UI under program Grants of International Publication Indexed for Final Project Student 2018 (PITTA 2018 contract No. 2238/UN2.R3.1/HKP.05.00/2018).

References

[1] E. Basuki 2016 Paduan Logam untuk Aplikasi temparatur tinggi dan penghematan Energi Bandung ITB,
[2] F. Abe 2001 *Material Science and Enginering* Vol. **A, 319-321**, pp. 770-773,
[3] H. Harada 2003 *High Temperature Materials for Gas Turbine* International Gas Turbine Congress, Tokyo,.
[4] C. R. Mechi RF 1983 Historical Development Physical Metalurgy North Holland
[5] D. J. Fischer JJ 1991 *Heat Treat,* vol. **23**, p. 15–6,.
[6] R. Mitra 2018 Structural intermetallics and intermetallic-matrix composites: An introduction, India: Elsevier Ltd. All , , pp. 1-18
[7] R. G. Frommeyer 2008 *SteelRes.Int.*, vol. **79**, pp. 507-513,.
[8] P. R. M. I. Baker 1997 *International Materials Review*, vol. **42**, pp. 181-205,.
[9] U. Prakash 2018, "Intermetallic matrix composites 2," India,.
[10] N. Wright U 1998 *Materials Science and Engineering* A, vol. **258**, pp. 285-290,
[11] C. K. J. D. C 1996 *Intemetallics*, vol. **4**, pp. 339-355
[12] Benjamin 1976 *Mechanical alloying* Sci Am.
[13] Benjamin 1992 *Novel powder processing*, vol. **7**, p. 155–68,.
[14] N. A.-A. C. Suryanarayana 2012 *Progress in Materials Science*, vol. **58**, p. 383–502,
[15] S. C 1998 *ASM handbook*, vol. **7** p. 80–90.,.
[16] L. K. S. N. A. Deepankar Panda 2015 4th International Conference on Materials Processing and Characterization, India,.
[17] M. P., P. K. V. S. D. A. Hotarˇ2012, *Corrosion Science* , vol. **63**, pp. 71-81,.
[18] H. GAJ 1987 *Metals Mater*, vol. **3**, p. 457–62.
[19] A. H. P. K. M. P. J. Mlnaˇr´ık 2015 *Corrosion Science*,.
[20] G. J. T. Karl Dawson 2017 Journal of Nuclear Materials.
[21] G. Jangg 1989,DGM Informationgesellschaft, p. 39–52,
[22] F.-l. K. R. S. D. H. N. Y. Y. Jia Li a 2009, *Materials Science and Engineering,* vol. **A 526**, pp. 50-55.
[23] H. Wentou, Yabuuchi K. 2016 *Nuclear Materials and Energy*, p. 1-6

The Ion Transport Characteristics of Chitosan Composite Membranes 'Ch-AgNP' with Emphasis on the Amount of Filler

Ni Nyoman Rupiasih*, W. G. Suharta, Made Sumadiyasa, I Ketut Putra and Ni Made Ayu Mega Trisnayani Putri

Department of Physics, Faculty of Mathematics and Natural Sciences, Udayana University, Kampus Bukit Jimbaran, Bali 80362, Indonesia

*Corresponding author: rupiasih@unud.ac.id

Abstract. In this study, we report about the characteristics of ion transport of chitosan composite membranes "chitosan-silver nanoparticle". The chitosan composite membranes were prepared by a casting method using chitosan as a matrix, silver nanoparticle (AgNP) as filler, and acetic acid 1% as a solvent. The various amount of filler added e.g. 10, 100, 250, 500, 750, and 1000 µg. These membranes called chitosan composite membrane Ch-AgNP10, Ch-AgNP100, Ch-AgNP250, Ch-AgNP500, Ch-AgNP750, and Ch-AgNP1000. A membrane without AgNP has also been prepared named chitosan membrane (Ch membrane) as a comparison. The ion transport experiments conducted by using a cell model consisting of two chambers, chamber 1 and 2, and the area of the membrane was 9.616 cm^2. The driving force used is a gradient concentration (dC/dx). The electrolyte solutions used were KCl and $CaCl_2$ with concentrations of 0.1, 1, 10, 100, 250, 500, 750, and 1000 mM. The voltage difference (V) is measured using Ag/AgCl electrodes which connected to a voltmeter. All measurements were done at room temperature, 28.1°C. The result showed that the obtained J_{diff} -V curves of chitosan composite membranes are similar to J_{diff} -V curves of Ch membrane, where the diffusion current densities of chitosan composite membranes increased with increasing the concentration gradient of solution. It also depends on the electrolyte solutions. The diffusion current density of chitosan composite membranes had affected by the amount of AgNP added that is increased as increasing the mass of AgNP added.

Keywords: chitosan membrane, chitosan composite membranes, silver nanoparticles, electrolyte solution, ion transport, diffusion current density

1. Introduction

Membrane technology is widely used in the field of industry, biology, and health so that more research on membrane phenomenon both biological membrane and artificial membrane. Membrane characterization is one of the many research subjects to learn more about membrane phenomena.

Chitosan is a natural polymer composed of glucosamine and N-acetyl glucosamine units linked by 1-4 glucosidic bonds. It is obtained by alkaline deacetylation of chitin which extracted from the exoskeleton of crustaceans such as shrimps and crabs, as well from the cell walls of some fungi [1].

Content from this work may be used under the terms of the Creative Commons Attribution 3.0 licence. Any further distribution of this work must maintain attribution to the author(s) and the title of the work, journal citation and DOI.
Published under licence by IOP Publishing Ltd

Chitosan can be processed into several products including fine powders, membranes, sponges, fibers, and gels [2].

Some applications of chitosan membranes involve for active transport of chloride ions in aqueous solution [3], as a chelating agent of heavy metal ions [4], as a carrier and/or a selective barrier controlling the transport rate of the substances involved. Chitosan membrane has also been used to recover silver from X-ray film processing wastes [5], and as an anion-exchange membrane directly [6]. Also it has been reported, chitosan-polytetrafluoroethylene composite membrane applied for separation of methanol/toluene mixture by pervaporation [7], and the polyvinylidene fluoride (PVDF)/chitosan (CS) membranes is a potential solution on the low flux and fouling of common PVDF microfiltration membranes [8]. These characteristics make chitosan has considerable application in various industrial areas such as pharmaceutical and biomedical engineering, food industry, agriculture and wastewater treatment [9-12].

Many studies with various methods used by researchers, in particular, to improve the application of technology for industrial purposes. In various fields, membrane is an option in filtering, separating, and purifying chemically sensitive substances and environment.

In this research the focus is physical membrane characterization through ion transport properties which can be seen from the diffusion current density value of the membrane. By measuring this parameter, it will be able to explain the mechanism of external ion transport through the membrane especially on chitosan composite membranes so that through this research is expected to obtain important information in the progress of membrane technology.

2. Experimental methods

2.1. Materials

The chitosan composite membranes were prepared by a casting method using chitosan as a matrix, silver nanoparticle (AgNP) as filler, and acetic acid 1% as a solvent. The chitosan powder used has a degree of deacetylation (DD) of 87.4%, average molecular weight of 900,000, and the solubility in acetic acid 1% is 99.4%. The sodium hydroxide and acetic acid were analytical grades and were used directly without any purification. Demineralized water was used in preparing solutions. Silver nanoparticles used has characteristics of particles size about 10-30 nm, *Surface Plasmon Resonance* (SPR) wavelength of 423 nm and crystal structure of *face center cubic* (FCC) with lattice parameter, a, of 4.03 Å [13]. The various amount of AgNP added e.g. 10, 100, 250, 500, 750, and 1000 μg. These membranes called chitosan composite membrane Ch-AgNP10, Ch-AgNP100, Ch-AgNP250, Ch-AgNP500, Ch-AgNP750, and Ch-AgNP1000. A membrane without AgNP has also been prepared named Ch membrane as a comparison.

2.2. The ion transport measurement

The ion transport measurements were done using a cell membrane model which composed of two compartments, compartment 1 and 2, and the area of the membrane was 9.616 cm^2. The driving force used is a gradient concentration of the solution (dC/dx) between the two chambers. The electrolyte solutions used were KCl and $CaCl_2$ with concentrations of 0.1, 1, 10, 100, 250, 500, 750 and 1000 mM. The potential difference through the membrane was measured using Ag/AgCl electrodes which connected to a voltmeter. All measurements were conducted at room temperature of 28.1 °C.

3. Results and discussion

3.1. The characteristics of the voltage difference of chitosan composite membranes (Ch-AgNP)

The average voltage differences (further it calls as the voltage differences (V)) of the membranes measured are shown in Tables 1 and 2, for KCl and $CaCl_2$ solutions respectively. In accordance with the Nernst equation (1), if the concentrations in the two chambers are equal then the voltage difference equals to zero. But experimentally at the same concentrations, there is a voltage difference measured.

This can be described as the voltage difference generated by the membrane itself, in this case the chitosan membrane is a positive membrane, so it formed polarization of charges on both sides of the membrane surfaces e.g. Cl^- ion at the both sides membrane surfaces [14].

$$V = 2.303\frac{k_B T}{ze}\log\left(\frac{C_1}{C_2}\right) \quad (1)$$

Where k_B is Boltzmann constant (1.38×10^{-23} JK^{-1}), T is absolute temperature, e is electron charges (1.602×10^{-19} C), z is ion valence, C_1 and C_2 are concentration of the solution in chamber 1 and 2.

Table 1. The voltage difference data of Ch membrane and chitosan composite membranes (Ch-AgNP) measured in KCl solution.

Concentration of solution (mM)		**V ± sd (x 10^{-3} V)**						
Chamber 1	Chamber 2	Ch-membrane	Ch-AgNP10-membrane	Ch-AgNP100-membrane	Ch-AgNP250-membrane	Ch-AgNP500-membrane	Ch-AgNP750-membrane	Ch-AgNP1000-membrane
0.1	0.1	0.50 ± 0.10	0.87 ± 0.06	1.10 ± 0.10	1.30 ± 0.10	1.43 ± 0.06	1.53 ± 0.06	1.70 ± 0.10
1	0.1	0.80 ± 0.04	1.30 ± 0.10	1.43 ± 0.06	1.63 ± 0.06	1.73 ± 0.06	1.87 ± 0.06	2.13 ± 0.12
10	0.1	1.20 ± 0.06	1.67 ± 0.06	1.77 ± 0.06	1.97 ± 0.06	2.07 ± 0.06	2.17 ± 0.06	2.47 ± 0.06
100	0.1	1.33 ± 0.06	1.93 ± 0.06	2.10 ± 0.10	2.30 ± 0.10	2.33 ± 0.06	2.50 ± 0.10	2.80 ± 0.10
250	0.1	1.67 ± 0.06	2.27 ± 0.06	2.47 ± 0.06	2.67 ± 0.06	2.70 ± 0.10	2.83 ± 0.06	3.13 ± 0.06
500	0.1	1.93 ± 0.06	2.53 ± 0.06	2.73 ± 0.06	2.93 ± 0.06	3.03 ± 0.06	3.20 ± 0.10	3.50 ± 0.06
750	0.1	2.23 ± 0.06	2.83 ± 0.06	3.10 ± 0.10	3.30 ± 0.10	3.40 ± 0.10	3.60 ± 0.10	3.90 ± 0.06
1000	0.1	2.53 ± 0.06	3.10 ± 0.10	3.50 ± 0.06	3.70 ± 0.10	3.80 ± 0.10	4.03 ± 0.06	4.30 ± 0.06

Table 2. The voltage difference data of Ch membrane and chitosan composite membranes (Ch-AgNP) measured in $CaCl_2$ solution.

Concentration of solution (mM)		**V ± sd (x 10^{-3} V)**						
Chamber 1	Chamber 2	Ch-membrane	Ch-AgNP10-membrane	Ch-AgNP100-membrane	Ch-AgNP250-membrane	Ch-AgNP500-membrane	Ch-AgNP750-membrane	Ch-AgNP1000-membrane
0.1	0.1	0.60 ± 0.10	0.93 ± 0.06	1.17 ± 0.06	1.33 ± 0.06	1.57 ± 0.06	1.77 ± 0.06	1.90 ± 0.10
1	0.1	0.97 ± 0.06	1.33 ± 0.06	1.53 ± 0.06	1.63 ± 0.06	1.87 ± 0.06	2.07 ± 0.06	2.37 ± 0.06
10	0.1	1.23 ± 0.06	1.77 ± 0.06	1.97 ± 0.06	2.07 ± 0.06	2.17 ± 0.06	2.33 ± 0.06	2.63 ± 0.06
100	0.1	1.57 ± 0.06	2.10 ± 0.10	2.33 ± 0.06	2.43 ± 0.06	2.57 ± 0.06	2.67 ± 0.06	3.00 ± 0.10
250	0.1	1.87 ± 0.06	2.37 ± 0.06	2.77 ± 0.06	2.87 ± 0.06	2.93 ± 0.06	3.10 ± 0.10	3.57 ± 0.06
500	0.1	2.27 ± 0.06	2.63 ± 0.06	3.13 ± 0.06	3.23 ± 0.06	3.33 ± 0.06	3.47 ± 0.06	4.00 ± 010
750	0.1	2.53 ± 0.06	2.97 ± 0.06	3.47 ± 0.06	3.57 ± 0.06	3.63 ± 0.06	3.37 ± 0.06	4.47 ± 0.06
1000	0.1	2.87 ± 0.06	3.23 ± 0.06	3.83 ± 0.06	3.97 ± 0.06	4.03 ± 0.06	4.27 ± 0.06	4.80 ± 0.10

The graphs of the voltage differences (V) vs. $\log(C_1/C_2)$ are presented in Figure 1 and 2. Figures 1 and 2 show graph of the voltage differences vs. $\log(C_1/C_2)$. It shows that the voltage difference increased as increasing the concentration ratio (C_1/C_2). At low concentration ratios, 0.1: 0.1,1: 0.1, 10: 0.1, and 100: 0.1, the voltage differences increase linearly, but at high concentration ratios, 250: 0.1, 500: 0.1, 750: 0.1, and 1000: 0.1, the increase in voltage difference is exponential relation. These results are in accordance with Nernst equation (1) that in aqueous solution (solution with low concentration), the membrane voltage difference is directly proportional to logarithmically with the ion concentration ratio ($\log(C_1/C_2)$). This is observed both on chitosan membrane (Ch) and chitosan composite membranes and is observed in both types of electrolyte solutions (KCl and $CaCl_2$). Figures 1 and 2 also show that the addition of AgNP with various masses affect the voltage difference characteristics of the chitosan composite membranes, that is, by increasing the mass of AgNP added, the greater the membrane voltage difference obtained.

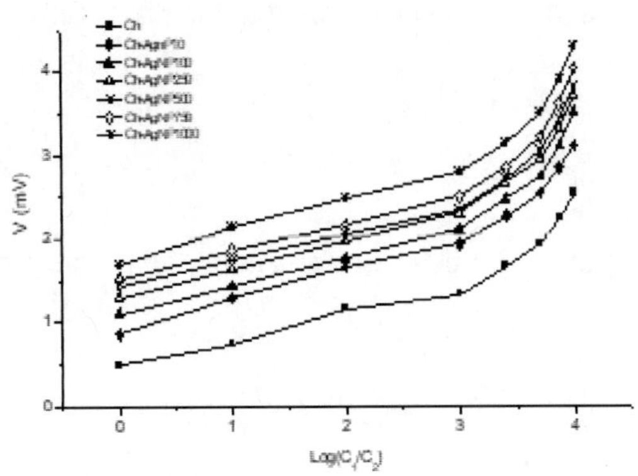

Figure 1. Graph of voltage difference (V) vs. $\log(C_1/C_2)$ of chitosan membrane and chitosan composite membranes (Ch-AgNP) in KCl electrolyte solution.

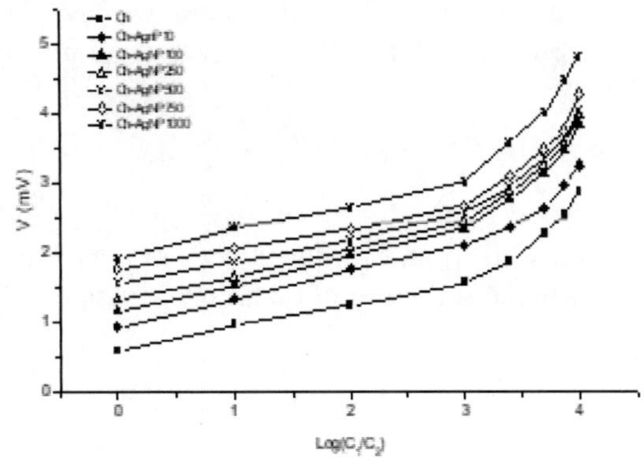

Figure 2. Graph of voltage difference (V) vs. $\log(C_1/C_2)$ of chitosan membrane and chitosan composite membranes (Ch-AgNP) in $CaCl_2$ electrolyte solution.

At low concentration regions, by doing linear regression obtained gradients of each curve as shown in Table 3. Table 3 shows that the gradient of the Ch membrane is smaller than that of the chitosan composite membranes. While the variation of AgNP added from 10 - 1000 μg affect the gradients around 1%-7% in KCl solution and around 1.2%-7.6% in $CaCl_2$ solution randomly. These results show that the solutions affect the ion transport process (gradient) of the membranes.

Table 3. The gradient of each curve of 'membrane voltage difference vs. $\log(C_1/C_2)$' at concentration ratios (C_1/C_2) is 0.1: 0.1, 1: 0.1, 10: 0.1, and 100: 0.1.

Membranes	Gradient	
	in KCl solution	in $CaCl_2$ solution
Ch	0.293	0.317
Ch-AgNP10	0.357	0.393
Ch-AgNP100	0.333	0.393
Ch-AgNP250	0.333	0.373
Ch-AgNP500	0.303	0.330
Ch-AgNP750	0.32	0.329
Ch-AgNP1000	0.363	0.357

3.2. The characteristic of the diffusion current density of chitosan composite membranes (Ch-AgNP)
The diffusion current density (J_{diff}) of each membrane can be calculated using equation (2) and the results are presented in Figure 3 and 4.

$$J_{diff} = V \frac{x^2 . z . e}{k_B . T . t . \left(\log\left(\frac{C_1}{C_2} \right) \right)^2} \frac{\Delta C}{\Delta x} \tag{2}$$

Where x is the distance of the voltage difference measured and t is time.

Figures 4 and 5 show that the diffusion current densities increase linearly as increased concentration gradient on two chambers. This condition is observed both, in the Ch membrane and in chitosan composite membranes. The similar study of Ch membrane also has been reported in the ref. [15]. This is in accordance with first Fick's Law (equation (3)) [14].

$$J_{diff} = -D\frac{dC}{dx} \tag{3}$$

Where D is diffusion coefficient and (dC/dx) is concentration gradient. So, by taking the gradient of each curve, the properties of diffusion current densities of both membranes can be explained as shown in Table 4.

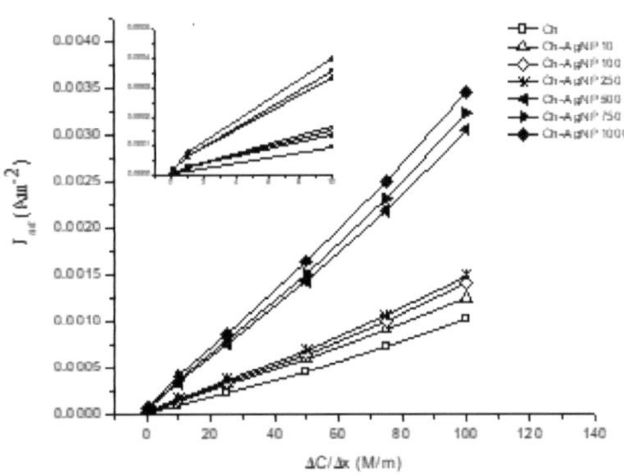

Figure 3. Graph of the diffusion current density (J_{diff}) with concentration gradient ($\Delta C/\Delta x$) of Ch membrane and chitosan composite membranes (Ch-AgNP) in KCl solution.

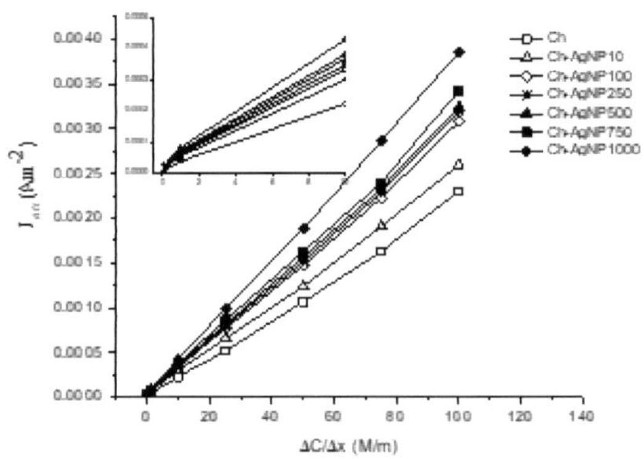

Figure 4. Graph of the diffusion current density (J_{diff}) with concentration gradient ($\Delta C/\Delta x$) of Ch membrane and chitosan composite membranes (Ch-AgNP) in $CaCl_2$ solution.

Table 4 shows the gradients of each curve (which according to Fick I law is called diffusion coefficient (D)) of Ch membrane is smaller than the chitosan composite membranes which occurs on both solutions. This means that the number of ions diffusing per unit area on Ch membrane is smaller than that of the chitosan composite membranes. It also appears that the gradient increases with increasing AgNP content. From the gradients ratio (the last column on Table 4), it showed that the diffusion current density in aqueous $CaCl_2$ solution (low concentration solution) on the Ch membrane and the chitosan composite membranes with AgNP of 10 to 250 μg were more than 2-fold compared with KCl solution. While on the composite membranes with AgNP of 500 - 1000 μg, the diffusion current density on $CaCl_2$ solution slightly larger about 5% - 13% compared with KCl solution. These results indicate that the addition of AgNP has affected the characteristics of the diffusion current density of the chitosan composite membranes. It is also observed that the solutions greatly affect the ion transport characteristics, both on Ch membrane and chitosan composite membranes. The similar study also has been reported in the ref. [16, 17].

Table 4. Gradient of the J_{diff} - ($\Delta C/\Delta x$) graph of each membrane and the gradient ratio.

Membranes	Gradient ($A.m^{-1}M^{-1}$)		gradient in $CaCl_2$: gradient in KCl
	KCl solution	$CaCl_2$ solution	
Ch	0.99×10^{-05}	2.24×10^{-05}	2.26
Ch-AgNP10	1.22×10^{-05}	2.54×10^{-05}	2.08
Ch-AgNP100	1.36×10^{-05}	3.01×10^{-05}	2.21
Ch-AgNP250	1.45×10^{-05}	3.11×10^{-05}	2.14
Ch-AgNP500	2.97×10^{-05}	3.16×10^{-05}	1.06
Ch-AgNP750	3.15×10^{-05}	3.31×10^{-05}	1.05
Ch-AgNP1000	3.37×10^{-05}	3.80×10^{-05}	1.13

From the voltage differences (V) data in Tables 1 and 2, and the diffusion current densities (J_{diff}) on Figure 3 and 4, graph made to represent the relation between J_{diff} and V as shown in Figure 5 and 6. Figures 5 and 6 showed that the diffusion current density increases exponentially with an increase in the potential differences. These are in accordance with the potential difference which is a function of log(C1/C2), Figure 1 and 2 so that the diffusion current densities are also very dependent on a log(C1/C2). At initial, the diffusion current density increases very slowly (image insert), then at a certain potential difference the diffusion current density increases very sharply. The similar result also has reported in the previous study in the ref. [15]. Where for the composite membranes, the curve begins to rise sharply it shifts toward a larger potential difference and a larger the diffusion current density the curve begins to parallel to each other. It shows that the diffusion current density in $CaCl_2$ solution shifts to the right side, toward a larger potential difference when compared to the diffusion current density in KCl solution. These results indicated that the AgNP addition and the use of different types of solutions have an effect on the Jdiff -V characteristics on both membranes, Ch membrane, and chitosan composite membranes.

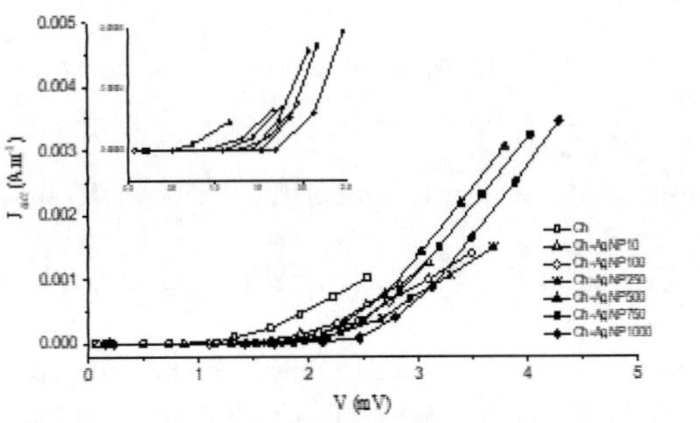

Figure 5. Graph of the diffusion current density (J_{diff}) with concentration gradient ($\Delta C/\Delta x$) of Ch membrane and chitosan composite membranes (Ch-AgNP) in KCl solution.

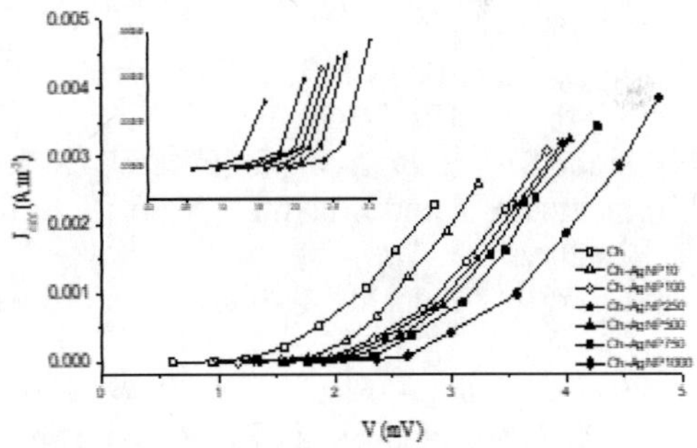

Figure 6. Graph of the diffusion current density (J_{diff}) with concentration gradient ($\Delta C/\Delta x$) of Ch membrane and chitosan composite membranes (Ch-AgNP) in $CaCl_2$ solution.

4. Conclusion

This study described the characteristics of ion transport of chitosan composite membranes "chitosan-silver nanoparticle (Ch-AgNP)". The J_{diff} -V curves were determined in KCl and $CaCl_2$ solutions. The obtained J_{diff} -V curves of chitosan composite membranes are similar to J_{diff} -V curves of Ch membrane, where the diffusion current densities of chitosan composite membranes increased with increasing the concentration gradient of the solution. It also depends on the electrolyte solutions. The diffusion current density of chitosan composite membranes had affected by the amount of AgNP added that is increased as increasing the mass of AgNP content.

Acknowledgement

This work is financially supported by the Ministry of Research, Technology and Higher Education Republic of Indonesia, under the Competence Research Grant of Udayana University 2018 are highly acknowledged.

References

[1] No H K and Meyers S P 1997 *Preparation of chitin and chitosan* in *Chitin Handbook* Muzzarelli R A A and Peter M G (Eds.) (European Chitin Society, Grottammare, Italy) p. 475-489
[2] Mohamed E I Badawy and Entsar I Rabea 2011 *International Journal of Carbohydrate Chemistry* 460381 p. 29
[3] Ying Wan, Katherine A. M. Creber, Brant Peppley and V. Tam Bui 2003 Synthesis, Characterization and Ionic Conductive Properties of Phosphorylated Chitosan Membranes *Macromol. Chem. Phys.* **204** p. 850-858
[4] Barbara Krajewska 2001 Diffusion of metal ions through gel chitosan membranes *Reactive & Functional Polymers* **47** p. 37-47
[5] N Nyoman Rupiasih, Rendra Rustam Purnomo and Made Sumadiyasa 2016 Preparation and Application of Chitosan Membranes to Filter Silver from X-ray Film Processing Wastes *J. Phys.: Conf. Ser.* **710** 012009
[6] Zeng Xianfang and Ruckenstein Eli 1998 Cross-linked macroporous chitosan anion-exchange membranes for protein separations *Journal of Membrane Science* **148** (2) p. 195-205
[7] Siddhartha Moulik, Bukke Vani, Sajja S Chandrasekhar and S Sridhar 2018 Chitosan-polytetrafluoroethylene composite membranes for separation of methanol and toluene by pervaporation *Carbohydrate Polymers* **193** p. 28-38
[8] Cynthia Noemi Baeza Elizalde, Samer Al-Gharabli, Joanna Kujawa, Musthafa Mavukkandy, Shadi W Hasan, Hassan A Arafat 2018 Fabrication of blend polyvinylidene fluoride/chitosan membranes for enhanced flux and fouling resistance *Separation and Purification Technology* **190** p. 68-76
[9] Rabea E I, Badawy M E T, Stevens C V, Smagghe G and Steurbaut W 2003 *Biomacromolecules* **4** (6) p. 1457-1465
[10] Hirano S 1996 *Biotechnology Annual Review* **2** p. 237-258
[11] Shigemasa Y and Minami S 1996 Biotechnology and Genetic Engineering Reviews **13** p. 413-420
[12] Shahidi F, Arachchi J K V and Jeon Y J 1999 *Trends in Food Science and Technology* **10** (2) p. 37-51
[13] *Wendri Nyoman, Rupiasih Ni Nyoman and Sumadiyasa Made* 2017 Biosintesis Nanopartikel Perak Menggunakan Ekstrak Daun Sambiloto: Optimasi Proses Dan Karakterisasi *Indonesian Journal Material Science* **18** (4) p. 162-166
[14] Bertil Hille 2001 Ion Channels of Excitable Membranes 3rd ed. Sinauer Associates Inc. U.S.A
[15] Rupiasih N Nyoman, Yayuk Eka Puspita and Sumadiyasa Made 2015 Transport phenomena in chitosan synthetic membranes with emphasis on the effect of variations in the ratio of matrix/solvent *J. Phys.: Conf. Ser.* **622** 012004
[16] Jae-Hwan Choi, Hong-Joo Lee and Seung-Hyeon Moon 2001 Effects of Electrolytes on the Transport Phenomena in a Cation-Exchange Membrane *Journal of Colloid and Interface Science* **238** p. 188-195
[17] Rupiasih N N 2016 Effects of electrolytes on ion transport in Chitosan membranes *J. Phys.: Conf. Ser.* **776** 012045.

Characteristics of Corncob-Originated Activated Carbon Using Two Different Chemical Agent

Nur Aini Rahma[1], Asih Kurniasari[1], Yoyok Dwi Setyo Pambudi[1], His Muhammad Bintang[1], Anne Zulfia[2], Chairul Hudaya[1*]

[1] Department of Electrical Engineering, Faculty of Engineering, Universitas Indonesia, Depok 16424, Indonesia
[2] Department of Metallurgical and Materials Engineering, Faculty of Engineering, Universitas Indonesia, Depok 16424, Indonesia
[3] Energy System Engineering, Faculty of Engineering, Universitas Indonesia, Depok 16424, Indonesia

[*]Corresponding author: c.hudaya@eng.ui.ac.id

Abstract. The research and development of biomass-based activated carbon (AC) has attracted much attention from researchers due to the abundant resource of biomass, including corncob waste. The urgency to find alternative and innovative applications for simple, inexpensive carbon material can be obtained by synthesizing the corncob waste which is abundant renewable resource and suitable for carbon properties. The use of chemical agent during activation process is of important to produce the desired AC, including high surface area and excellent electrical conductivity. Among the various chemical agents, KOH and $ZnCl_2$ have been widely applied for synthesizing AC. This study aims to find out the characteristics of corncob-originated activated carbon (CAC) using these two chemical agents. Step by step of activating carbon from corncob will be determined briefly. Corncob was dried and chopped. Then it was carbonized. After that, the carbon result was soaked in each chemical agent solution, KOH and $ZnCl_2$, in different molarity for carbon chemical activation. For physical activation, impregnated carbon was carbonized again in high temperature under inert gas atmosphere until AC was obtained. We employed scanning electron microscopy (SEM), X-Ray diffraction (XRD), and Raman Spectroscopy measurement to characterize the CAC samples. The results showed that the application of KOH and $ZnCl_2$ at a different optimized process parameters exhibited the different results of surface morphology, structures, and crystallyte size. The crystallite size of the activated carbon using different chemical activating agents with varied concentrations is diverse enough. The XRD data revealed the average crystallite size of carbon with KOH as the activator is ~45 nm in three different conditions. However, in the case of $ZnCl_2$ as the activating agent, it shows the average size of ~65 nm. This number is significantly higher than the activated carbon impregnated with KOH . Visual observation of SEM images gives an impression on the carbon pore where CACK12 posess the highest pores among those analytes. The synthesized corncob activated carbon can be used in many functional application such as energy storage materials, agriculture, and adsorbents in industrial and environmental sectors.

Keywords: activated carbon, corncob, chemical Agent, KOH, $ZnCl_2$.

Content from this work may be used under the terms of the Creative Commons Attribution 3.0 licence. Any further distribution of this work must maintain attribution to the author(s) and the title of the work, journal citation and DOI.
Published under licence by IOP Publishing Ltd

1. Introduction

Due to abundant benefit, development of activated carbons (AC) which have large surface area and high microporosity is becoming a hot topic for many researcher at present. CO2 adsorbents [1, 2], electrode materials for energy storage [3-7]and catalyst [8] are some examples of AC's benefits. Many commercial AC are produced from non-eco-friendly materials with difficult and complicated process, also high cost production. The research and development of simple, low cost [9], and biomass based activated carbon [4, 6, 10] get many interest from researcher. Several studies have reported biomass based carbon such as coconut shell [11], sunflower seed [12], palm kernel shell [1] , fish waste [13], etc. One of potential biomass that can be turned into active carbon is corncob. Based on data from the Central Statistics Agency (Badan Pusat Statistik-BPS), corncob waste in Indonesia reached 7.5 million tons in 2014. During this time, corncob in Indonesia is only used as fodder. Through this study expected to increase the added value of corncob.

In this study, we demonstrated a simply way to produce a biomass based activated carbon that derived from corncob. The previous studies, researcher use only the innermost part of corncob [14], but in this experiment we used the whole part of corncob. Various strategies have been developed to fabricate AC which has high surface area and great conductivity, one of which is chemical or physical activation methods.

First physical activation aims to remove vaporous substance that can worsen char performance. In chemical activation, the char is mixed with a chemical activating agent such as KOH[15] , NaOH [16], HCl [10], $ZnCl_2$ [16-18] to eliminate the tar content and expand the pore size. Selection of chemical agent important because some organic matter can only make a bonding and reaction in certain chemicals to form excellent activated carbon.

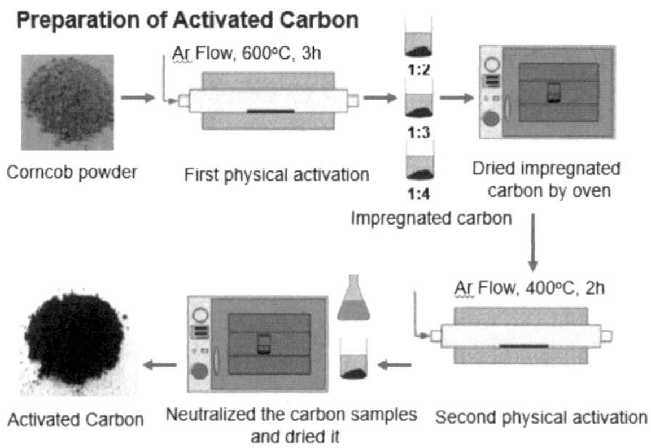

Figure 1. Schematic Diagram of Activated Carbon Preparation

After completion of chemical immersion, it is followed by carbonization under an inert gas at high temperature ranging 300-700°C [19] then washed to get neutralized charcoal. Carbonization temperature, carbonization time, gas flow rate, and impregnation ratio are crucial parameters in the AC surface area and pore formation [10, 19, 20]. Among chemical activation method, $ZnCl_2$ and KOH are widely used for activating carbon because $ZnCl_2$ and KOH can develop the pore size distribution, volume, and surface area with efficient and flexible method [2, 3]. This work aimed at characterizing AC with chemical activator KOH and $ZnCl_2$. Effect of various C: Solutions mass ratio also investigated to find the optimum activations.

2. Experimental methods

2.1. Active carbon preparation

Preparation of active carbon including three processes; pre-carbonization process, physical activation, and chemical activation. The corncob were sundried then grinded into powder. Then, the powder were carbonized in a horizontal tube furnace under argon flow atmosphere of 200 cc / min, the temperature was increased until 600°C for 1h and held in 600°C for 2h. After turning to room temperature, carbonized sample was mixed in chemical solution by magnetic stirrer for chemical activation process. Chemical solution that used in this experiment was $ZnCl_2$ and KOH solution with different mass ratio for variations. The char: activating agent mass ratio chosen in this study were 1: 2, 1: 3, and 1: 4. The bigger mass of chemical agent makes the molarity of chemical solution. The mixture was dried in oven for 12h then furnaced again under Ar flow of 300cc / min with increasing temperature 400°C for 1h and held in 400°C for 1h. The corncob originated activated carbon (CAC) obtained from $ZnCl_2$ chemical agents were washed with aquadest and CAC obtained from KOH solutions were washed with HCl and aquadest to neutralized the CAC. The washed CAC then dried in oven for 12h until final CAC were obtained. The CAC produced from $ZnCl_2$ activation denoted as CACZ12 (1:2), CACZ13 (1:3), and CACZ14 (1:4). The CAC produced from KOH activation denoted as CACK12 (1:2), CACK13 (1:3), and CACK14 (1:4).

2.2. *Material Characterization*

*2.2.1. Scanning Electron Microscopy (SEM).*The morphologies and microstructures of the samples were determined using scanning electron microscopy (NOVA-SEM, FEI Company). The operating voltage was 10 kV, and the magnifications of the images were 5000 and 10.000.

2.2.2. X-Ray Diffraction (XRD). X-ray diffraction analysis is used to exhibit the structures and composition of the samples. The crystallite size can be measured by XRD. X-ray diffraction (XRD) patterns were obtained on a Rigaku using Cu Kα radiation (with measurement range 10-80 degree, λ=1.540598 Å). Scan were recorded with scanning rate 0.04° s^{-1}.

2.2.3. Raman Spectroscopy. Raman spectra were recorded by micro Raman Spectrometer from Renishaw Company at an excitation wavelength of 532 nm.

3. Results and discussion

3.1. *SEM Analysis*

Error! Reference source not found. shows the SEM photographs of activated carbon from corncob that applying KOH and $ZnCl_2$ activation with different mass ratio. The pore width of CACK12 is around 5.71 μm, CACK13 is 7.43 μm, and CACK14 is 8 μm. While the pore size of carbon impregnated by $ZnCl_2$ as follows, CACZ12 is around 11.71 μm, CACZ13 is 10.85 μm, and CACZ14 is 6.57 μm. This result shows that all of the samples has the macroporous morphological structures as the pore size is larger than 50 nm. CACK12 has the clearest surface morphology like a coral with distributed pores, indicating the homogeneous cover of chemical agents. While CACZ14 displays like solid ball with small pores, indicating that the processing method faced some problems. The hole and granular of carbon displayed in the picture were the result of activation both physical and chemical activation as reported in other research [15]. It can be inferred that various pore size of carbon are depending on the impregnation ratio. It was observed from Fig.2 that the higher KOH molarity employed, the greater pore size of carbon resulted. Contrast with KOH activation, SEM photograph of $ZnCl_2$ shows that the smaller $ZnCl_2$ molarity, the greater CAC pore size obtained.

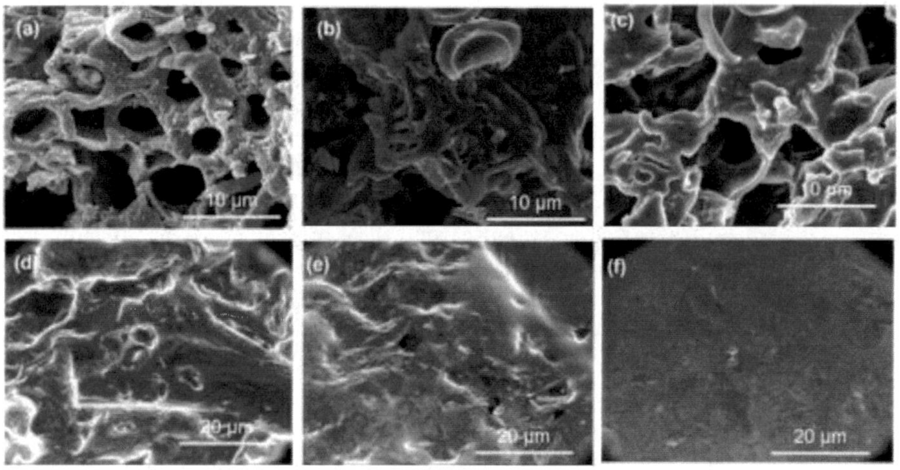

Figure 2. SEM Images of (a) CACK12, (b) CACK13, (c) CACK14, (d) CACZ12, (e) CACZ13, and (f) CACZ14

3.2. XRD Analysis

Giving a better understanding about the carbon structure, XRD test has been conducted to analyze more deeply about the structural materials using software Match! version 3.0. The plotting XRD data is provided in the Fig. 2. It is clearly indicated that broad peaks are centered at $2\theta = \sim 28^{\circ}$ and $2\theta = \sim 40^{\circ}$ that belongs to biomass-based carbon peaks. Based on our calculations, the structure of obtained carbon turns out to be amorphous that is reflected by the miller indices of (002) and (101) respectively [12], [14]. This is supported by our next peak observation at $\sim 28^{\circ}$ is due to (002) lattice plane of typical graphite structure of carbon-black. The XRD graph also shows that CACKs exhibited the uniform pattern changes. The peaks represent a graphitic crystal structure of the samples. Nevertheless, some other peaks were still present, associated with the chemical agents that might be there due to incomplete removal during washing processes to neutralize the yielded activated carbon.

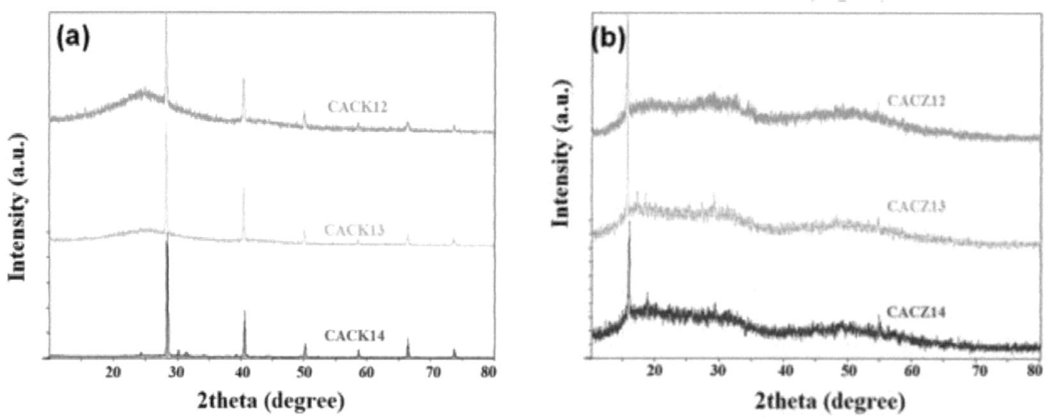

Figure 3. SEM Images of (a) CACK12, (b) CACK13, (c) CACK14, (d) CACZ12, (e) CACZ13, and (f) CACZ14

To further calculate the crystallite size we use the Scherrer's equation as following:

$$B = \frac{0.9 \text{ x } \lambda}{d \text{ x } cos\theta} \quad (1)$$

where λ: wavelength (1.54060 Å of Cu Kα1), θ: $2\theta/2$; d: Full Width at Half Maximum (FWHM) intensity of the peak (rad) ; Angle's unit conversion (degree to rad): Rad = π. degree/180.

The crystallite size of CACK12 is 51.21 nm, CACK13 is 41 nm, and CACK14 is 40.96 nm. While the crystallite size of carbon impregnated by $ZnCl_2$ as follows, CACZ12 is around 82 nm, CACZ13 is 66.8 nm, and CACZ14 is 42.65 nm. The absence of C-atom in Match! analysis determine that the carbon still make bonding with other compound.

3.3. Raman Spectra Analysis

Using OriginPro Software, the Raman spectra are deconvoluted into two peaks as shown in Error! Reference source not found.. From the graph we can obtain the value of D-Band and G-Band. D-Band is corresponding to an irregular mode or defects on $\boldsymbol{A_{1_g}}$ Symmetry. Defects are required to disseminate elastically for the momentum of conservation is fulfilled, observed on 1330 and 1360 cm^{-1}. While G-band is a tangential shear mode of the carbon atom corresponding to the stretching mode in the graphite plane,show the ideal mode of graphite on $\boldsymbol{E_{2_g}}$ Symmetry, which is observed on 1580 cm^{-1}. Table 1 show the measurement of G-band and D-band. The dominant peak of D-band than G-Band indicates the presence of amorphous carbon in the sample.

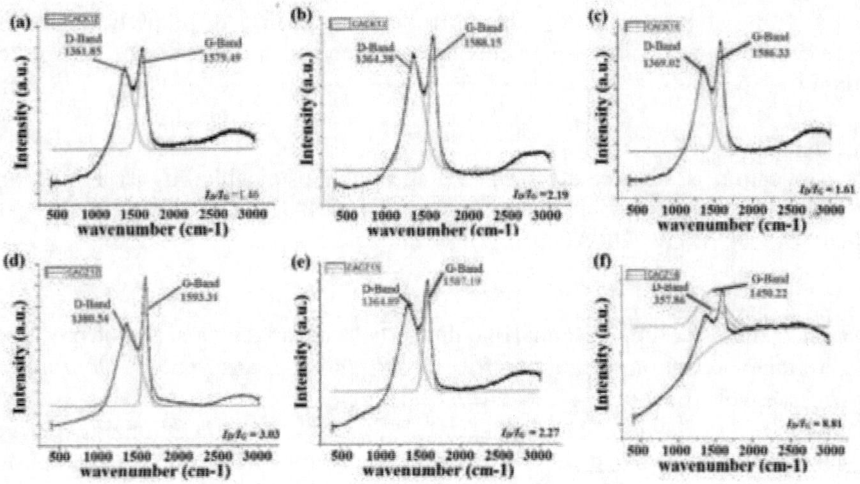

Figure 4. Raman Spectra of (a) CACK12, (b) CACK13, (c) CACK14, (d) CACZ12, (e) CACZ13,and (f) CACZ14

Quality of the sample is evaluated using intensity band ratio I_D/I_G. For high quality samples without defects and without amorphous carbon ratio is usually less than 2%. The samples intensity band ratio has been calculated as showed Table 1. Intensity band ratio is also used for assessing the graphitic in-plane microcrystalline size of carbon ($\boldsymbol{L_a}$), which can be formulated [3] :

$$\frac{I_D}{I_G} = \frac{C_\lambda}{L_a} \quad (2)$$

Where $\boldsymbol{C_\lambda}$ is 4.4 nm.

Comparing$\boldsymbol{L_a}$, values to the XRD crystal size calculation (1 Å = 0.1 nm), a little disparity has appeared, which CACK14 has the biggest distinction (2.01 nm). The CACZ12 precursor has the most similar crystal size, with the distinction is only 0.01 nm. This discrepancy can occur because the inaccuracy possibility of XRD graph analysis and $\boldsymbol{L_a}$calculation.

Table 1. Measurement of G-band and D-band

Samples	D-Band	G-Band	I_D/I_G	L_a (nm)
CACK12	1361.85	1579.49	1.46	3.01
CACK13	1364.38	1588.15	2.19	2.01
CACK14	1369.02	1586.33	1.61	2.73
CACZ12	1380.4	1595.31	3.03	1.45
CACZ13	1364.89	1587.19	2.27	1.98
CACZ14	1357.86	1450.22	8.81	0.49

4. Conclusion

The study of characterization of corncob-originated activated carbon (CAC) using two different chemical agent has been carried out. Among two chemical agents, KOH provides the best performance in terms of surface morphology, meanwhile the $ZnCl_2$ has the best impact in terms of structure and crystallite size. It can be concluded that $ZnCl_2$ is more suitable to be an activator agent for corncob biomass than KOH. Variation in mass ratio also generate diversity in CAC properties. It becomes a great challenge to find optimum chemical agent mixture and method to obtain excellent performance of activated carbon.

Acknowledgement

This article's publication is supported by USAID through Sustainable Higher Education Research Alliances (SHERA) Project for Universitas Indonesia's SMART CITY Center for Collaborative Research (No. 0144/UN2.R3.SC/HKP.05.01/ 2018).

References

[1] N. A. Rashidi and S. Yusup, "Potential of palm kernel shell as activated carbon precursors through single stage activation technique for carbon dioxide adsorption," *Journal of Cleaner Production,* vol. 168, pp. 474-486, 2017/12/01/ 2017.

[2] K. Chomiak, S. Gryglewicz, K. Kierzek, and J. Machnikowski, "Optimizing the properties of granular walnut-shell based KOH activated carbons for carbon dioxide adsorption," *Journal of CO2 Utilization,* vol. 21, pp. 436-443, 2017/10/01/ 2017.

[3] Z. J. Zhang, C. Dong, X. Y. Ding, and Y. K. Xia, "A generalized ZnCl2 activation method to produce nitrogen-containing nanoporous carbon materials for supercapacitor applications," *Journal of Alloys and Compounds,* vol. 636, pp. 275-281, 2015/07/05/ 2015.

[4] X. Xu, Z. Meng, X. Zhu, S. Zhang, and W.-Q. Han, "Biomass carbon composited FeS2 as cathode materials for high-rate rechargeable lithium-ion battery," *Journal of Power Sources,* vol. 380, pp. 12-17, 2018/03/15/ 2018.

[5] K. Ö. Köse, B. Pişkin, and M. K. Aydınol, "Chemical and structural optimization of ZnCl2 activated carbons via high temperature CO2 treatment for EDLC applications," *International Journal of Hydrogen Energy,* 2018/04/24/ 2018.

[6] F. Sun *et al.*, "Converting biomass waste into microporous carbon with simultaneously high surface area and carbon purity as advanced electrochemical energy storage materials," *Applied Surface Science,* vol. 436, pp. 486-494, 2018/04/01/ 2018.

[7] Q. Abbas, M. Mirzaeian, A. A. Ogwu, M. Mazur, and D. Gibson, "Effect of physical activation/surface functional groups on wettability and electrochemical performance of carbon/activated carbon aerogels based electrode materials for electrochemical capacitors," *International Journal of Hydrogen Energy,* 2018/05/30/ 2018.

[8] T. Tsoncheva *et al.*, "Activated carbon from Bulgarian peach stones as a support of catalysts for methanol decomposition," *Biomass and Bioenergy,* vol. 109, pp. 135-146, 2018/02/01/ 2018.

[9] G. G. Stavropoulos and A. A. Zabaniotou, "Minimizing activated carbons production cost," *Fuel Processing Technology,* vol. 90, no. 7, pp. 952-957, 2009/07/01/ 2009.

[10] M. Danish and T. Ahmad, "A review on utilization of wood biomass as a sustainable precursor for activated carbon production and application," *Renewable and Sustainable Energy Reviews,* vol. 87, pp. 1-21, 2018/05/01/ 2018.

[11] K. Sun *et al.*, "Microporous activated carbons from coconut shells produced by self-activation using the pyrolysis gases produced from them, that have an excellent electric double layer performance," *New Carbon Materials,* vol. 32, no. 5, pp. 451-459, 2017/10/01/ 2017.

[12] U. Moralı, H. Demiral, and S. Şensöz, "Optimization of activated carbon production from sunflower seed extracted meal: Taguchi design of experiment approach and analysis of variance," *Journal of Cleaner Production,* vol. 189, pp. 602-611, 2018/07/10/ 2018.

[13] A. B. Fadhil, A. I. Ahmed, and H. A. Salih, "Production of liquid fuels and activated carbons from fish waste," *Fuel,* vol. 187, pp. 435-445, 2017/01/01/ 2017.

[14] J. Guo, J. Zhang, F. Jiang, S. Zhao, Q. Su, and G. Du, "Microporous carbon nanosheets derived from corncobs for lithium–sulfur batteries," *Electrochimica Acta,* vol. 176, pp. 853-860, 2015/09/10/ 2015.

[15] S. Li, K. Han, J. Li, M. Li, and C. Lu, "Preparation and characterization of super activated carbon produced from gulfweed by KOH activation," *Microporous and Mesoporous Materials,* vol. 243, pp. 291-300, 2017/05/01/ 2017.

[16] M. Kılıç, E. Apaydın-Varol, and A. E. Pütün, "Preparation and surface characterization of activated carbons from Euphorbia rigida by chemical activation with ZnCl2, K2CO3, NaOH and H3PO4," *Applied Surface Science,* vol. 261, pp. 247-254, 2012/11/15/ 2012.

[17] R. G. Pereira *et al.*, "Preparation of activated carbons from cocoa shells and siriguela seeds using H3PO4 and ZnCL2 as activating agents for BSA and α-lactalbumin adsorption," *Fuel Processing Technology,* vol. 126, pp. 476-486, 2014/10/01/ 2014.

[18] S. Yorgun, N. Vural, and H. Demiral, "Preparation of high-surface area activated carbons from Paulownia wood by ZnCl2 activation," *Microporous and Mesoporous Materials,* vol. 122, no. 1, pp. 189-194, 2009/06/01/ 2009.

[19] O. Üner and Y. Bayrak, "The effect of carbonization temperature, carbonization time and impregnation ratio on the properties of activated carbon produced from Arundo donax," *Microporous and Mesoporous Materials,* vol. 268, pp. 225-234, 2018/09/15/ 2018.

[20] Ş. Karadirek and H. Okkay, "Statistical modeling of activated carbon production from spent mushroom compost," *Journal of Industrial and Engineering Chemistry,* vol. 63, pp. 340-347, 2018/07/25/ 2018.

IOP Conf. Series: Materials Science and Engineering **622** (2019) 012031 doi:10.1088/1757-899X/622/1/012031

The effect of temperature and reaction time on the shape of Ag nanowires synthesized using the polyol method

Fairuz Septiningrum[1], Liszulfah Roza[2], Vivi Fauzia[1,*]

[1] Department of Physics, Universitas Indonesia, Depok 16424, Indonesia
[2] Physics Education Program, Muhammadiyah Prof. Dr. Hamka University, East Jakarta 12130, Indonesia

[*]Corresponding author: vivi@sci.ui.ac.id

Abstract. Silver nanowires (Ag NWs) have attracted more attention in the last decade due to their potential applications in the next generation of flexible and wearable electronic devices. Several simple approaches were used to produce highly crystalline Ag NWs, but the synthesis parameter should still be precisely optimized to obtain high aspect ratio Ag NWs. In this study, we observed the morphological changes of Ag NWs during the synthesis to obtain the precise temperature and reaction time. $AgNO_3$ is used as a source of metal ions, polyvinylpyrrolidone (PVP) as a capping agent or stabilizing agent, NaCl as a control agent, and ethylene glycol (EG) as a reducing agent and solvent. Synthesis was performed with temperatures of 150°C and 170°C, and it was observed in reaction times of 10, 15, 40, 80, and 120 min. The results show that Ag nanoparticles were formed in the first 10 min of the reaction then continue to form Ag NWs up to 80 min. After that, the diameter of Ag NWs starts to expand from 7.1 nm to 8.8 nm. The highly crystalline Ag NWs with an aspect ratio of 191.7 ± 9.8 were obtained in the synthesis performed at temperature of 170°C in 120 min.

Keywords: Ag NWs, polyol, aspect ratio, nucleation and growth.

1. Introduction

Silver nanowires (Ag NWs) play an important role as an interconnection component in nanoscale electronic devices due to having the highest electrical conductivity (6.3×10^7 S/m) among all metals, as well as high transparency and flexibility [1–4]. These properties are beneficial for the next generation of flexible and wearable electronic devices, such as transistors, touch screens, light-emitting diodes, solar cells [1,2], transparent heaters [3], and wearable antennas [4].

There are many methods used in the synthesis of Ag NWs, such as electrochemical, hydrothermal, ultrasonic reduction, and template methods [5,6]. Template methods are divided into hard-template and soft-template methods. The most widely used soft-template method is the polyol method, which is a chemical approach in which metal ions are dissolved in a polyol solution and heated to a certain temperature. This method has some advantages in terms of cost, yield, and simplicity, making it suitable for mass production [6,7].

Content from this work may be used under the terms of the Creative Commons Attribution 3.0 licence. Any further distribution of this work must maintain attribution to the author(s) and the title of the work, journal citation and DOI.
Published under licence by IOP Publishing Ltd

To obtain Ag NWs with a high aspect ratio (the ratio of the length to the diameter of NWs), many researchers have observed the effect of synthesis parameters, such as temperature and reaction time, injection rate, stirring rate, concentration of the capping agent polyvinylpyrrolidone (PVP), the metal precursor $AgNO_3$, and the control agent NaCl [7–12]. Many previous studies have shown that the temperature plays a key role in increasing the reducing power of ethylene glycol to glycolaldehyde [6]. Therefore, the temperature of reaction becomes a important factor influencing the morphology of Ag NWs. In the early investigations, most of AgNW synthesis with a polyol methods were conducted at 160 °C for 1 h. Until now, the temperature and the reaction time were varied in a wide range from 100°C to 185°C with a reaction time up to 12 h [6]. The differences findings obtained with regard to the influence of temperature and reaction times may suggest that the temperature and the reaction times are not the independent parameter influencing the shape and size of Ag NWs and that other factors may also work [6]. In this study, we observed the morphological changes of Ag NWs during the synthesis to obtain the precise temperature and reaction time that can produce the Ag NWs with a high aspect ratio.

2. Materials and methods

Ag nitrate ($AgNO_3$) and PVP (Mw = 55.000) were purchased from Sigma-Aldrich, sodium chloride (NaCl) was purchased from Emsure, and all materials were used without further purification. The $AgNO_3$ is used as a source of metal ions, PVP as a capping agent or stabilizing agent, NaCl as a control agent, and ethylene glycol (EG) as a reducing agent and solvent [6,10,11,13]. In a typical experiment, 0.66 g of PVP and 2.2 mg of NaCl were dissolved in 35 ml of EG to form solution A by stirring at 500 rpm at T = 120°C and then cooling to 60°C. Meanwhile, 0.2 g of $AgNO_3$ was dissolved in 15 ml of EG to form solution B. After, solution A was heated slowly to a certain temperature and, simultaneously, solution B was injected into solution A using a burette. Then, the reaction was kept constant until 120 min. In this study, synthesis was performed with temperatures of 150°C and 170°C, and it was observed at reaction times of 10, 15, 40, 80, and 120 min. For the purification process, the Ag NW solution was then centrifuged at 4,000 rpm for 20 min in ethanol and repeated three times.

The morphology of Ag NWs was characterized by using field emission scanning electron microscopy (FESEM) with the Hitachi S-4800 instrument and transmission electron microscopy (TEM) with the FEI Tecnai G20 F20 instrument. The microstructural characterization was performed with x-ray diffraction (XRD) using the RIGAKU SMARTLAB 3 kV instrument and the optical characterization was performed using the Thermo Fisher Scientific Genesys 10S Ultraviolet Visible instrument.

3. Results and discussion

3.1. *Effect of Temperature*

Figure 1 shows the morphology of Ag NWs grown at temperatures of 150°C and 170°C with a reaction time of 40 min. It is clear that at a temperature of 150°C, there are Ag nanoparticles besides Ag NWs. It can be interpreted that NWs did not grow perfectly, resulting in the low yield. In addition, the aspect ratio of Ag NWs was calculated and presented in Table 1. The aspect ratio of Ag NWs grown at 170°C is slightly higher (141.0 ± 8.7) than that of the samples grown at 150°C (136.0 ± 11.8). This may be because the thermal energy at 150°C is not enough to reduce the Ag^+ ions to Ag atoms, so the number of Ag atoms is lesser in the self-assembly of Ag NWs. When the temperature is higher, the molecules and ions are more energetic and the more Ag atoms are available to form longer Ag NWs [7,14].

Table 1. The sizes of Ag NWs in various temperatures

Temperature (°C)	Diameter (nm)	Length (μm)	Aspect ratio
150	46.3 ± 1.8	6.3 ± 0.3	136.0 ± 11.8
170	47.5 ± 0.8	6.7 ± 0.3	141.0 ± 8.7

The morphology of Ag NWs is also confirmed by TEM images, as shown in Figure 2. Both samples are still capped by PVP, which may be due to an inadequate centrifugation process for the purification of Ag NWs. The presence of PVP is also evidence that PVP was absorbed on the Ag surface with the Ag-O bond to assist in the anisotropic growth of Ag NWs [15]. TEM images (Figure 2a and 2c) show that the lattice distances of Ag NWs grown at 150°C and 170°C are 0.24 nm and 0.23 nm, respectively. These spacings represent the lattice distances of the {111} plane, which is in agreement with the calculation from the XRD spectrum, as shown in Figure 3. The Fast Fourier Transform (FFT) pattern in the inset of Figure 2a shows the bright dots identifying that Ag NWs are single crystal materials.

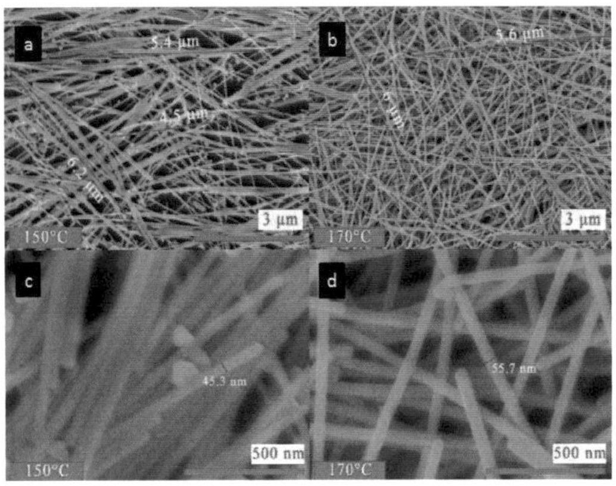

Figure 1. FESEM images of Ag NWs synthesized at temperatures of (a, c) 150°C and (b, d) 170°C.

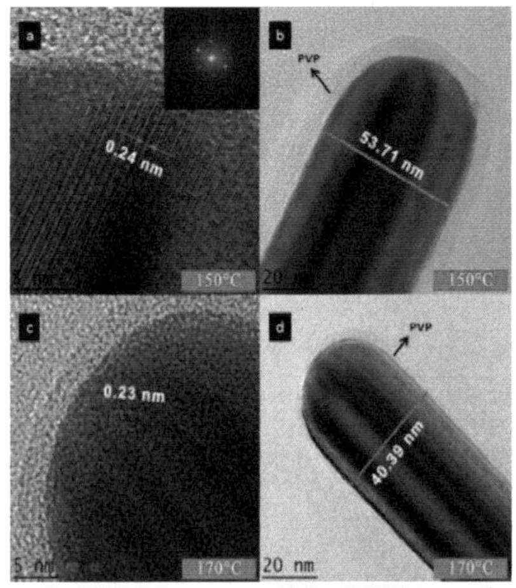

Figure 2. TEM images of Ag NWs synthesized at temperatures of (a, b) 150°C and (c, d) 170°C. Inset (a) FFT pattern of Ag NWs

The XRD pattern of Ag NWs is shown in Figure 3. Based on the Ag standard JCPDS 96-110-0137, it is known that Ag NWs are a single phase with a face center cubic (FCC) structure. Figure 3 shows that the {111} plane has the highest intensity, indicating that Ag atoms tend to grow dominantly in the {111} plane and are grown anisotropically to form Ag NWs. The preferred orientation of the {111} plane was then used for the calculation of lattice parameters and *d*-spacing. It was found that both samples have similar lattice parameters ($a = b = c = 4.09$ Å) and *d*-spacing (= 2.36–2.42 Å).

The UV-VIS absorption spectra of Ag NWs grown at T = 150°C and 170°C are shown in Figure 4. Both samples show the absorption peaks at wavelengths of 345 nm, 360 nm, and 430 nm. The first two peaks are coming from the transverse mode of the localized surface plasmon resonance (LSPR) of Ag NWs due to the vibration of surface plasmons in the shortest distance [9,15,16]. Meanwhile, the longitudinal mode cannot be detected in the UV-Visible range due to the very high aspect ratio of Ag NWs. The peak around 430 nm represents the existence of Ag nanoparticles [8,16]. The sample grown at T = 150°C has a higher intensity at 430 nm that is consistent with the FESEM image in Figure 1a, showing that many Ag nanoparticles were formed.

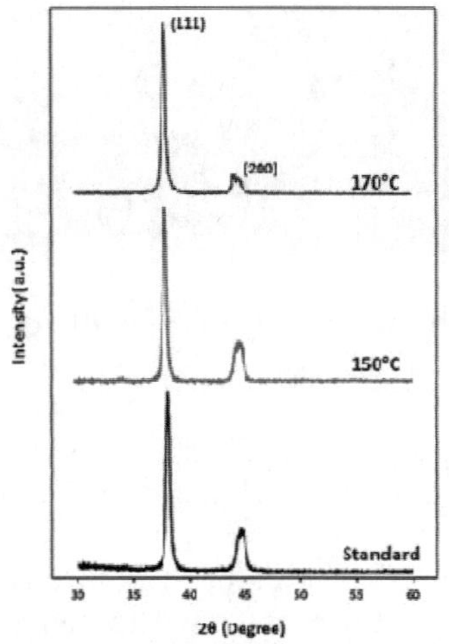

Figure 3. XRD pattern of Ag NWs Ag NWs grown at 150°C and 170°C (left).

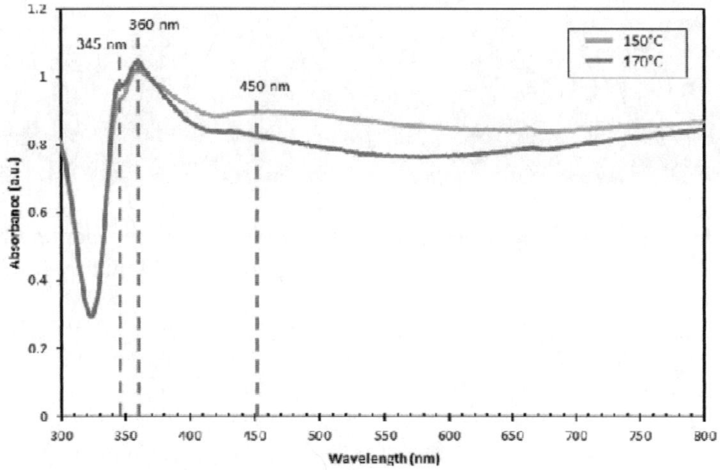

Figure 4. The absorption spectra of Ag NWs grown at T = 150°C and T = 170°C (right).

3.2. Effect of Reaction Time

To obtain the optimum reaction time of Ag NWs, the synthesis process was carried out at T = 170°C with observation times that varied from 10 to 120 min. The solutions at each observation time are shown in Figure 5. During the injection process of Ag nitrate, the solution changes colour from yellow to brownish yellow, red, greenish red, green, and greyish green when the synthesis process was completed. These colour changes indicate changes in the nanostructure's shape during the synthesis process, and the greyish green colour indicates that the Ag NWs were formed [15].

The morphological change of Ag NWs was observed via FESEM images, as shown in Figure 6. In Figure 6a, it shows that at t = 10 min, the nanostructures that formed are nanoparticles. This agrees with the red colour solution, as shown in Figure 5. The change process from nanoparticles to NWs occurs when the reaction is more than 10 min. This can be seen visually, where the colour of the solution turns to green within 15 min. Figures 6b–d show the formation of Ag NWs in 40–80 min. It is clear that the Ag nanoparticles exist beside the Ag NWs. At t = 120 min, Ag NWs are grown longer (with an average length of 8.8 ± 0.3 μm and larger (with an average diameter of 45.9 ± 0.8 nm), while the number of Ag nanoparticles is significantly reduced. Table 2 summarizes the calculation of the average diameter, length, and aspect ratio of Ag NWs with different reaction times. The highest aspect ratio of Ag NWs (191.7 ± 9.8) was reached at a 120-min reaction time. This is still relatively low if it is compared with the previous study using the polyol method (T = 140°C for 90 min) that produced Ag NWs with a diameter of 44 nm and length of 35 μm (the aspect ratio of 495) [14].

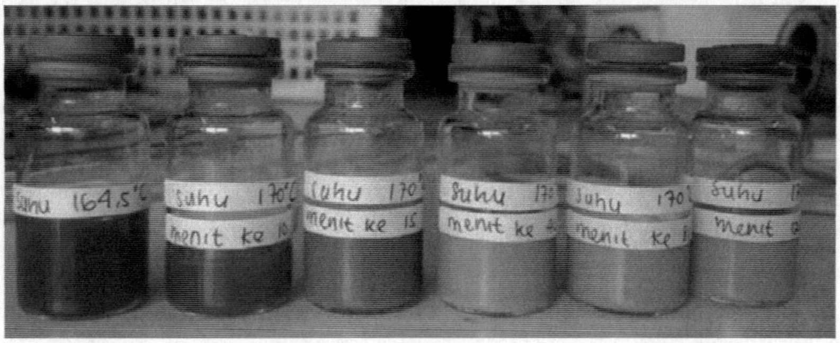

Figure 5. The colour changes of the Ag NW solution during the synthesis process.

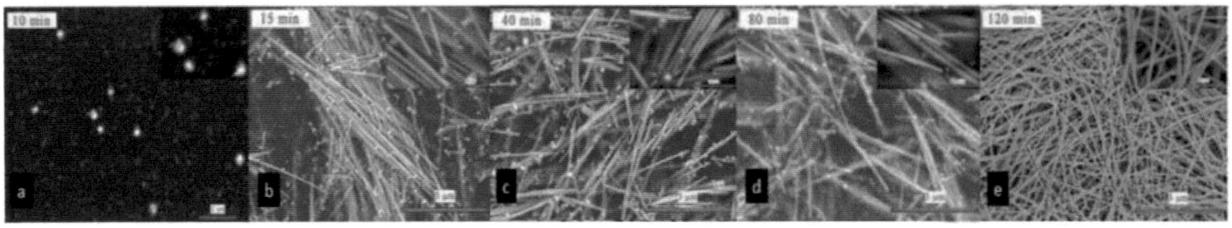

Figure 6. FESEM images of Ag NWs synthesized with reaction times of (a) 10, (b) 15, (c) 40, (d) 80, and (e) 120 min.

The optical analysis of Ag NWs was also observed via UV-VIS spectroscopy. Figure 7 shows the absorption peak of the solution synthesized for 10 min at a wavelength of 460–470 nm, and it is the typical absorption peak of Ag nanoparticles [8,16]. This indicates that the product of synthesis is still Ag nanoparticles. In the process, the sample changes to red. This is consistent with the FESEM image in Figure 6a, which shows the presence of Ag nanoparticles. For reaction times of 10 and 15 min, the absorption peaks are at a wavelength of 450 nm with a low intensity. This is due to the LSPR peaks of Ag nanoparticles. When the reaction time is more than 40 min, there are three peaks at wavelengths of 350 nm, between 380–400 nm, and between 430–450 nm. These results indicate that the Ag NWs are successfully formed, but the Ag nanoparticles are still present. The peak at 350 nm is constant, but the two other peaks tend to shift to a longer wavelength with an increase in reaction time. This may be due to the increase in the diameter of Ag NWs. These results agree with the shapes of the Ag NWs, as demonstrated in Figure 6, which shows that the lengths and diameters of the Ag NWs produced are increased with an increase in the reaction time. Based on these results, the reaction time of 120 min can produce Ag NWs with more homogenous diameters and lengths. This property is beneficial for the optoelectronic device application because the resistance of the Ag NW network could decrease with a higher transparency [17].

Table 2. The sizes of Ag NWs in different reaction times

Reaction Time (min)	Diameter (nm)	Length (μm)	Aspect ratio
15	37.6 ± 0.8	4.3 ± 0.2	114.4 ± 7.7
40	40.2 ± 1.0	7.1 ± 0.4	176.6 ± 14.3
80	43.3 ± 1.3	7.1 ± 0.5	164.0 ± 16.5
120	45.9 ± 0.8	8.8 ± 0.3	191.7 ± 9.8

3.3. Growth Mechanism of Ag NWs

The mechanism of Ag NW formation is explained as follows. There are three steps in the formation of Ag NWs, i.e., the nucleation, seeding, and growth of Ag NWs. At the beginning of the synthesis process, the heated EG acts as a solvent and reductant and it decomposes into glycol aldehyde. A temperature of more than 150°C is required as the thermal energy to decompose the EG [2,17,23]. At this stage, chemical reactions occur as follows [15]:

$$2HOCH_2 - CH_2OH \text{ (EG)} \rightarrow 2CH_3CHO \text{ (glycol aldehyde)} + 2H_2O \quad (1)$$

When the $AgNO_3$ is injected into the glycol aldehyde solution, the Ag^+ ions are reduced to Ag atoms as follows [15]:

$$2CH_3CHO + 2AgNO_3 \rightarrow CH_3CO - COCH_3 + 2Ag + 2HNO_3 \quad (2)$$

The continuous injection of Ag nitrate results in the solution reaching a supersaturated state, where the concentration of the solute is more than the solubility limit of the solvent [6]. In such a stage, the nucleation of Ag atoms occurs to form the crystal nucleus [6]. The halide ions from NaCl also act as the nucleating substrate for Ag^+ ions to form the seeds AgCl for the future growth of Ag NWs. Typically, this process is the heterogeneous nucleation following the chemical reaction [19,20]:

$$NaCl + AgNO_3 \rightarrow AgCl + NaNO_3 \quad (3)$$

Continuous heating of the solution leads to the growth of Ag nanoparticles, while the other small Ag nanoparticles will shrink and dissolve through the Ostwald ripening mechanism [21]. In this stage, the presence of structural defects leads to the formation of a certain morphology, such as multiple-twinned decahedral particles (MTPs), which have the lowest surface energy [15,21]. The MTPs become the seeds of Ag NWs, as shown in Figure 8 [8].

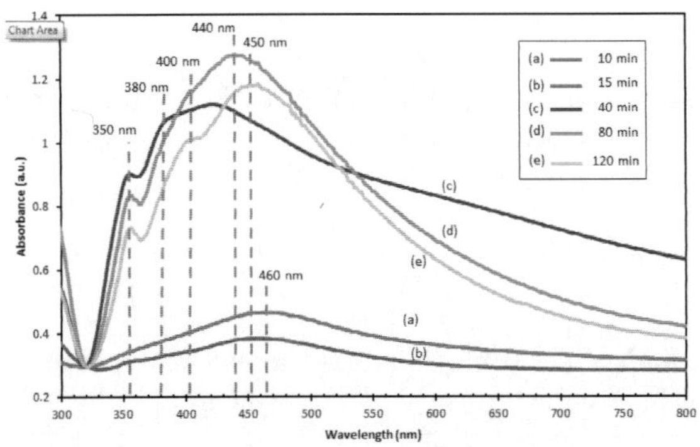

Figure 7. The absorption spectra of Ag NWs in different reaction times

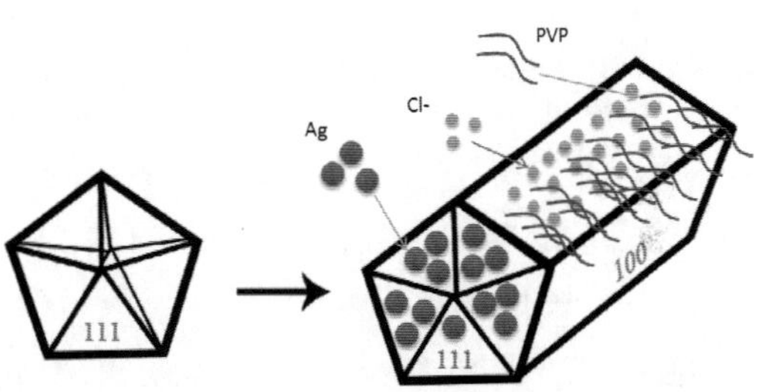

Figure 8. The growth of Ag NWs from MTP seeds with the guidance of PVP and NaCl molecules [8].

Meanwhile, the PVP and Cl-anions adsorb and bind preferentially on the {100} plane of Ag NWs (Figure 8); hence, the Ag atoms continually deposit along the {111} plane and inhibit the lateral {111} plane growth [15,22–24]. Other previous results stated that the atom density of {111} plane of Ag NWs is higher than the atomic density of the {100} plane, so the {100} plane of Ag NWs becomes the active sites for binding the oxygen atoms of the PVP with the Ag atoms [10,11,15]. Coskun et al. pointed out that the addition of NaCl also could inhibit the agglomeration of Ag nanoparticles and decrease the reduction rate of Ag^+ ion to obtain higher thermodynamic stability for higher yields of Ag NWs [7,19].

4. Conclusion

In this study, Ag NWs were successfully synthesized with the polyol method. The growth temperature of 170°C could produce a higher aspect ratio of Ag NWs, as the thermal energy is enough to reduce the Ag^+ ions. Hence, the number of Ag atoms is much higher for the self-assembly of longer Ag NWs. With the assistance of PVP and NaCl, single crystalline Ag NWs grow anisotropically in the {111} plane until the reaction time reaches 120 min. Ag nanoparticles were formed in the first 10 min of the reaction, then continue to form Ag NWs up to 80 min. After that, the diameter of AgNWs starts to expand from 7.1 nm to 8.8 nm. At $t = 120$ min, Ag NWs were obtained with homogenous diameters and lengths but still have a relatively low aspect ratio (191.7 ± 9.8).

Acknowledgement

This research was financially supported by Hibah Publikasi Internasional Terindeks untuk Tugas Akhir Mahasiswa (PITTA) 2018 2329/UN2.R3.1/HKP.05.00/2018 from Universitas Indonesia.

IOP Conf. Series: Materials Science and Engineering **622** (2019) 012031 doi:10.1088/1757-899X/622/1/012031

References

[1] Shinde M A, Mallikarjuna K, Noh J and Kim H 2018 Highly stable silver nanowires based bilayered flexible transparent conductive electrode *Thin Solid Films* **660** 447–54

[2] Yang H, Chen T, Wang H, Bai S and Guo X 2018 One-pot rapid synthesis of high aspect ratio silver nanowires for transparent conductive electrodes *Mater Res Bull* **102** 79–85

[3] Han D, Li Y, Jiang X, Zhao W, Wang F, Lan W, Xie E and Han W 2018 A facile method to prepare transparent and stretchable epidermal thin film heaters *Compos Sci Technol* **168** 460–6

[4] Razaq A, Ali A, Unsa K and Asim S 2018 Next Generation Flexible Antennas for Radio Frequency Applications *Trans Electr Electron Mater* **19** 311–8

[5] Xiang X, Gong W, Kuang M and Wang L 2016 Progress in application and preparation of silver nanowires *Rare Met*

[6] Zhang P, Wyman I, Hu J, Lin S, Zhong Z, Tu Y, Huang Z and Wei Y 2017 Silver nanowires: Synthesis technologies, growth mechanism and multifunctional applications *Mater Sci Eng B* **223** 1–23

[7] Coskun S, Aksoy B and Unalan H E 2011 Polyol Synthesis of Silver Nanowires : An Extensive Parametric Study 4963–9

[8] Sun Y, Yin Y, Mayers B T, Herricks T and Xia Y 2002 Uniform silver nanowires synthesis by reducing AgNO3 with ethylene glycol in the presence of seeds and poly(vinyl pyrrolidone) *Chem Mater* **14** 4736–45

[9] Shobin L R and Manivannan S 2014 One pot rapid synthesis of silver nanowires using NaCl assisted glycerol mediated polyol process *Electron Mater Lett* **10** 1027–31

[10] Lin J Y, Hsueh Y L and Huang J J 2014 The concentration effect of capping agent for synthesis of silver nanowire by using the polyol method *J Solid State Chem* **214** 2–6

[11] Lin J Y, Hsueh Y L, Huang J J and Wu J R 2015 Effect of silver nitrate concentration of silver nanowires synthesized using a polyol method and their application as transparent conductive films *Thin Solid Films* **584** 243–7

[12] Zhang Y, Guo J, Xu D, Sun Y and Yan F 2017 One-Pot Synthesis and Purification of Ultralong Silver Nanowires for Flexible Transparent Conductive Electrodes *ACS Appl Mater Interfaces* acsami.7b07146

[13] Basarir F, Irani F S, Kosemen A, Camic B T, Oytun F, Tunaboylu B, Shin H J, Nam K Y and Choi H 2017 Recent progresses on solution-processed silver nanowire based transparent conducting electrodes for organic solar cells *Mater Today Chem* **3** 60–72

[14] Ding H, Zhang Y, Yang G, Zhang S, Yu L and Zhang P 2016 Large scale preparation of silver nanowires with different diameters by a one-pot method and their application in transparent conducting films *RSC Adv* **6** 8096–102

[15] Mao H, Feng J, Ma X, Wu C and Zhao X 2012 One-dimensional silver nanowires synthesized by self-seeding polyol process *J Nanoparticle Res* **14**

[16] Lai X, Feng X, Zhang M, Hong R, Chen Y, Li A, Deng X and Hu J 2014 Large-scale synthesis

and surface plasmon resonance properties of angled silver/silver homojunction nanowires *J Nanoparticle Res* **16**

[17] Li Y, Ximin Y, Hongwei Y, Yunxiu C, Shuailong G and Chuan W 2019 One-Step Synthesis of Silver Nanowires with Ultra-Long Length and Thin Diameter to Make Flexible Transparent Conductive Films *Materials (Basel)* **12** 401

[18] V. Nghia N, N. K. Truong N, M. Thong N and P. Hung N 2012 Synthesis of Nanowire-Shaped Silver by Polyol Process of Sodium Chloride *Int J Mater Chem* **2** 75–8

[19] Jiu J, Tokuno T, Nogi M and Suganuma K 2012 Synthesis and application of Ag nanowires via a trace salt assisted hydrothermal process *J Nanoparticle Res* **14** 975

[20] Jiu J, Sugahara T, Nogi M and Suganuma K 2013 Ag nanowires: Large-scale synthesis via a trace-salt-assisted solvothermal process and application in transparent electrodes *J Nanoparticle Res* **15** 1–13

[21] Wu Z, Yang S and Wu W 2016 Shape control of inorganic nanoparticles from solution *Nanoscale* **8** 1237–59

[22] Sun Y, Mayers B, Herricks T and Xia Y 2003 Polyol synthesis of uniform silver nanowires: A plausible growth mechanism and the supporting evidence *Nano Lett* **3** 955–60

[23] Hofmeister H 2009 Shape variations and anisotropic growth of multiply twinned nanoparticles *Zeitschrift fur Krist* **224** 528–38

[24] Sun Y and An C 2011 Shaped gold and silver nanoparticles *Front Mater Sci* **5** 1–24

IOP Conf. Series: Materials Science and Engineering **622** (2019) 012032 doi:10.1088/1757-899X/622/1/012032

Optimization of Fabricating Fluorine-doped Tin Oxide by Using Spray Pyrolysis Methods

S Saehana[1*] Darsikin[1], Z Arifin[2], and Nasar[1]

[1] Department of Physics Education, Tadulako University, Jln. Soekarno-Hatta km. 09 Kel. Tondo, Palu, Indonesia
[2] Mechanical Engineering Departement, Sebelas Maret University, Jl. Ir. Sutami 36A Kentingan Surakarta 57126, Solo, Indonesia

*Corresponding author: sahrulsaehana@gmail.com

Abstract. In this study, Fluorine-doped Tin Oxide (FTO) was fabricated and synthesized by using spray pyrolysis methods for solar cells applications through the variation of the concentration and deposition time of precursor solution of $SnCl_2$:F. The glass substrate furthermore was heated at a temperature of 450°C prior to deposition and the sheet resistance was about 56 to 137.25 Ω/cm^2. X-Ray Diffraction, Scanning Electron Microscopy, UV-Vis Spectroscopy, and four-point probe were employed to investigate the characteristic of the FTO. This product was further applied as an electrode in the dye-sensitized solar cell (DSSC) and a photovoltaic effect was observed.

Keywords: FTO, spray pyrolysis, DSSC application.

1. Introduction

Fluorine-doped Tin Oxide (FTO), as a type of Transparent Conducting Oxide (TCO), is widely applied in the photovoltaic area and some of its advantages include stability under atmospheric conditions, chemically inert, mechanically hard, high-temperature resistant, high tolerance to physical abrasion, and less expensive than indium tin oxide [1-3].

Fluorine-doped Tin Oxide (FTO) with high performance (low resistance and high transmittance) is required in many application [4-8]. There is a possibility of producing this compound through the use of different methods such as spray pyrolysis [7-9], sol-gel [10,11], chemical vapor deposition [12] and DC magnetron sputtering [13]. However, some of these methods are complicated and, as a result, the spray pyrolysis method was used due to its requirement of low-cost materials and effectiveness at low temperature [8].

In the previous study [14], several parameters were reported to be affecting the performance of the FTO, and they include the concentration of the solution and the time of deposition. In this paper, FTO was produced by using the spray pyrolysis method and an investigation was conducted on the effect of the concentration and deposition time on its performance. The combination of these parameters in

Content from this work may be used under the terms of the Creative Commons Attribution 3.0 licence. Any further distribution of this work must maintain attribution to the author(s) and the title of the work, journal citation and DOI.
Published under licence by IOP Publishing Ltd

producing high-performance FTO was chosen as optimum because of their usefulness in producing DSSC solar cells.

Furthermore, in this study, Dye-Sensitized Solar Cells (DSSC) was produced through the application of the FTO as an electrode. A comparison was made between this approach and the use of commercial FTO and the performance for both was calculated under solar lighting.

2. Materials and methods

The materials used for the fabrication of the FTO glass include the solution of Tin(II) Chloride ($SnCl_2.2H_2O$, Merck, Germany), Ammonium Fluoride (NH_4F, Merck, Germany), and Ethanol (96%, Merck, Germany). 1 gram each of $SnCl_2$ and NH_4F materials were fed into the ethanol solvent and stirred till it became homogeneous. Furthermore, spray pyrolysis method through the application of OMRON NE-C28 Nebulizer was used to deposit the solution on a glass substrate with a wide scale of about 1cm x 1cm and thickness of 2mm. The concentration of the solution and time deposition of nanoparticle were investigated to obtain the optimum resistance. Furthermore, the sample was heated at 450°C for 30 minutes and a conductive transparent thin layer was formed. The sheet resistance and transmittance were measured by using a four-point probe (PROHEX digital multitester) and UV-vis spectroscopy (8452 UV-Visible spectrophotometer - Agilent). Scanning Electron Microscopy (JEOL JSM-6360LA) was used to investigate the morphology of the FTO film and XRD diffraction (XRD, PW1710) applied to characterize its crystallization. However, in the optimization process, the solution concentration and deposition time were varied to obtain the optimum performance of FTO and after the production, it was used as an electrode in the DSSC solar cell. It is important to point out that the Ruthenium complex was used as a dye, Titanium Dioxide as a photo electrode, Tri-Iodide solution as an electrolyte, and platinum as a catalyst.

3. Results and discussion

FTO was fabricated by spray pyrolysis method as shown in Figure 1 [14,15]. This involves the production of nanoparticles by spraying droplets from the solution onto the heated surface (Fig 1). When a droplet contacts the hot surface, it vaporized and the lagging oxide particles became attached to the surface [9]. However, several parameters were observed to be affecting FTO resistance and they include solution concentration, deposition time, temperature, and distance between the nebulizer and the substrate [14].

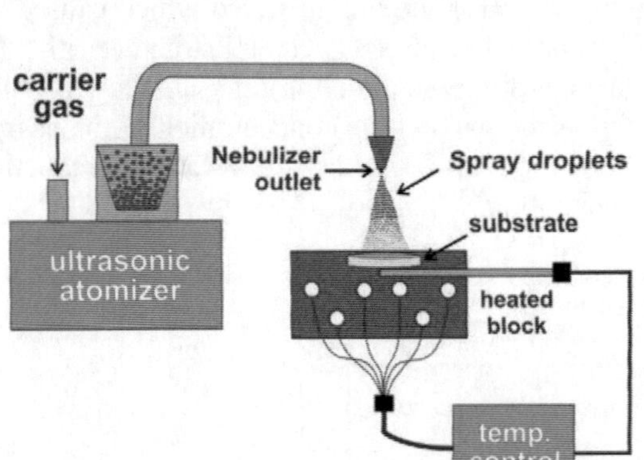

Figure 1. FTO fabrication process using spray pyrolysis method [14].

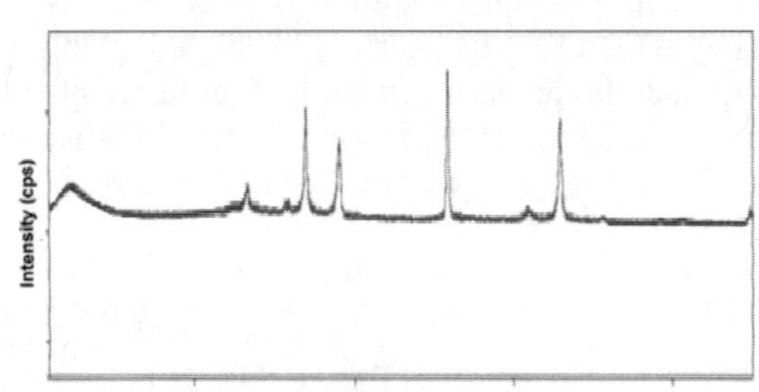

Figure 2. XRD pattern of FTO films after annealing at temperature 400 °C (fabricated in 0.3 M for 10 minutes)

In order to understand the crystallization and morphology of the FTO film, it was fabricated using 0.3 M concentration, deposited for 10 minutes, and later characterized by using X-Ray Diffraction

Spectroscopy (XRD). Moreover, Scanning Electron Microscopy (SEM) was conducted for two different FTOs fabricated from different concentration and deposition time.

Figure 2 shows the XRD pattern in the films after annealing at 450 °C for 30 min. The crystalline structure was found to be tetragonal SnO_2 (JCPDS no. 770451 and space group: F42/mnm(136)). However, the density and size of the FTO nanoparticle grain were found to be increasing with the concentration of the solution as shown in Figure 3.

Figure 3. Scanning Electron Microscopy of FTO fabricated in different conditions: (a) 0.3 M of solution concentration and deposition for 40 minutes and (b) 0.7 M of solution concentration and deposition for 40 minutes.

The transmittance of FTO at visible wavelength was observed by using UV-vis spectroscopy (8452, UV-Visible spectrophotometer - Agilent) as shown in Figure 4. The transmittance of two FTOs with different concentration and deposition were compared and was discovered to be reducing with increased value. This is associated with the increase in particle size due to the high value of these parameters as depicted in Figure 3. Therefore, to fabricate high performing FTO, the combination of solution concentration and deposition time was required.

Figure 4 shows the transmittance values decreasing with increase in concentration and time deposition. This is associated with the increasing amount of fluorine administered which caused an increment in the concentration. This makes several F atoms enter the SnO_2 crystal structure either by filling in the O atomic vacancy or by occupying the position of the Sn atom through substitution. This, consequently, led to the increment in the number of free electrons and the concentration of the charge carrier and a reduction in the amount of light passed on by the glass FTO [20]. Therefore, the reduction in transmittance may be attached to the increasing concentration.

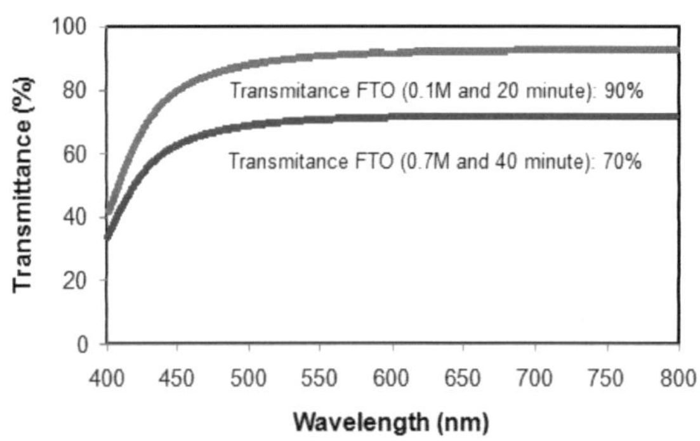

Figure 4. The optical transmittance of FTO fabricated from: (a) 0.1 M of solution concentration for 20 minutes; (b) 0.7 M of solution concentration for 40 minutes.

In this study, to obtain a high performing FTO, optimization was conducted through the testing of a few solution concentration and deposition time. However, the effect of both parameters on sheet resistance and transmittance are shown in Tables 1 and 2. The sheet resistance was determined by employing a 4-point probe method with PROHEX digital multi tester. It was, therefore, predicted that high solution concentration containing a high amount of fluorine and more F^- anions substituted with O^{2-} anions created more free electrons and decreased the resistance value [20].

Table 1. FTO performance in 20 minutes of deposition time.

No	Solution Concentration (M)	Sheet Resistance (Ω/cm^2)	Transmittance (%)
1	0.1	1,000	91.91
2	0.3	630.50	83.20
3	0.5	181.25	90.24
4	0.7	56.00	80.77

Table 1 shows the value of solution concentration was varied and deposition time kept constant at about 20 minutes. For 0.7 M, the sheet resistance and transmittance were found to be about 56 Ω/cm^2 and 89.77, respectively. In Table 2, the deposition time less than 20 minutes produced high resistance while those above produced low resistance with transmittance value below 80%.

From the results obtained, as shown in Table 1 and 2, both parameters (solution concentration and deposition time) affect the resulting sheet resistance of FTO. Through the combination of the two parameters, a value between 56 to 137.25 Ω/cm^2 was obtained with the lowest found at a concentration of 0.7 M and 20-minute deposition time. This was predicted to be related to the loading level of nanoparticle formation on the substrate [16].

Table 2. Performance optimization of FTO in 0.7 M of concentration.

No	Deposition Time (Minute)	Sheet Resistance (Ω/cm^2)	Transmittance (%)
1	5	10.000	79.38
2	7	1.382	85.87
3	10	578	84.21
4	15	129	79.99
5	20	56	80.77
6	25	26	72.17
7	30	29	77.65
8	35	31	78.89
9	40	15	64.50

From the results obtained, as shown in Table 1 and 2, both parameters (solution concentration and deposition time) affect the resulting sheet resistance of FTO. Through the combination of the two parameters, a value between 56 to 137.25 Ω/cm^2 was obtained with the lowest found at a concentration of 0.7 M and 20-minute deposition time. This was predicted to be related to the loading level of nanoparticle formation on the substrate [16].

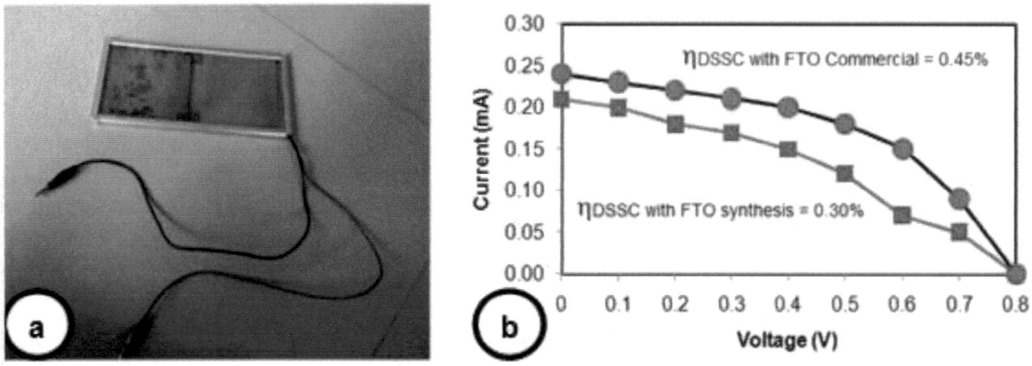

Figure 5. (a) DSSC solar cell with optimized FTO; (b) I-V Characteristics of DSSC applying synthesis FTO (blue square) and its comparison with commercial FTO (red circle).

4. Conclusion

In this study, the conductive thin layer of FTO produced using the spray pyrolysis method has a low sheet resistance between 56 to 137.25 Ω/cm^2 and transmittance over 80%. In the fabrication process, the varied solution concentration and deposition time were tested to get the optimum value of 0.7 M and 20 minutes, respectively. The FTO obtained was used as a DSSC component and the phenomenon of photovoltaics was observed.

Acknowledgement

This work was supported by the Riset Terapan Unggulan Perguruan Tinggi Decentralization Research Grant from the Research, Technology and Higher Education of the Ministry of the Republic of Indonesia for 2018 fiscal year. The researcher also appreciates Dr. Eng. Agus Purwanto, M.Si for supplying materials needed for the research.

References

[1] Chiba, Y., et. al. 2006 *Japan. J. Appl. Phys.***45**(25) L638–L640.

[2] Dai, S., et. al. 2005 *J. Sol. Energy Mat. Sol. Cells***85** 447–455.

[3] Gratzel, M., 2001 *J. Sol-Gel Sci. Tech.***22** pp. 7–13.

[4] Noh E., et. al. 2013 *Sci. World J.***2013** pp. 1-8.

[5] Chuang S-H, Tsung C-S, Chen C-H, Ou S-L, Horng R-H, Lin C-Y, and Wuu D-S 2015 *ACS Appl. Mater. Interfaces* **7** (4) pp 2546–2553.

[6] Chou, T.P., et. al. 2008 *J. Nanophotonics* **2** pp. 023511.

[7] Purwanto A, Widiyandari H, and Jumari A 2012 *Thin Solid Films***520** (6) pp. 2092–2095.

[8] Russo B, and Cao G.Z 2008 *Appl. Phys. A***90**pp. 311–315.

[9] Suyitno, Arifin Z, Santoso A.A, Setyaji A.T, and Ubaidillah 2014 *Appl. Mech. Mat.* **575** 689-695.

[10] Senthilkumar V, Vickraman P, and Ravikumar R 2010 *J. Sol-Gel Sci. Tech.***53**(2) 316–321.

[11] Shi, X.H. and K.J. Xu 2017 *Mat. Sci. Semiconductor Process.* **58** 1–7.

[12] Chavarría-Castillo K.A et. al. 2016 *Mat. Research* **19** pp. 97-102.

[13] Banyamin Z.Y, Kelly PJ, West G, and Boardman J 2014 *Coatings***4** 732-746.

[14] Darsikin, Z Arifin, Nasar and Saehana S 2018 *Mat. Sci. Eng.* **395** 1-5.

[15] Montero, Chialvo, M.R.G and Chialvo, A.C 2009 *J. Mater. Chem.***19** 3276–3280.

[16] Supriyono, Surahmana H, Krisnandia Y.K, and Gunlazuardia 2015 *J. Procedia Environ. Sci.***28** 242–251

[17] Saehana S, Darsikin, Yuliza E, Khairurrijal and Abdulah M 2012 *J. Sol. Energy Eng.***136** 044504-1.

[18] Saehana S, Muslimin, and Abdulah M 2014 *J. Renew. Sustain. Energy.***6** 023109.

[19] Saehana S, Arifin A, Khairurrijal, Abdulah M 2012 *J. Appl. Phys.***111** 123109.

[20] Widhiyanuriyawan D, Maksum I.M, Bintarto R, and Arifin Z 2019 Mat. Sci. Eng. **494** 012066.

IOP Conf. Series: Materials Science and Engineering **622** (2019) 012033 doi:10.1088/1757-899X/622/1/012033

Influence of Inorganic Binder Composition on the Structure-Property Relationship of Monolithic Dye-sensitized Solar Cells with Carbon-based Counter Electrode

N M Nursam[1,*], Shobih[1], E S Rosa[1], J Hidayat[1], and Z Mubarak[2]

[1]Research Center for Electronics and Telecommunications, Indonesian Institute of Science, Kampus LIPI Building 20 Level 4, Cisitu, Bandung – Indonesia 40135

[2]Department of Physics, Hasanuddin University, Jl. Perintis Kemerdekaan KM. 10 Makassar 90245, Indonesia

Email: natalita.maulani.nursam@lipi.go.id

Abstract. In order to fabricate a low-cost dye-sensitized solar cell (DSSC), carbon has become a highly preferred catalytic material as counter-electrode, particularly for DSSC with monolithic structure. This paper presents novel synthesis method of carbon-based composite pastes using two types of carbon material, i.e. carbon nanopowder and activated carbon. The concentrations of the inorganic binder added to the composite pastes were varied to investigate their effect on the physical properties of the counter electrode and the electronic properties of the constructed monolithic DSSC. The inorganic binder used in this work was titanium dioxide nanoparticles, Evonik P25. After optimization, power conversion efficiency of 0.221% was achieved by the monolithic DSSC with counter electrode composite comprising activated carbon and titanium dioxide with weight concentration of 0.5 g and 0.25 g, respectively. Characterizations using gas sorption technique showed that the shape of the hysteresis curves obtained for all composites resembled the isotherm curve Type II and H3, indicating the presence of micropores. Furthermore, higher concentration of titanium dioxide nanoparticles as binder led to counter electrode with lower surface area. The solar cell efficiency, however, was found to be not only correlated to the surface area or the binder composition, but it was also determined by the type of the carbon material.

1. Introduction

Dye-sensitized solar cell (DSSC) has been considered as a promising alternative to replace solid-state silicon based solar cells due to its lower production cost. In addition to its cost, the attractive and colorful appearance, flexibility, transparency, as well as its short energy payback time have also became selling points that boost interest on the further development of DSSC. The performance of DSSC has progressed considerably since its breakthrough in 1991 [1-3]. However, much effort is still required especially for allowing DSSC to enter commercial market and industrialization as a future renewable power source.

One way to promote the use of DSSC could be carried out either by increasing the power conversion efficiency or by further lowering its production cost. At the moment, competing with the efficiency of silicon based solar cell that has already reached more than 20% appears to be rather difficult to achieve in a relatively short time frame. Thus, lowering the manufacturing cost of DSSC seems to be more reasonable solution. DSSC is generally fabricated on top of two glass substrates coated with a layer of

Content from this work may be used under the terms of the Creative Commons Attribution 3.0 licence. Any further distribution of this work must maintain attribution to the author(s) and the title of the work, journal citation and DOI.
Published under licence by IOP Publishing Ltd

transparent conducting oxide (TCO) that are sandwiched together [4, 5]. With regard to the production cost, the TCO glass holds a significant portion from the total material cost for fabricating DSSC, i.e. approximately 40-60% [6]. Not to mention, having two substrates often imposed on alignment issue particularly when it comes to the manufacturing of dye-sensitized solar modules in large sizes. DSSC with monolithic design offers an attractive solution for these challenges. The use of a single substrate in a monolithic structure could substantially reduce the cost for TCO substrates, thereby cutting back the material cost as well as simplifying the fabrication process [7, 8].

Further cost reduction could also be achieved by replacing the counter electrode material. It has been reported that the use of platinum as counter electrode could be replaced with inexpensive carbon based materials with comparable catalytic performances [8-11]. Furthermore, platinum iodide or PtI_4 could also be formed in the triiodide-based electrolyte solution due to corrosion [12]. Not only serves as low-cost replacement for platinum due to its abundant availability, carbon also has several excellent properties, such as high conductivity and catalytic activity toward the redox couples existing in the electrolyte [13]. There are several requirements for obtaining carbon based composite material with suitable properties for DSSC counter electrodes. The carbon based composite materials with large surface area is typically preferred, since it could allow more contact with electrolyte. In addition, since a good bonding between carbon layer and glass substrate is required, a certain amount of binder is usually added to strengthen the mechanical properties of the film and prevent any cracking. Titanium dioxide nanoparticle is an inorganic binder that is commonly used for this purpose.

In this paper, the effect of the inorganic binder concentration on the physical properties of the counter electrode and the electronic properties of monolithic DSSC will be investigated by varying the concentrations of the inorganic binder added to the composite pastes. To date, the addition of titanium dioxide nanoparticles has been mostly aimed at improving the mechanical properties of the counter electrode films, however, further investigation on how it affect the electrical properties of the solar cells has not yet been reported. Herein, the effect of the binder concentration will also be correlated to type of carbon material, i.e. carbon nanopowder and activated carbon. The results of this work could be used as a future benchmark to formulate low-cost counter electrode materials for high performing monolithic DSSCs.

2. Experimental Section

2.1. Materials

Fluorine-doped transparent tin oxide coated glass (FTO) with a conductivity of ~15 Ω/sq (TEC-15), TiO_2 opaque paste (18NR-AO), Ru-based dye sensitizer (Z907), iodide-based electrolyte (EL-HPE), thermoplastic sealant (Surlyn®), and hermetic sealing compounds were purchased from Greatcell, Australia. Carbon nanopowder and ZrO_2 were obtained from Aldrich and Solaronix, respectively. Isopropyl alcohol, ethanol, acetone, and ethyl-cellulose were obtained from Merck. The rest of the materials used were activated carbon, colloidal graphite (Polaron Instruments, Inc.), TiO_2 powder (P25, Evonik, Germany), and α-terpineol (Aldrich). The resistivity of deionized (DI) water used throughout the experiments was higher than 15 MΩ.

2.2. Synthesis of Carbon Pastes

The carbon-based composites for the counter electrodes (CE) were prepared as follows: first, graphite paste was dried and grounded before the dried powder was collected. The carbon composite paste was made by grinding using mortar and pestle 2 g of graphite powder, 0.5 g of carbon nanopowder or activated carbon (alternately), 0.3 g of ethyl-cellulose, 4.25 g of terpineol, and various amount of P25. Herein, the amount of P25 was varied to 0.25 g and 2 g for CE with carbon nanopowder (namely $CNP_{0.25}$ and CNP_2, respectively) and activated carbon ($AC_{0.25}$ and AC_4, respectively). The obtained pastes were collected after being grounded constantly for at least 1 h and then stored at room temperature in closed containers.

IOP Conf. Series: Materials Science and Engineering **622** (2019) 012033 doi:10.1088/1757-899X/622/1/012033

2.3. Solar Cell Fabrication
FTO substrate glass were initially cut to a size of 20 mm × 15 mm and then sanded on certain areas to remove the FTO layer in order to isolate the anode and cathode sides. The FTO substrates were subsequently washed in acetone, soapy water, DI water, and isopropyl alcohol for 10 minutes each using ultrasonic cleaner bath (Branson 3200, Emerson Electric Co, USA). The layers on monolithic DSSC were deposited by screen printing method. TiO_2 paste was printed twice on the FTO substrate glass and dried at 120 °C for 10 min after each deposition before sintered at 500 °C in conveyor furnaces (Lindberg, USA) for 30 min. The sintered films were then immersed in 40 mM of $TiCl_4$ aqueous solution to 70 °C for 30 min as a post-treatment, followed by another annealing under the same conditions as the TiO_2 sintering. When the substrate was cool, Zr-Nanoxide paste (Solaronix, Switzerland) was then printed twice on top of TiO_2 layer and dried at 120 °C for 10 min after each deposition before sintered at 400 °C for 30 min in a conveyor furnace. Carbon composite paste was subsequently printed on top of the ZrO_2 layer and dried in conveyor oven at 120 °C for 30 min and sintered at 400 °C for 30 min (Radiant Technology Corp, USA). The photograph of the cells with complete layers is shown in Figure 1. The samples were then soaked in dye solution (Z907 Ru-dye, 0.25 mM in ethanol) for 20 h in dark. The dyed samples were then rinsed in ethanol, dried and then assembled using thermoplastic sealant and pressed for 120 °C for 30 s. Non-conductive cover slips were used to encapsulate the top part of the monolithic cells. Electrolyte (EL-HPE, Dyesol) was then dropped in the gap between the FTO and cover slip using syringe before the gap was sealed with hermetic sealing compound and then dried naturally.

2.4. Characterizations
The thickness of the counter electrodes was measured using thickness gauge (Mitutoyo), while the sheet resistance was measured using four-point probe with Hewlett-Packard 3468A multimeter and 6186C DC current source. The surface morphology of the carbon composite pastes was observed using field-emission scanning electron microscopy (FESEM, STEM JEOL JB-4610F). Meanwhile, the surface area and pore size analysis were conducted via nitrogen sorption technique (Quantachrome Nova Analyzer), wherein all samples were dried into powders and degassed while heated at 150 °C prior to the measurement. The electrical parameters of the monolithic DSSC were analyzed by measuring the current-voltage characteristics under sun simulator (Oriel 9112) with an intensity of 50 mW/cm^2 equipped with AM1.5G filter using National Instrument I-V measurement system. The photoconversion efficiency of the cells was determined using the following equation:

$$\eta = \frac{V_{OC} I_{SC} FF}{P_{in}} \times 100\%$$

where V_{OC} and I_{SC} are the open circuit voltage and the short circuit current, respectively, FF is the fill factor, and P_{in} is the power input received during the illumination.

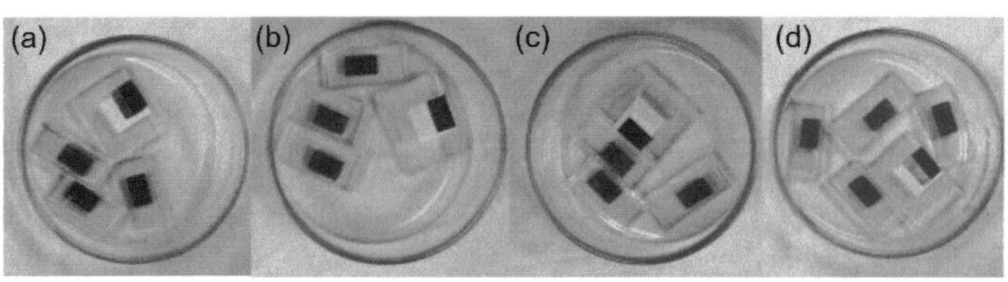

Figure 1. Photograph of the constructed cells with various counter electrodes: (a) $CNP_{0.25}$, (b) CNP_2, (c) $AC_{0.25}$, and (d) AC_2.

IOP Conf. Series: Materials Science and Engineering **622** (2019) 012033 doi:10.1088/1757-899X/622/1/012033

3. Results and Discussion

3.1. Physical Properties of Carbon Composite Pastes

Nitrogen gas sorption measurements were performed to characterize the surface and microstructure of the carbon composite pastes. Prior to each measurements, all samples were degassed to achieve physical desorption from any residual liquid or moisture. Figure 2 shows that all samples exhibited gas sorption isotherms with similar shape, regardless of the carbon types and the binder concentrations. Herein, all of the isotherm profiles resemble Type II isotherm curve with H3 hysteresis loop, which are commonly associated with the characteristics of microporous materials having pore width below 2 nm according to the classifications from the International Union of Pure and Applied Chemistry (IUPAC) [14]. There was also a clear trend that the increase in the binder concentration produced higher gas uptake or more complete pore filling compared to the samples with lower binder concentration, particularly at high pressure ratio range (P/P_0~1.0). This phenomenon could be an indication that the carbon composites with higher P25 content had bigger cavities, possibly as the result of large aggregates, which will be further analysed via morphological analysis later in this work. Meanwhile, pore size determination using Barrett-Joyner-Halenda (BJH) [15] method also indicated that the increase in binder concentration also led to slightly larger pore diameter. For samples with carbon nanopowder the pore diameter for sample $CNP_{0.25}$ and CNP_2 was 2.43 and 14.58 nm, respectively, while the pore diameter for samples with activated carbon was 2.43 and 4.79 nm for $AC_{0.25}$ and AC_2, respectively. In terms of surface area, all of the carbon composites did not particularly show any profoundly high surface areas. The surface area in this work was determined using Brunauer-Emmett-Teller (BET) method [16], giving an approximate values of 49.4, 3.42, 54.22, and 50.52 m^2/g for sample $CNP_{0.25}$, CNP_2, $AC_{0.25}$, and AC_2, respectively.

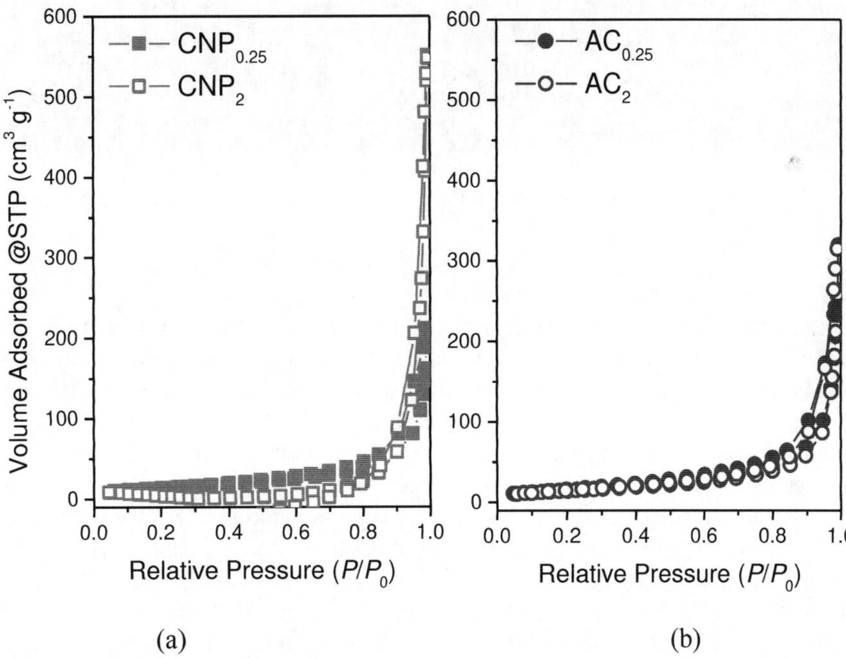

Figure 2. Nitrogen gas sorption isotherms for carbon composite paste containing (a) carbon nanopowder and (b) activated carbon with P25 binder concentrations of 0.25 and 2 g, as indicated by the index.

Figure 3 shows the surface morphology of the carbon composite as counter electrode films. It can be clearly seen that P25 existed as smaller particles, which dispersed around carbon aggregates that have significantly larger size, i.e. in the order of micrometre. The higher the P25 concentration added into the composite, the more dispersion of nanoparticles could be observed, regardless the type of carbon used (see Figure 2b and 2d). By comparing Figure 2a and 2c, there was an indication that the composites with activated carbon tends to form larger agglomeration with a size of tenths of micrometre compared to that of carbon nanopowder, wherein the basic form of the activated carbon itself was existed as thin flakes. Similarly, larger agglomeration in the activated carbon containing composites remained appear at higher P25 concentration (Figure 2d).

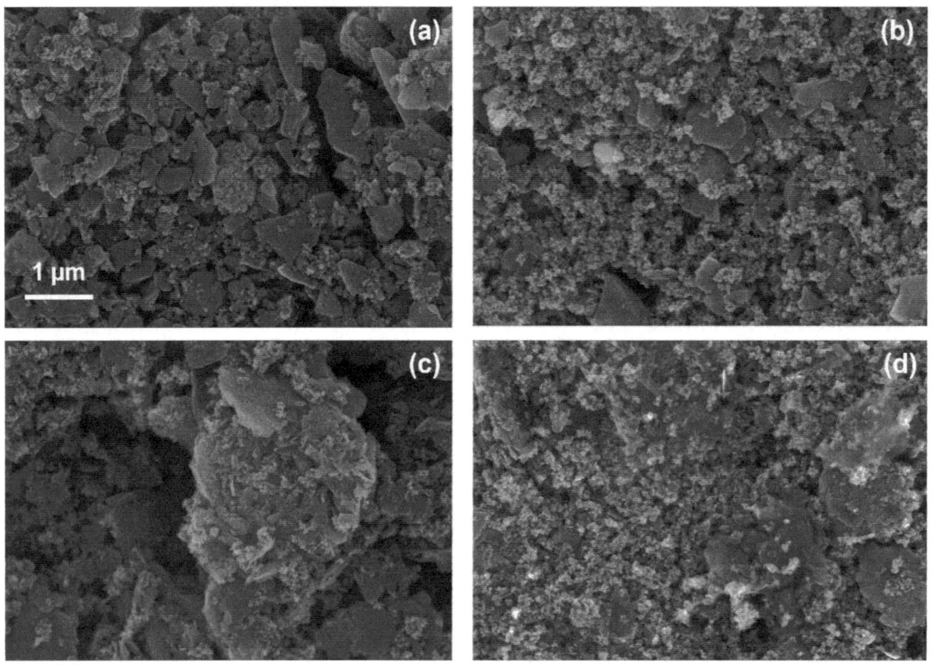

Figure 3. SEM images of (a) $CNP_{0.25}$, (b) CNP_2, (c) $AC_{0.25}$, and (d) AC_2.

3.2. Electrical Characteristics

Prior to characterizing the electrical parameters, the thickness and the sheer resistivity of the counter electrodes containing various carbon composites were initially measured and the results are summarized in Table 1. It is clear that the thickness of the counter electrode was consistently increased with the increase in the P25 concentration, regardless of the carbon type used in the composite. The thickest film was obtained on the counter electrode containing carbon nanopowder and P25 with a ratio of 1:2 (i.e. CNP_2). The sheet resistivity, however, did not suggest any particular trends with regard to the P25 concentration nor the carbon type. For the case of carbon nanopowder, the increase in the P25 concentration seems to slightly improve the film conductivity by lowering the sheet resistance. In contrast, the opposite effect occurred when it came to the composite with the activated carbon. This phenomenon might be related to the nature of the activated carbon that tend to form larger aggregates than carbon nanopowder, as previously discussed according to the SEM observation. Such behaviour may have led to the decrement in the conductivity, particularly at high P25 concentration, since the P25 nanoparticles could easily blend into the agglomeration of the activated carbon. Nevertheless, it was clear that the overall carbon-containing counter electrodes in this work has improved the conductivity of the counter electrodes, as indicated by their lower sheet resistance values compared to the blank uncoated FTO substrate.

Table 1. Thickness and sheet resistivity data of counter electrodes with various carbon composite films.

Sample	**Thickness (μm)**	**Sheet resistance (Ω/sq)**
$CNP_{0.25}$	16.4 ± 4.0	11.1 ± 0.7
CNP_2	17.8 ± 4.3	10.8 ± 0.3
$AC_{0.25}$	10.4 ± 1.3	10.7 ± 0.4
AC_2	14.4 ± 3.5	11.5 ± 0.2
Blank FTO	-	14.1 ± 1.2

The current-voltage (*I-V*) characteristics of the solar cells constructed in this work were measured under simulated solar irradiation with an intensity of 50 mW/cm^2 or equals to 0.5 Sun. The AM1.5G filter was also used during the irradiation, mimicking the standard solar spectrum at the earth's surface. Additionally, it should be noted that all of the measured parameters during the *I-V* measurements hereafter were normalized to the size of the cell active area that was fixed at 0.25 cm^2. Figure 4 shows the *I-V* characteristics of DSSC with various carbon composites, while Table 2 summarizes the electrical parameters obtained from the *I-V* measurements. The characteristic of the *I-V* curves shown in Figure 4 describes an unideal *I-V* curve, since the shape is rather straight with overall *FF* values still less than 0.5. Such *I-V* curve profile could be attributed to an incomplete filling of electrolyte [17], which is rather commonly observed in DSSC with monolithic structure due to the solid interfaces existed between the multiple layers of photoelectrode, spacer and counter electrode [8].

The best photovoltaic performance was achieved by sample $AC_{0.25}$ with photoconversion efficiency of 0.221%. Despite still suffered from low *FF*, sample $AC_{0.25}$ consistently produced the highest J_{SC}, V_{OC} and P_{max} among all samples, thus demonstrating the best photovoltaic performance. Interestingly, the electrical performance obtained was in good agreement with the previous trend observed on the sheet resistivity data (Table 1). In this case, the increase in the P25 binder concentration was advantageous for the samples containing carbon nanopowder, but vice versa for samples containing activated carbon. For instance, sample $CNP_{0.25}$ and AC_2 herein showed rather similar performance with almost comparable photoconversion efficiency, while their sheet resistance values were also rather proportional (see Table 1). However, although sample CNP_2 and $AC_{0.25}$ had comparably low resistances, their photovoltaic performance were significantly different. This phenomenon suggests that there is another factor that determine the electrical performance in addition to the sheet resistance or conductivity. According to the gas sorption data, it was suspected that the lower performance exhibited by CNP_2 was caused by its significantly low BET surface area (i.e. 3.42 m^2/g). The relatively low BET surface area obtained on sample CNP_2 could arise from the TiO_2 P25 nanoparticles blocking the large pores or voids between the particles of the carbon nanopowder composite. This phenomenon was plausible considering that the SEM results have suggested that the carbon nanopowder composite exhibited smaller particle size with better blending with the P25 binder compared to the activated carbon composite. Thus, the superior performance exhibited by sample $AC_{0.25}$ was likely related to a combination between low sheet resistivity (i.e. 10.7 Ω/sq) and high BET surface area (i.e. 54.22 m^2/g).

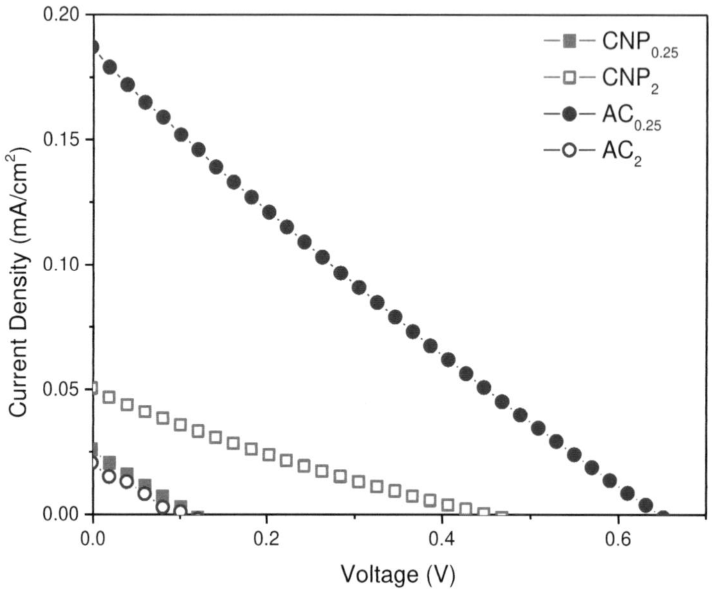

Figure 4. Current-voltage (*I-V*) curves of monolithic DSSC fabricated with various carbon composites containing various P25 concentrations as the counter electrode.

Table 2. Electrical parameters of monolithic DSSC with various carbon composites containing various P25 concentrations as the counter electrode.

Sample	J_{SC} (mA/cm^2)	V_{OC} (V)	P_{max} (mW)	*FF*	*Efficiency* (%)
$CNP_{0.25}$	2.06×10^{-2}	0.10	6.86×10^{-4}	0.33	0.005
CNP_2	4.96×10^{-2}	0.45	4.83×10^{-3}	0.23	0.039
$AC_{0.25}$	17.90×10^{-2}	0.63	2.76×10^{-2}	0.24	0.221
AC_2	1.51×10^{-2}	0.10	5.13×10^{-3}	0.34	0.004

4. Conclusions

Monolithic DSSCs with carbon composite based counter electrodes were developed. The type of carbon used as the main material in the composite and the concentration of TiO_2 P25 that was used as binder were varied in order to analyze their influence on the photovoltaic performance of the resulting DSSC. Herein, composites containing carbon nanopowder and activated carbon were compared. With regard to the presence of binder, it was imperative that the a good blending between the carbon content and P25 binder needs to be fully achieved in order to gain an improved photovoltaic performance. Since the composites containing activated carbon existed as large aggregates and did not mix well with P25, increasing the concentration of P25 added into the composite reduced the conductivity and deteriorated the photovoltaic performance. The opposite effect was observed on the samples containing carbon nanopowder. The best performing DSSC was obtained when carbon composite containing activated carbon and TiO_2 P25 with weight concentration of 0.5 g and 0.25 g, respectively. The main contributor for its excellent performance was suspected due to the low resistivity and high BET surface area.

Acknowledgement
This research was funded by INSINAS research program 2017-2018 from the Ministry of Research, Technology and Higher Education of the Republic of Indonesia, entitled "Design, Fabrication and Up-scaling of Monolithic Dye-sensitized Solar Modules for Indoor Applications". We also thanked members of the Materials and Devices for Solar Cells research group at the Research Center for Electronics and Telecommunication, Indonesian Institute of Sciences (P2ET LIPI) for the support. Research Center for Physics (P2F LIPI) is also acknowledged for performing the SEM and gas sorption measurement.

References
[1] O'Regan B and Gratzel M 1991 *Nature* **353** 737
[2] Kroon J M, Bakker N J, Smit H J P, Liska P, Thampi K R, Wang P, et al. 2007 *Prog. Photovoltaics Res. Appl.* **15** 1
[3] Simon Mathew, Aswani Yella, Peng Gao, Robin Humphry-Baker, Basile F. E. Curchod, Negar Ashari-Astani, et al. 2014 *Nature Chemistry* **6** 242
[4] Fakharuddin A, Jose R, Brown T M, Fabregat-Santiago F and Bisquert J 2014 *Energy Environ. Sci.* **7** 3952
[5] Hagfeldt A, Boschloo G, Sun L, Kloo L and Pettersson H 2010 *Chem. Rev.* **110** 6595
[6] Kalowekamo J and Baker E 2009 *Sol. Energy* **83** 1224
[7] Nursam N M, Anggraini P N, Shobih and Hidayat J 2017 Low-cost monolithic dye-sensitized solar cells fabricated on single conductive substrate. In: *2017 International Conference on Radar, Antenna, Microwave, Electronics, and Telecommunications (ICRAMET),* pp 164
[8] Nursam N M, Istiqomah A, Hidayat J, Anggraini P N and Shobih 2017 *Jurnal Elektronika dan Telekomunikasi* **17** 30
[9] Lee W J, Ramasamy E, Lee D Y and Song J S 2008 *Sol. Energy Mater. Sol. Cells* **92** 814
[10] Maiaugree W, Lowpa S, Towannang M, Rutphonsan P, Tangtrakarn A, Pimanpang S, et al. 2015 *Sci. Rep.* **5** 15230
[11] Jayaraman T, Raja S A, Jagannathan M and Thandavarayan M 2015 *ChemElectroChem* **2** 928
[12] Olsen E, Hagen G and Eric Lindquist S 2000 *Sol. Energy Mater. Sol. Cells* **63** 267
[13] Kouhnavard M, Ahmad Ludin N, Vazifehkhah Ghaffari B, Sopian K, Abdul Karim N and Miyake M 2016 *Int. J. Photoenergy* **2016** 7
[14] Schneider P 1995 *Appl. Catal., A* **129** 157
[15] Barrett E P, Joyner L G and Halenda P P 1951 *J. Am. Chem. Soc.* **73** 373
[16] Brunauer S, Emmett P H and Teller E 1938 *J. Am. Chem. Soc.* **60** 309
[17] Nursam N M, Hidayat J, Muliani L, Anggraeni P N, Retnaningsih L and Idayanti N 2017 *IOP Conference Series: Materials Science and Engineering* **214** 012007

Synthesis of Geopolymer from Blends of Tropical Biomass Ashes

Tjokorde Walmiki Samadhi*, Winny Wulandari, Afiana Suroto and Dendi Satriya Agung

Chemical Engineering, Faculty of Industrial Technology, Institut Teknologi Bandung, Jl. Ganesha 10, Bandung 40132, Indonesia

*Corresponding author: twsamadhi@che.itb.ac.id

Abstract. Thermal conversion of biomass to produce energy inevitably generates ash as residual matter. Therefore, sustainable utilization of biomass should also consider reuse measures for the ash. This research examines the conversion of locally available biomass ashes in Indonesia into geopolymer. This is an environmentally more benign alkali-aluminosilicate alternative to the ordinary Portland cement produced by low-temperature reactions between aluminosilicates and concentrated alkali solution. Biomass ash sources in this study are corn stover, coconut shell, and sugarcane bagasse. Biomass ash blends are prepared according to a ternary simplex centroid mixture experiment design. These blends are reacted with activator solutions containing NaOH and Na-silicate and blended with sand aggregate to produce geopolymer mortar specimens. Early compressive strengths of geopolymer mortars range from 7.2 to 10.4 MPa, which surpasses the SNI 15-2049-2004 national standard for Portland cement with low heat of hydration. Statistical analysis of the experimental data indicates that the early strength as a function of biomass ash blend formulation is represented by a quadratic mixture model. Bagasse ash produces the highest strength. The quadratic terms consist of bagasse-corn and bagasse-coconut antagonistic blending terms. Morphology of the geopolymer mortar fracture surface indicates good bonding with the sand aggregate. Extensive acicular crystal growth within the amorphous geopolymer gel phase is observed in lower-strength formulations, which points to the formation of zeolite-like phase in the geopolymerization process. While the correlation of biomass ash type to the structure of the resulting geopolymer is yet to be established, this work has clearly identified the technical feasibility of producing geopolymer with satisfactory strength from tropical biomass ashes.

Keywords: geopolymer, biomass ash, cement

1. Introduction

The contribution of biomass-based fuels to the Indonesian national energy resources mix is expected to rise in the coming decade as the country strives to fulfil its greenhouse gases emission reduction by 26% from 2009 to 2020 as stipulated in the UN Framework Convention on Climate Change (UNFCCC). To achieve this voluntary reduction, biomass-based fuel is projected to contribute 5% to the national primary energy resource mix by 2025 through the Presidential Decree No.5 / 2006.

Content from this work may be used under the terms of the Creative Commons Attribution 3.0 licence. Any further distribution of this work must maintain attribution to the author(s) and the title of the work, journal citation and DOI.
Published under licence by IOP Publishing Ltd

A problem that can be expected with the rising share of biomass-based energy is the concurrent increase in the generation of biomass combustion ashes by thermal conversion processes. As a figure of reference, it has been estimated that biomass-based electrical power generation in 2018 produces roughly 10 million tons of biomass ash per year globally [1]. To further ensure that biomass-based energy is a viable fundamental solution for global climate control, utilization of this waste material needs to be thoroughly established.

This work explores the preliminary feasibility of the conversion of several biomass ashes abundant in tropical climes such as Indonesia into geopolymer. Geopolymer, or more generally referred to as alkali-activated cement, is a largely amorphous cementitious material obtained by complex reactions between aluminosilicate solids and concentrated alkali solution. The typical range of the overall reaction mixture Si/Al ratio is 2-4 range [2], although values several orders of magnitude higher than this range have been investigated as well [3]. These reactions involve the initial dissolution of silica network structure of the aluminosilicate raw materials, followed by the formation of alkali-aluminosilicate gel phase. As the reaction proceeds, the gel phase solidifies into a mechanically strong random 3D network of polymeric sialate groups. Geopolymers have been identified as favourable materials for Portland cement substitute in concrete constructions, hazardous waste encapsulation, and other applications [4].

A wide range of aluminosilicate solids has been known to be amenable for geopolymerization, including waste materials such as coal fly ash, metallurgical slag, and even construction demolition waste [5,6]. This ability to reuse waste materials is further bolstered by the lower energy consumption in geopolymer production. Geopolymerization reactions can proceed at ambient or near ambient conditions, compared to 1400 °C sintering temperature required to produce ordinary Portland cement (OPC). It has been argued that geopolymers, when produced in the vicinity of point of use, may reduce the greenhouse gas emissions by 64% relative to OPC [7].

Studies on the utilization of biomass ashes for geopolymer synthesis is still rather limited. Of the obviously wide array of biomass refuse available, perhaps only rice husk ash has been studied in detail in geopolymer synthesis [8]. A key problem in the geopolymerization of rice husk ash is its extremely high Si/Al ratio, which necessitates the addition of alumina-rich precursors. Earlier works in the Author's research group has indicated the feasibility of using bamboo ash for geopolymer synthesis, both in neat form or as an auto thermally-prepared mixture of bamboo ash and calcined kaolin [9].

This study aims to expand the knowledge base of the geopolymerization of biomass ashes, particularly those associated with tropical biomass resources. Prospect for further development is evaluated based on mortar compressive strength as one of key engineering properties of cementitious materials. Since an extensive description on the geopolymerization of biomass ashes is not yet available in the literature, this work is expected to contribute significantly to the science of this highly interesting and diverse class of engineering material.

2. Materials and methods

Biomass selected based on their local abundance were coconut shell (CO), corn stover (ST), and sugarcane bagasse (BG). Coconut shell was purchased in the form of charcoal in a local market in Bandung. Corn stover was obtained as fresh food market waste in Bandung. Sugarcane bagasse was obtained from a sugar mill in Central Java. Corn stover and bagasse are sun-dried for 7 days, and charred in a fixed bed combustor. The chars are ground using mortar and pestle and stored in airtight plastic bags. The coconut shell charcoal is then ground and bagged.

Biomass char ashing temperatures were determined by thermogravimetric analysis (TGA) in ambient air atmosphere. These were found to be 475 °C for corn stover, 400 °C for bagasse, and 730 °C for coconut shell. The oxide compositions of biomass ashes measured by x-ray fluorescence (XRF) are summarized in Table 1.

IOP Conf. Series: Materials Science and Engineering **622** (2019) 012034 doi:10.1088/1757-899X/622/1/012034

Table 1. Oxide Compositions of Biomass Ashes

Oxide	Content, %-mass		
	corn stover (ST)	sugarcane bagasse (BG)	coconut shell (CO)
SiO_2	25.58	63.44	66.58
Al_2O_3	14.26	9.44	7.65
Fe_2O_3	15.48	4.99	4.44
MgO	2.78	4.40	4.32
CaO	9.35	4.03	1.63
Na_2O	4.33	0.15	<0.001
K_2O	16.69	2.80	8.88
TiO_2	0.67	0.15	0.06
P_2O_5	3.52	1.49	4.44

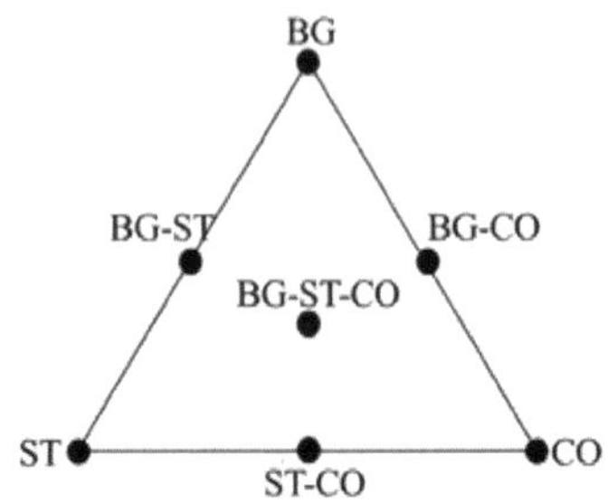

Figure 1. Simplex centroid mixture design to measure the effect of ash blend formulation on biomass ash geopolymer compressive strength (BG = sugarcane bagasse, ST = corn stover, CO = coconut shell)

To compare the strength of geopolymers produced using the three ashes, as well as to measure the effect of ash blend formulation on the strength, a three-component simplex centroid mixture experiment design was run. As described in Figure 1, this design included three pure-component runs, one balanced-composition centre point run, and three binary composition runs. Ash blend formulations in this design are summarized in Table 2. Due to the preliminary nature of this research, geopolymer mortar compressive strength after 14 days of curing was selected as the sole response variable. This key engineering property was deemed adequate in identifying biomass ashes with potentials for further investigation.

To prepare the mortar specimens, biomass ash was dry-mixed with graded sand classified according to ASTM C778-06 standard method, at a sand to ash mass ratio of 2.75:1. The dry mix was tumbled in a jar mill for 3 minutes to ensure its homogeneity. The activator solution was prepared by dissolving technical-grade NaOH flakes and sodium silicate in water at a water to solid mass ratio of 2.2:1. Sodium silicate to hydroxide mass ratio was kept at 2.0:1 for all runs.

The dry ash-sand mix and activator solution was then mixed in a planetary mixer for 4 minutes. The activator solution to ash mass ratio was adjusted from batch to batch to produce a visually similar consistency throughout all runs. The wet mortar was then cast into a 5×5×5 cm steel mold and heat-cured in an electric oven for 24 hours. Afterwards, the cast was removed from the oven and kept at ambient conditions for further curing. After 7 days, the mortar was removed from the mold. After a total period of 14 days since casting, the mortar specimens were subjected to compressive strength (crushing strength) measurement using a uniaxial load tester.

3. Results and discussion

Table 2 compiles the compressive strengths of the biomass ash geopolymer mortars. Pure-component and centre point runs were replicated twice, while binary-component blends were run once.

The compressive strengths varied from 7.18 to 10.46 Mpa. These values compared very well with the ASTM C270 standard for intermediate-strength cement mortar minimum compressive strength of 5.25 Mpa, and also with results described by the authors in a preceding publication [10]. Sugarcane bagasse (BG) exhibits the highest compressive strength among the three biomass ashes, while coconut shell produces the lowest strength.

Results of ANOVA treatment of compressive strength data in Table 2 using the special cubic mixture model are summarized in Table 3. Significances of the model terms are indicated by the P-values. The ANOVA table clearly identifies the corn stover – coconut shell binary mixing term as the only non-significant model component at a confidence level of 95%. The special cubic model produced a very good fit to the measurement data, as indicated by the coefficient of correlation.

Table 2. Compressive Strength of Biomass Ash Geopolymer Mortars After 14-Day Curing

Ash blend composition	Si/Al	Activator /ash	Rep.	Compressive strength, Mpa
ST	12.3	1.15	I	7.92
			II	8.79
CO	3.2	1.50	I	7.26
			II	7.25
BG	9.5	1.15	I	10.25
			II	10.46
ST+CO	5.3	1.32	I	7.75
ST+BG	10.7	1.15	I	8.00
CO+BG	4.9	1.32	I	7.18
ST+CO+BG	6.3	1.25	I	8.09
			II	8.16

Table 3. ANOVA Treatment Results of the Measured Compressive Strength Data

Term	Effect coefficients	P-value	F_0
BG	10.34	0.00	617.59
ST	8.35	0.00	1344.95
CO	7.25	0.00	455.54
BG·ST	-5.00	0.00	568.50
BG·CO	-6.48	0.00	958.68
ST·CO	-0.22	0.75	1.01
BG·ST·CO	20.88	0.003	200.80
S	R^2	R^2(adj)	R^2(pred)
0.0604	99.81%	99.65%	99.25%

By eliminating the ST·CO binary mixing term, an empirical model describing the correlation between ash blend composition and geopolymer mortar compressive strength was obtained (Equation (1) - independent variables are ash contents in %-mass).

$$CS(MPa) = 10.3BG + 8.4ST + 7.2CO - 5.2BG \cdot CO + 21.2BG \cdot ST \cdot CO \quad (1)$$

Consistent with the numbers in Table 2, the empirical model produces the highest strength when the geopolymer is formulated using only sugarcane bagasse. No optimum ash blend formulation is found within the simplex experimental region. Compositional shift towards coconut shell ash reduces the strength. The antagonistic effect of mixing bagasse and coconut shell ashes produces the lowest strength along the binary mixing line, at an approximate BG:CO mass ratio of 1:3.

Figure 2. Scanning electron micrograph of coconut ash geopolymer at 1000x magnification.

As an example of typical fresh fracture surface microstructure of the geopolymer mortars obtained in this work, Figure 2 presents the SEM micrograph of the coconut shell ash geopolymer mortar. The amorphous geopolymer gel phase cover almost the entire graph, with the lower right hand side indicating the surface of the aggregate phase. Formation of acicular crystals from the amorphous gel is clearly indicated. Kumar and Kumar [11] attributed this acicular phase to the Na-zeolite crystals formed during heat-curing of the amorphous geopolymer gel phase. Li and co-workers [12] in their research on the geopolymerization of paper sludge ash by sodium silicate – hydroxide activator solution, argued that such acicular and prismatic forms are attributed to the ettringite (calcium aluminum sulfate) phase. Since the coconut shell ash contains nearly negligible calcium (see Table 1), it is hypothesized that the acicular phase observed in this work is Na-zeolite.

The higher strength of the bagasse ash is likely to be due to its high potassium content. It is well-known that potassium is has a higher reactivity compared to sodium towards the silicate groups in the initial stages of geopolymerization [13]. Duxson and co-workers [14] measured the compressive strength of metakaolin geopolymers as a function of Si/Al and activator solution Na/K ratios. A significant interaction between the two variables was indicated, with the mixed-alkali activator producing higher strength than either potassium or sodium alone at Si/Al ratio of 1.4 to 1.9.

4. Conclusion

The preliminary technical feasibility of utilizing ashes from several locally abundant tropical biomass, namely coconut shell, corn stover, and sugarcane bagasse, has been proved. These waste materials are capable of producing geopolymer mortars with compressive strength which satisfies ASTM C270 standards via activation by NaOH-Na_2SiO_3 solution, at Si/Al ratios in the 3-12 range. Measurement of the effect of ash blend formulation on geopolymer mortar compressive strength indicated the absence of an optimum formulation. Rather, the highest strength is achieved when using bagasse ash alone. Acicular crystals observed in some of the specimens suggests the formation of Na-zeolite from the amorphous gel phase due to heat curing. The correlation between phase development during curing and the measured compressive strength is subject to further study.

Acknowledgement

Financial support of this research was provided by the RI Ministry of Research, Technology and Higher Education through the PDUPT 2018 program, administered by the ITB Institute for Research and Community Services (LPPM-ITB).

References

[1] Lamers F, Cremers M, Matschegg D, Schmidl C, Hannam K, Hazlett P, Madrali S, Dam B, Roberto R, Mager R, Davidsson K, Bech N, Feuerborn H, Saraber A, *Options for Increased Use of Ash from Biomass Combustion and Co-firing,* IEA Bioenergy Report, 2018

[2] He P, Wang M, Fu S, Jia D, Yan S., Yuan J., Xu J., Wang P and Zhou Y 2016 *Ceramics International* **42** 1416

[3] Fletcher RA, MacKenzie KJD, Nicholson CJ and Shimada S 2005 *Journal of the European Ceramic Society* **25** 1471

[4] Salih MA, Ali, AAA, Farzadnia N 2014 *Construction and Building Materials* **65** 592

[5] Zhang Z, Wang H, Zhu Y, Reid A, Provis JL and Bullen F 2014 *Applied Clay Science* **88-89** 194

[6] Vasquez A, Cardenas V, Robayo RA and de Gutierrez RM 2016 *Advanced Powder Technology* **27** 1173

[7] McLellan BC, Williams RP, Lay J, van Riessen A and Corder GD 2011 *Journal of Cleaner Production* **19** 1080

[8] Hwang C and Huynh T 2015 Construction and Building Materials **101** 1

[9] Purbasari A, Samadhi TW and Bindar Y 2017 *Advanced Science Letters* **23** 5723

[10] Samadhi TW, Wulandari W, Prasetyo MI, Fernando MR 2017 *IOP Conference Series: Materials Science and Engineering* **248** 012008

[11] Kumar S and Kumar R 2011 *Ceramics International* **37** 533

[12] Li Z, Ohnuki T and Ikeda K 2016 *Materials* **9** 633

[13] Duxson P, Lukey GC, Separovic F and van Deventer JSJ 2005 *Industrial Engineering Chemistry & Research* **44** 832

[14] Duxson P, Mallicoat SW, Lukey GC, Kriven WM and van Deventer JSJ 2007 *Colloids & Surfaces A* **292** 8

AUTHOR INDEX